ADVANCES IN
Chromatography

VOLUME 26

ADVANCES IN
Chromatography

VOLUME 26

Edited by

J. Calvin Giddings

Executive Editor

UNIVERSITY OF UTAH
SALT LAKE CITY, UTAH

Eli Grushka

THE HEBREW UNIVERSITY OF JERUSALEM
JERUSALEM, ISRAEL

Phyllis R. Brown

UNIVERSITY OF RHODE ISLAND
KINGSTON, RHODE ISLAND

MARCEL DEKKER, Inc.   New York and Basel

Library of Congress Cataloging in Publication Data
Main entry under title:

Advances in chromatography. v.1-
 1965-
New York, M. Dekker
  v. illus. 24 cm.
 Editors: v.1- J.C. Giddings and R.A. Keller.
 1. Chromatographic analysis-Addresses, essays, lectures.
1. Giddings, John Calvin, [date] ed. II. Keller, Roy A., [date]
ed.
QD271.A23 544.92 65-27435
ISBN 0-8247-7664-X

MARCEL DEKKER, INC.
270 Madison Avenue, New York, New York 10016

Current printing (last digit):
10 9 8 7 6 5 4 3 2 1

PRINTED IN THE UNITED STATES OF AMERICA

Contributors to Volume 26

Raymond Annino Principal Research Scientist, Corporate Research, The Foxboro Company, Foxboro, Massachusetts

Lars G. Blomberg Department of Analytical Chemistry, University of Stockholm, Arrhenius Laboratory, Stockholm, Sweden

Joseph A. Caruso Department of Chemistry, University of Cincinnati, Cincinnati, Ohio

Emilio Gelpí Department of Neurochemistry, CSIC, Centro de Investigacion y Desarrollo, Barcelona, Spain

Yoshio Kato Central Research Laboratory, Toyo Soda Manufacturing Co., Ltd., Yamaguchi, Japan

Hermann J. Möckel Hahn-Meitner-Institut Berlin GmbH, Bereich Strahlenchemie, Berlin, Federal Republic of Germany

Ahmad H. Mohamad Department of Chemistry, University of Cincinnati, Cincinnati, Ohio

Roger M. Smith Department of Chemistry, Loughborough University of Technology, Loughborough, Leicestershire, England

Ian D. Watson Drug Investigation Unit, Department of Biochemistry, Royal Infirmary, Glasgow, Scotland

Contents of Volume 26

v

Contents of Other Volumes

ADVANCES IN
Chromatography

VOLUME 26

1

RPLC Retention of Sulfur and Compounds Containing Divalent Sulfur

Hermann J. Möckel *Hahn-Meitner-Institut Berlin GmbH, Berlin, Federal Republic of Germany*

I. INTRODUCTION

Interest in sulfur and its compounds has grown remarkably during the past two or three decades. For a long time sulfur had been used to a steadily increasing extent, but analytical research in the sulfur field was not very common. Now, people have started thinking about sulfur and the damage to the natural environment caused by sulfur products from industry, traffic, energy production, and other aspects of daily life. Sulfur dioxide has been recognized as one of the main pollutants, and attention is now being focused on its precursors and reaction products. Indeed, most sulfur compounds and even elemental sulfur must be regarded as hazardous if they are handled carelessly.

The interest in sulfur has also led to new activities in the analytical chemistry of compounds of sulfur, not only in environmental pollution control but also in a variety of disciplines in science and engineering. Consequently, there is much literature on the analysis of sulfur and surveys are available [1—5]. However, systematic investigations on the separation of complex mixtures containing sulfur and/or sulfur compounds were almost entirely missing. During the past 10 years some progress has been made, mainly by the application of high-performance liquid chromatography (HPLC). This chapter is intended to outline the present status and to present the main characteristics of the liquid chromatographic behavior of sulfur and a variety of its compounds.

II. SPECIAL PROBLEMS IN SULFUR ANALYSIS

Sulfur and many of its compounds have some properties which make their analytical chemistry comparatively difficult. Frequently, one has to work on complex mixtures of isomers, homologs, or other sets of similar compounds. The determination of the total sulfur content, e.g., by combustion to sulfate, is well established. However, if single components must be identified, it is obvious that this problem can be solved only by using chromatographic techniques.

Many sulfur compounds tend to undergo thermally or photochemically initiated interconversion reactions. A mixture of dimethyl and diethyl disulfides yields considerable amounts of methylethyl disulfide when it is heated or simply exposed to sunlight for some time [6—8]. Organic polysulfides with long sulfur chains decompose to compounds of shorter chain lengths when they are heated [9,10]. Numerous similar effects are known. Various sulfur compounds are sensitive to catalytic activity, mainly surface catalysis. Thiols and polysulfides decompose at metal surfaces. Most of the various molecular forms of elemental sulfur are rapidly converted to S_8 at the surface of silica or alumina [11].

Separation techniques for these fairly labile compounds should be designed so that the impact of heat and light on the sample components is kept as small as possible. Use of metal surfaces and strong adsorbents should be avoided whenever possible.

Liquid-liquid partition and reversed bonded phase liquid chromatography (RPLC) appear to meet the above demands best, since these systems are operated close to room temperature and the heats of sorption are low. It is somewhat surprising that very little work has been done in this direction [12–15].

The investigations described in this chapter have been performed almost exclusively in the author's laboratory. For convenience, we decided from the beginning to do all experiments on bonded alkyl silica, mostly octadecylsilica (ODS), phases. Although binary or ternary liquid partition systems might lead to better separations in some cases, the aim of the project was to obtain comparable information on the retention behavior of a representative selection of sulfur compounds.

III. QUOTATION OF DATA

The experimental data as read from the integrator printout are raw retention times, given in minutes with two digits after the decimal point. They include the column dead time t_m and the extra column dead time t_{extr}. The capacity factor is calculated from Eq. (1):

$$k' = \frac{t_{raw} - t_m - t_{extr}}{t_m} \tag{1}$$

The determination of t_{extr} is trivial. For the determination of the column dead time t_m we have chosen the "linearization of $\ln k'$" method. This method, first used in gas chromatography (GC) [16–18] and later proposed for RPLC [19–22], is based on the fact that the $\ln k'$ values of the members of a homologous series, e.g., the n-alkanes, depend linearly on the carbon number n_C. Therefore

$$\ln k' = a(alk) + b(alk)n_C \tag{2}$$

This linear relation holds only within certain limits of solute molecular size [23,24] and for n-alkanes on ODS phases from C_5 to C_{17}. After combining Eqs. (1) and (2), t_m is changed by computer until

the fit of the experimental data to linearity is a maximum, i.e., until
the least-squares fit correlation coefficient r comes as close as possible to the value 1.0. A simplified calculation has been proposed [21,
22,25] which is most useful for determining a(alk) and b(alk) in Eq.
(2) daily according to a fixed schedule. Any significant changes
indicate system malfunctions, for instance, small temperature fluctuations, may cause minor changes in the above values which can be
used to normalize measured retention data to standard conditions.
Under regular operating conditions, over a 1-month period, relative
standard deviations of 3% for a(alk) and 1% for b(alk) can be expected [26].

All sulfur compounds in this chapter can be described as thia
derivatives generated by substituting CH_2 groups in alkanes or
cycloalkanes by sulfur atoms. Retention changes caused by such
a thia substitution can best be visualized if the retention of the substitution product is compared to that of the original hydrocarbon.
In the retention system we have chosen, the retention index I_K is
equivalent to the GC Kovats index [27]. Thus we define the index
of an n-alkane as

$$I_K(alk) = 100n_C \tag{3}$$

Then, using Eqs. (2) and (3), the index of some sulfur-containing
solute A, which is not an n-alkane, is determined as follows:

$$I_K(A) = \frac{100}{b(alk)} [\ln k'(A) - a(alk)] \tag{4}$$

The retention change associated with the substitution then is

$$\Delta I_K = I_K(A) - 100n_{skeleton} \tag{5}$$

where $n_{skeleton}$ gives the number of atoms other than H.

As long as solute A and the alkanes show practically the same
solubility dependence on the composition of the eluent, $I_K(A)$ is invariant toward eluent composition. However, if the solubility of A
increases faster than that of the alkane having $n_{skeleton}$ carbon
atoms, $I_K(A)$ decreases with eluent composition change. An I_K increase is observed if the ratio of solubility changes is reversed. By
observing ΔI_K values as a function of the eluent composition, useful
information about changes in the solute-eluent interaction may be
gained.

IV. SIMPLE WORKING MODEL

All the solutes that will be discussed can be classified as low-polarity compounds, singly substituted compounds being not as polar as compounds with either an ether or OH group. Consequently, the useful eluent composition range extends from about 75% MeOH + 25% alkane through 100% MeOH to about 75% MeOH + 25% H_2O. Under these operating conditions the following general observations were made for alkanes and thia-alkanes:

1. The ln k' values of more than 50 representative compounds on an OS phase, plotted against the respective ln k' values on an ODS phase, yield a straight line, as shown in Fig. 1.
2. While the ln k' values on an ODS phase are generally higher, probably due to a higher phase ratio ϕ, the corresponding I_K values on OS and ODS are not distinguishable.

From this we conclude that in the system investigated differences in retention or selectively originate mainly in the mobile phase, while the stationary phase acts more as a nonspecific receptor, much as described in Locke's original RPLC selectivity model [28].

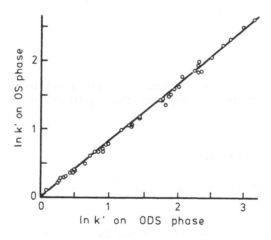

Fig. 1   Comparison of ln k' values of various hydrocarbons and thia substitution products on an OS phase and an ODS phase. Columns, 20 cm × 3 mm Nucleosil 10μ C_8 and 20 × 3 cm MicroPAK CH10 (10μ C_{18}); eluent, 70% MeOH + 30% H_2O; flow rate, 1 ml/min.

As far as solubility in the eluent is concerned, we must consider two opposing effects, which have been treated in detail by Horvath in his fundamental work on retention and solvophobicity [29,30]. First, we must deal with the free energy for the formation of a cavity within the polar eluent to accommodate the solute molecule. This energy, which depends on the molecular surface area SA of the solute molecule and the cohesive energy of the eluent, must be delivered to the system in the dissolution process and is regained in sorption, thus enhancing retention. Second, there is some interaction free energy between the solute residing in the cavity and the surrounding eluent molecules. This energy is gained in the dissolution process and must be resupplied in sorption, thus decreasing retention.

Therefore, for our very simple retention model, we make the following assumptions:

1. Different solute molecules with identical solvophobically effective surface areas SA demand the same cavity formation free energy.
2. Such solute molecules, in sorption at the "stationary phase," which is not specified in detail, produce about the same amount of free energy.
3. Equal surface area solutes can be separated if their interaction energies with the eluent are different (different chemical functionality, solvation states, etc.).
4. Solute molecules with the same functionality can be separated if their molecular surface areas are different (e.g., homologous series).

This model is simplified and should not be applied to other solutes or eluents. However, for our purpose it explains the experimental findings in a satisfactory way.

V. THERMODYNAMIC CONSIDERATIONS

The application of equilibrium thermodynamics in RPLC is doubtful for several reasons:

First, an RPLC system with a bonded alkylsilica phase is subject to more or less spontaneous changes and thus cannot be an equilibrium system [31].

Second, the volumes of all phases involved in the system are essential quantities in thermodynamics. The "volume of the stationary phase," however, is still waiting for a satisfactory definition. It certainly is not a volume in the sense of a three-dimensionally homogeneous liquid [32].

Third, the RPLC retention mechanism is not unambiguously clear, despite the fact that many sophisticated treatments of this subject have been published. It is highly probable that, depending on the eluent water content, several effects contribute to retention, and a dual mechanism has been proposed [50]. Application of classical thermodynamics to a poorly defined process will, at best, be an approximation.

On the other hand, excellent van't Hoff plots are obtained from the temperature dependence of ln k' for a great variety of compounds, in particular for the members of homologous or quasi-homologous series. Values of ΔH derived in this way are of the order of magnitude expected for partition systems.

To circumvent these problems, we will apply some simple formalisms of classical thermodynamics, keeping in mind that the resulting ΔH, ΔG, and ΔS values are not identical to equilibrium values. How large the discrepancy is may depend on the particular system and can hardly be judged at the present status of knowledge.

From the thermodynamic definition of k'

$$\ln k' = \frac{-\Delta G}{RT} + \ln \phi \qquad (6)$$

(where ϕ is the unknown phase ratio) we obtain with Eq. (2) for a homologous series

$$\ln k' = \frac{-\Delta G^*}{RT} - \frac{\Delta \Delta G}{RT} n \ln \phi \qquad (7)$$

where $\Delta \Delta G$ is the sorption "free energy" increment for one propagation step in the series. Since the experimental results show the dependence for the sorption enthalpy in Eq. (8):

$$\Delta H = \Delta H^* + \Delta \Delta H n \qquad (8)$$

one may conclude that Eq. (9) also holds:

$$\ln k' = \frac{-\Delta H^*}{RT} - \frac{\Delta \Delta H}{RT} n + \frac{\Delta S^*}{R} + \frac{\Delta \Delta S}{R} n + \ln \phi \qquad (9)$$

The ΔH terms contain contributions from cavity formation, solute-eluent interaction, and solute-stationary phase interaction. The cavity ΔH contribution is proportional to the solute molecular surface area. The same should hold for the solute-stationary phase interaction. The solute-eluent interaction may comprise Keesom, Debye, and London-type contributions, of which the London forces mostly outweigh the others. On a per mole basis, the London-type interaction again is proportional to the size of the solute molecule. However, in the vicinity of polar methanol and water molecules, local polar centers may be formed around heteroatomic bonds, which might considerably increase the solute-eluent interaction. It will be shown that remarkable ΔI_K values can arise in this way. The entropy terms incorporate a term for mixing (dilution), for ordering of eluent molecules around the cavities, and for the sorbed, most highly ordered state. Absolute values for "ΔS" cannot be given since $\Delta S^*/R + \ln \phi$ cannot be separated as long as ϕ is not defined. This seems to be the weakest point in every "thermodynamic" treatment of RPLC retention.

While the $\Delta\Delta H$, $\Delta\Delta G$, and $\Delta\Delta S$ values are attributed to the addition of a building unit of the homologous series, the starred quantities, e.g., ΔH^*, comprise contributions due to nonlinearity for the first members of a series and to the presence of functional groups [33–35].

VI. ELEMENTAL SULFUR AND POLYSULFANES

A. Sulfur Homocycles

General Remarks

The first sulfur ring separated from reaction mixtures by the RPLC technique was the S_8 [36]. In these experiments, which were planned for the quantitative analysis of S_8 in aqueous suspensions, two small peaks were observed to elute faster than S_8. Later we found that they represented S_6 and S_7 rings. Further investigations [8,11,37,38] revealed the existence of many hitherto unknown sulfur homocycles, all of which could be separated on ODS phases. RPLC is still the best and sometimes the only way to separate complex mixtures of sulfur rings.

Sulfur rings containing other than eight sulfur atoms are encountered more often than commonly thought; however, their presence is not recognized. They are all referred to as "elemental sulfur," which, traditionally, is assumed to be S_8. The physical properties, such as molecular and crystalline structure, color, and density, of the various sulfur homocycles are more or less different [39,40,51–55]. Schmidt and Siebert [40] pointed out that their chemical reactivity may differ by several orders of magnitude. If, for example, a reaction produces "elemental sulfur" consisting mainly of S_6 rings instead of S_8 rings, the chemical behavior of the solution may be different than expected. Therefore, from a chemical point of view, when referring to "elemental sulfur" the ring sizes n_S should be indicated.

The thermodynamically stable molecule S_8 is soluble in most organic solvents [1]. The RP chromatogram of a freshly prepared solution shows the solvent peak and that of S_8, which, with ODS phases and pure MeOH eluent, appears at $I_K = 1240 \pm 10$. After standing for a day or more, the solution always contains S_6 and S_7 and sometimes also S_9 to S_{12}. The mechanism of formation is unknown, although some photochemistry is probably involved [36].

If solutions of S_8 in nonpolar solvents like cyclohexane are ultraviolet-irradiated, larger amounts of all ring sizes from S_6 to about S_{20} are produced [55].

The sulfur formed when sodium thiosulfate solutions are acidified contains more S_6 and S_7 than S_8. Even medium-sized rings ($n \leqslant 15$) are produced with appreciable yields [38].

Sulfur melts contain many different homocycles, some of which have been isolated on a preparative scale for further investigation [39].

One could mention many more examples of the formation of sulfur rings with $n \neq 8$, several of which are of more than purely academic interest. They are, in fact, so common that it is difficult to obtain pure S_8 solutions!

Retention Effects

Figure 2 shows the separation of S_6 to S_{15} with pure MeOH eluent. Table 1 gives typical retention data from three runs. The sample was prepared by photolysis of S_8 in cyclohexane. The solubility of

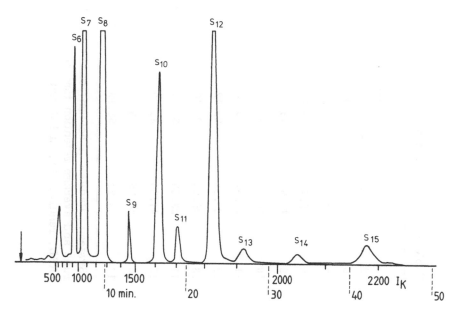

Fig. 2   Chromatogram of sulfur homocycles S_6 to S_{15}. Column, 10 cm × 8 mm RadPAK A (10μ C_{18}); eluent, MeOH; flow rate, 1 ml/min.

Table 1   Typical Retention Data for Sulfur Rings[a]

n_s	$t_{ms}(1)$	$t_{ms}(2)$	$t_{ms}(3)$	$\overline{t_{ms}}$	Percent (RSD)[b]	ln k'	I_K
6	6.37	6.44	6.34	6.383	0.8	0.3837	924
7	7.51	7.59	7.46	7.52	0.9	0.6498	1063
8	9.75	9.72	9.68	9.717	0.4	1.0221	1258
9	12.95	13.07	12.88	12.96	0.7	1.4008	1457
10	16.57	16.72	16.48	16.59	0.7	1.7019	1614
11	19.01	19.16	18.88	19.017	0.7	1.8624	1698
12	23.22	23.42	23.09	23.243	0.7	2.0922	1819
13	26.95	27.16	26.86	26.99	0.6	2.2594	1906
14	33.53	33.77	33.36	33.553	0.6	2.4981	2031
15	41.95	42.20	41.79	41.98	0.5	2.7393	2158
16	53.26	53.61	53.20	53.357	0.4	2.9934	2291
17	65.21	65.89	65.00	65.367	0.7	3.2059	2402

[a]System: RadPAK A column; eluent, MeOH; flow rate, 1 ml/min. Retention times (t_{ms}) for three runs and average retention time are in minutes.
[b]RSD, relative standard deviation.

larger sulfur rings in MeOH is so low that retention times become excessive and detection problems arise. Admixing of an alkane or cycloalkane to the MeOH eluent lowers the polarity and thus decreases sulfur retention. Figure 3 shows a chromatogram of "S_x" from a sulfur melt on an ODS phase with an eluent containing 75% MeOH + 25% cyclohexane. Twenty peaks are seen, which suggest that they originate from the sulfur homocycles S_6 to S_{25}. The chromatogram appears to represent a homologous series, although some slight irregularities disturb the typical peak sequence pattern.

In Figures 4 and 5 the ln k' values derived from the above chromatograms are plotted against the sulfur atom number or ring size n_S. The triangles in Fig. 5 are derived from the already identified and characterized [39] sulfur homocycles S_6, S_7, S_8, S_{10}, S_{12}, and S_{20}, which served as reference compounds. The perfect coincidence of the data for S_x and the reference compounds permits the conclusion that the other S_x data points represent new sulfur rings of the indicated size.

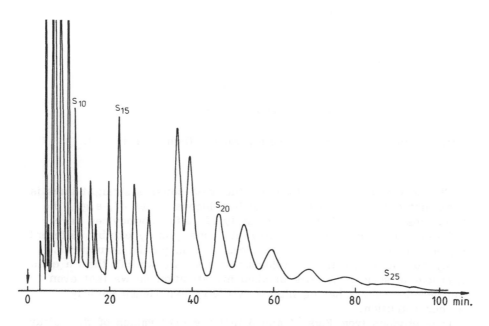

Fig. 3   Chromatogram of sulfur homocycles S_6 to S_{25}. Column, 10 cm × 8 mm RadPAK A; eluent, 75% MeOH + 25% cyclohexane; flow rate, 1 ml/min.   (From Ref. 34.)

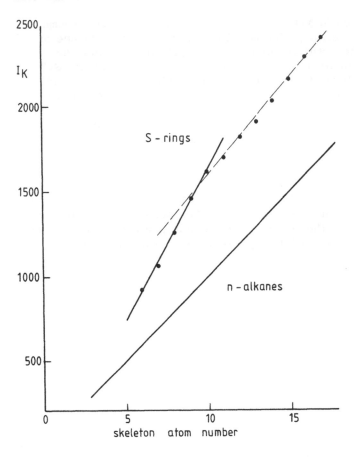

Fig. 4   Plot of ln k' versus sulfur rank.   Data from Fig. 2.

The scatter of data points about the least-squares straight lines is somewhat more pronounced than in homologous series of, e.g., the *n*-alkanes. The members of the family of sulfur homocylces have different structures and symmetry and consequently exhibit different dipole moments or dipole moment increments in partial structures. They do not form a true homologous series, and we call them a quasi-homologous series. A more detailed investigation shows that even numbered members having higher symmetry experience slightly higher retention.

It is obvious from Figs. 4 and 5 that the ln k' values of the sulfur homocycles cannot all be represented by only one linear function of n_S. At least two straight lines, intersecting near S_{10}, are necessary.

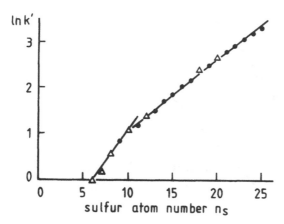

Fig. 5   Plot of ln k' versus sulfur rank.   Data from Fig. 3 [34].

The position of this discontinuity does not change with eluent composition or with bonded phase chain length. Two facts are noted: (a) the values of ln k' of S_6 to S_9 are lower than expected from the straight line of the larger rings, and (b) the slope of the function of the small rings is larger.

The ln k' of sulfur rings on a C_{18} and a C_8 column with pure MeOH eluent can be described [35] by Eqs. (10) to (13):

$$\ln k' = -1.75 + 0.3541 n_S \qquad r = 0.998 \ (S_6 \text{ to } S_{10} \text{ on ODS}) \qquad (10)$$

$$\ln k' = -0.4824 + 0.2236 n_S \qquad r = 0.995 \ (S_{10} \text{ to } S_{17} \text{ on ODS}) \qquad (11)$$

$$\ln k' = -1.7827 + 0.2039 n_S \qquad r = 0.988 \ (S_6 \text{ to } S_{10} \text{ on OS}) \qquad (12)$$

$$\ln k' = -1.1295 + 0.1348 n_S \qquad r = 0.999 \ (S_{10} \text{ to } S_{17} \text{ on OS}) \qquad (13)$$

Inspection of the respective retention indices I_K reveals an unexpected effect, which can be seen from Fig. 6. The I_K values for some sulfur homocycles and *n*-alkanes are shown as a function of the respective skeleton atom number. The I_K values of the sulfur ring are always distinctly greater than those of the corresponding *n*-alkanes, despite the fact that the respective S_n and C_n have similar values for the molecular surface areas.

The S_6 ring has $I_K = 900$ or $\Delta I_K = +300$; S_{10} has $I_K = 1600$ or $\Delta I_K = +600$. The regression for this n_S range yields

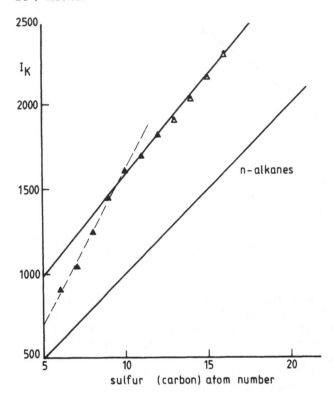

Fig. 6 Retention indices of sulfur rings and *n*-alkanes as a function of skeleton atom number. Column, Chrompack C_{18} glass cartridge; eluent, MeOH; flow rate, 1 ml/min.

$$I_K(S_6-S_{10}) = -156.0 + 177.4n_S \qquad r = 0.998 \qquad (14)$$

Thus the index increase is 177.4 units per sulfur atom. The difference of each respective *n*-alkane increases by $\Delta\Delta I_K = 77$ units on the average.

From S_{10} to the larger rings we find

$$I_K(S_{10}-S_{17}) = +443.3 + 115.0n_S \qquad r = 0.998 \qquad (15)$$

which means that the difference of ΔI_K for the respective *n*-alkane increases by only $\Delta\Delta I_K = 15$ units per sulfur atom.

The ln k' values of sulfur homocycles on ODS phases are greater than those on the OS phase. The I_K values, however, are identical on both phases, which indicates that retention changes with bonded-phase chain length are the same for *n*-alkanes and sulfur rings.

The high retention index of sulfur rings depends strongly on the eluent composition. In RPLC, selectivity with respect to the separation of two solutes can in most cases be increased by adding water to the methanol eluent. The reverse is observed for the separation of a sulfur ring and an *n*-alkane having the same number of skeleton atoms. Figure 7 shows the separation of *n*-hexane and cyclohexasulfur. The respective ln k' values are plotted as function of the eluent water content. In pure MeOH, we find that $\ln k' (S_6) > \ln k' (C_6)$. With increasing eluent water content, $\ln k'(C_6)$ increases faster than $\ln k'(S_6)$. With 72% MeOH + 28% H_2O, the S_6 and *n*-hexane co-elute, and with more water in the eluent $\ln k'(S_6) < \ln k'(C_6)$, so the elution order has reversed.

Analogous observations were made with other sulfur rings and alkanes. Figure 8 shows some examples. It is seen that the indices

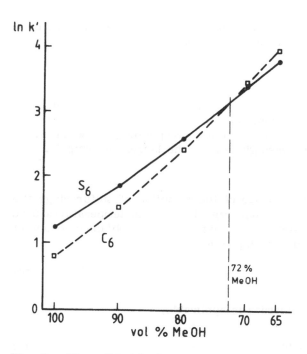

Fig. 7   Plog of ln k' of cyclohexasulfur and *n*-hexane versus MeOH volume percentage in eluent.   Column, 10 cm × 8 mm RadPAK A; flow rate, 1 ml/min.

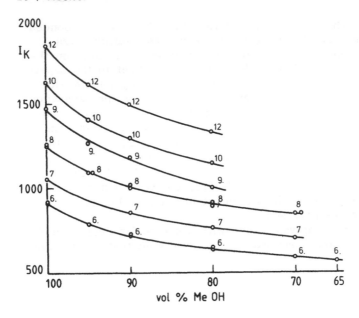

Fig. 8   Retention indices of some sulfur homocycles as function of eluent MeOH volume percentage.   Conditions as in Fig. 7.

of S_6, S_8, S_9, S_{10} and S_{12} decrease with increasing eluent H_2O content.   Apparently the eluent H_2O content necessary to reduce the I_K of a sulfur ring to the value of the corresponding n-alkane, i.e., to $n_S \times 100$, must be increasingly higher the larger the ring is.

Heats of Sorption

From the temperature dependence of the retention of sulfur rings the sorption enthalpies have been determined.   Column temperatures used were 15 to 50°C in 5°C steps.   The experimentally obtained values of ln k' were plotted according to Eq. (16):

$$\ln k' \;=\; \alpha + \beta \frac{1}{T} \tag{16}$$

Since, as far as thermodynamics is applicable,

$$\ln k' \;=\; \frac{-\Delta H}{RT} + \frac{\Delta S}{R} + \ln \phi \tag{17}$$

we obtain the relations

$$\beta = -\frac{\Delta H}{R} \qquad\qquad (18)$$

$$\alpha = \frac{\Delta S}{R} + \ln \phi \qquad\qquad (19)$$

An example is given in Fig. 9, which shows $\ln k'$ of S_8 as a function of $1/T$ for four eluent compositions. We find straight lines with slopes that increase with increasing eluent water content. Table 2 shows the ΔH data for some sulfur rings, together with n-alkane data. The values are averages of two independent data sets, the deviation from the mean being typically 5%. Values for ΔG and ΔS are given only to indicate trends and to compare sulfur rings with alkanes. The figures were calculated using the indicated $\ln \phi$. The phase volume ratio ϕ, in turn, was calculated from the column total liquid volume, the interstitial volume, and the column dead volume [24]. Since this procedure is not generally accepted [32] and the

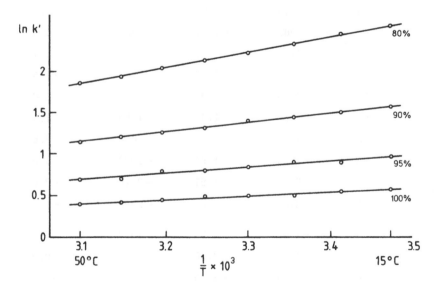

Fig. 9   Linear plot of $\ln k'$ of S_8 versus reciprocal temperature $1/T \times 10^3$. Eluent MeOH content was 100, 95, 90, and 80%; the rest was H_2O. Column, 20 cm × 3 mm Nucleosil 10μ C_{18}.

Table 2 Heats of Sorption and Other Parameters Relevant to Retention of Sulfur Homocycles and n-Alkanes

Solute	MeOH (%)	ΔH (cal/mol)	ln φ	ΔG (cal/mol)	ΔS (cal/mol)
C_5	100	−190	−1.53	−540	1.2
C_6	100	−410	−1.53	−630	0.7
C_9	100	−1090	−1.53	−910	−0.6
C_{10}	100	−1300	−1.53	−1000	−1.0
C_{14}	100	−2170	−1.53	−1360	−2.7
S_6	100	−410	−1.53	−910	1.7
S_8	100	−920	−1.53	−1220	1.0
S_9	100	−1280	−1.53	−1430	0.5
S_{10}	100	−1550	−1.53	−1560	0.0
S_{12}	100	−1850	−1.53	−1770	−0.3
C_5	95	−610	−1.97	−930	1.1
C_6	95	−900	−1.97	−1060	0.5
C_9	95	−1770	−1.97	−1450	−1.1
C_{10}	95	−2060	−1.97	−1580	−1.6
C_{14}	95	−3220	−1.97	−2090	−3.8
S_6	95	−860	−1.97	−1300	1.5
S_8	95	−1510	−1.97	−1690	0.6
S_9	95	−1920	−1.97	−1920	0.0
S_{10}	95	−2280	−1.97	−2110	−0.6
S_{12}	95	−2730	−1.97	−2380	−1.2
C_5	90	−1160	−2.35	−1350	0.6
C_6	90	−1510	−2.35	−1510	0.0
C_9	90	−2570	−2.35	−2010	−1.9
C_{10}	90	−2920	−2.35	−2170	−2.5
C_{14}	90	−4330	−2.35	−2930	−5.0
S_6	90	−1470	−2.35	−1710	0.8

Table 2 (Continued)

Solute	MeOH (%)	ΔH (cal/mol)	$\ln \phi$	ΔG (cal/mol)	ΔS (cal/mol)
S_8	90	−2250	−2.35	−2180	−0.2
S_9	90	−2750	−2.35	−2490	−0.9
S_{10}	90	−3180	−2.35	−2670	−1.7
S_{12}	90	−3690	−2.35	−2990	−2.3
C_5	80	−2510	−3.21	−2350	−0.5
C_6	80	−2930	−3.21	−2590	−1.1
C_9	80	−4200	−3.21	−3310	−3.0
C_{10}	80	−4620	−3.21	−3540	−3.6
C_{14}	80	−6310	−3.21	−4520	−6.0
S_6	80	−2690	−3.21	−2660	−0.1
S_8	80	−3630	−3.21	−3290	−1.2
S_9	80	−4220	−3.21	−3550	−2.2
S_{10}	80	−4740	−3.21	−3900	−2.8
S_{12}	80	−5630	−3.21	−4360	−4.3

actual meaning of ϕ in RPLC is still not clear, the calculation of the sorption entropy according to Eq. (20):

$$\Delta S = \frac{\Delta H + R(\ln k' - \ln \phi)}{T} \tag{20}$$

yields only relative data. Furthermore, it should be kept in mind that the thermodynamic meaning of such data is somewhat obscure since they were obtained from a nonequilibrium system.

Nevertheless, it is seen that sorption entropy becomes more negative with increasing solute molecular size and with increasing eluent water content. For solutes having equal skeleton atom numbers, the sorption entropy of an *n*-alkane is always more negative than that of the respective sulfur rings.

The sorption enthalpies of sulfur rings are more negative than those of the corresponding alkanes. The dependence of ΔH on the molecular size of the solute according to Eq. (8) is shown in Fig. 10.

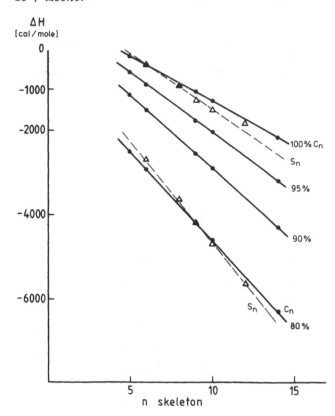

Fig. 10 Sorption of enthalpies ΔH of *n*-alkanes and sulfur rings versus skeleton atom number. Eluent MeOH content as indicated.

The coefficients of Eq. (8) are given in Table 3. The overall sorption enthalpies are low, on the order of a few kilocalories per mole. This is one important reason for the fact that even unstable sulfur compounds can be separated in water-lean RPLC systems. The homolog increments $\Delta\Delta H$ of sulfur atoms are somewhat more negative than those of CH_2 groups.

Correlation Between Solute Size and Retention

The sulfur homocycles are basically nonpolar solutes having similar chemical functionality. The RPLC retention of the members of such a quasi-homologous series is largely governed by the eluent-accessible solute surface area (ASA) [35,43].

Table 3 Coefficients of $\Delta H^* = \Delta\Delta H \cdot n_{skeleton}$ for Sulfur Rings and n-Alkanes

Solutes	MeOH (%)	ΔH^*	$\Delta\Delta H$
C_5-C_{14}	100	907	-220
C_5-C_{14}	95	840	-290
C_5-C_{14}	90	602	-352
C_5-C_{14}	80	-398	-422
S_6-S_{10}	100	1336	-288
S_6-S_{10}	95	1297	-356
S_6-S_{10}	90	1130	-429
S_6-S_{10}	80	423	-514

Surface area values can be calculated by a method developed by Pearlman [41,42]. The total molecular surface areas (TSA) of sulfur rings S_n, given in angstroms squared (Å^2), is

$$TSA(S_n) = -1.95 + 23.76 n_S \qquad r = 0.9998 \; (n_S \geqq 6) \qquad (21)$$

The n-alkane TSA follows Eq. (22):

$$TSA(C_n) = 33.06 + 20.47 n_C \qquad r = 0.999997 \; (n_C > 3) \qquad (22)$$

For $n \leqslant 10.6$ a sulfur ring has a smaller TSA than the respective reference alkane; for $n \geqslant 10.6$ the sulfur ring is larger.

As far as the absolute TSA values are concerned, an important parameter for the computation is the "radius" of the respective atom. The van der Waals radii used for this purpose are not very accurately defined. In particular, not much is known about the dependence of the atomic radius on the bond status. Consequently, the absolute TSA data may include some errors. Changes of TSA in a homologous series, however, are consistent within themselves.

It has been shown that the $\ln k'$ or retention indices I_K of linear and branched alkanes depend linearly on their respective TSA values. An equivalent relation is expected to hold for sulfur homocycles. The experimental results, however, show that there are two

I_K functions, Eqs. (14) and (15), with quite different slopes, while there is only one relation that describes TSA as a function of n_S. We must assume that only for the larger sulfur rings is the total surface area eluent-accessible and effective in producing retention (ASA = TSA). When proceeding from S_{10} to the larger rings, each additional sulfur atom is fully exposed to eluent and contributes a constant surface area increment of 23.76 Å^2. This, in turn, causes a constant ln k' increase.

In the smaller rings, the inner part of the surface is thought to be not accessible to the surrounding eluent (ASA < TSA); thus the "hidden" part does not contribute to retention. This would explain why ln k' values of the smaller rings are lower than expected. The accessible surface area increase per additional atom must be greater in the smaller rings since, as the ring is expanded, more of the inner surface becomes accessible to eluent. This explains the greater slope of the ln k' versus n_S functions for S_6 to S_{10}.

If the I_K data from Fig. 6 are plotted against the respective TSA instead of n_{skeleton}, we get Fig. 11. On this scale, S_6 has $I_K = 910$ while the corresponding hypothetical alkane would have $I_K = 510$. At the intersection point between S_9 and S_{10} the index difference is $\Delta I_K = 1550 - 940 = 610$ units. For the larger rings, ΔI_K remains almost constant; the straight lines for the larger S rings and the alkanes are nearly parallel. The small I_K increase toward the larger rings observed in Fig. 6 is due to the fact that, according to Eqs. (21) and (22), the S rings now have a larger surface area than the corresponding alkanes. The ratio of TSA increases per atom for the larger S rings and the alkanes is expected to be equal to the ratio of the respective I_K increases per atom. From Eqs. (14) and (3) we obtain 1.15 for this ratio and from Eqs. (21) and (22) we find 1.16, which are in good agreement.

The high retention indices of sulfur rings in Fig. 11 seem to be related to the presence of protons in alkanes. As a function of solute surface area, the same index increase has been observed for other proton-free solutes such as selenium rings, sulfur-selenium rings, As_4, and P_4 [44]. Obviously, the presence of protons in the alkane solute permits an additional interaction with the eluent. This stronger interaction is also reflected by the higher solubility of alkanes in methanol: 8.65×10^{-4} mol S_8/liter MeOH at 18.5°C [1] versus 1.12 mol n-octane/liter MeOH at 25°C [43]. It is evident, however, that only a small fraction of these solubility differences is effective in producing retention differences. Since with increasing eluent water content the ln k' values of n-alkanes increase faster than the ln k' values of the sulfur rings, one may conclude that the extra solute-solvent interaction involves mainly the methyl protons of the eluent. The fact that ΔI_K of sulfur homocycles in pure MeOH has a constant value of 610 units suggests that the proton

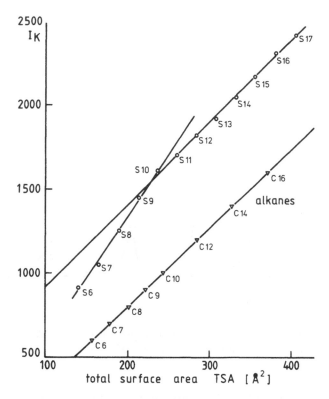

Fig. 11   Retention indices of S rings and *n*-alkanes as a function of the respective molecular surface area TSA.   Replotted from Fig. 6.

density of the solute per accessible surface area rather than the number of protons per solute molecule is responsible for this effect.

B.   Polysulfanes

The polysulfanes H_2S_n, which consist of hydrogen-terminated sulfur chains, are relatively unstable and very sensitive to catalytic activities.   In alcoholic (methanol to pentanol) solutions they can be kept for several hours and separated in RP systems [33].   After standing for about 1 day, they have decomposed to yield H_2S and sulfur rings S_{n-1}.   Figure 12 shows the chromatogram of H_2S_2 to H_2S_{14} in methanolic solution.   The presence of the sulfur homocycles S_6, S_7, and S_8 can also be seen.   If the ln k' values from Fig. 12 are plotted against n_S, we get curve b in Fig. 13.   Evidently the polysulfanes form a perfect homologous series.   After H_2S_n have decomposed to

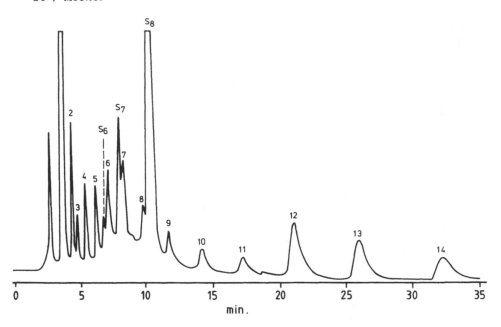

Fig. 12   Chromatogram of polysulfanes H_2S_2 to H_2S_{14} (figures on peaks give n_S).   Conditions as in Fig. 2.   (From Ref. 33.)

S_{n-1}, the ln k' values of the sulfur rings give curve a in Fig. 13. The point corresponding to $n_S = 5$ may result from cyclopentasulfur, a hitherto unknown compound which might have been formed according to $H_2S_6 \rightarrow H_2S + S_5$. From Fig. 13 it is seen that the ln k' (H_2S_n) values are similar to the ln k' (S_n) values and greater than the values for the n-alkanes (curve c). The slope of the H_2S_n curve is close to that of the sulfur rings with $n_S > 10$. At $n_S = 7$ curves a and b intersect; for $n_S < 7$ the ln k' values for the sulfur rings are lower. On addition of water to the methanol eluent, the retention index of polysulfanes decreases in almost the same way as that of sulfur homocycles.

From the above observations it must be concluded that the presence of two terminal hydrogen atoms has only a small effect on lowering retention. The lower ln k' for $n_S < 7$ supports the idea that part of the surface area of these smaller sulfur rings is not effective in producing retention. The surface of the chainlike H_2S_n is fully exposed to the eluent, and the corresponding ln k' function is linear down to $n_S = 2$.

In both polysulfane chains and sulfur homocycles, a sulfur atom has greater retention than a methylene group in n-alkanes or cyclo-alkanes. The retention difference between sulfur solutes and

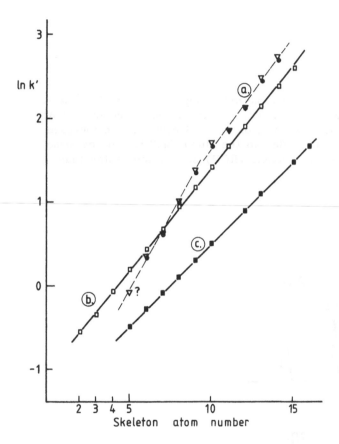

Fig. 13   Plot of ln k' of polysulfanes (b), sulfur homocycles (a),
and *n*-alkanes (c) versus skeleton atom number.   On trace a
triangles are derived from the decomposition products of polysulfanes.
Solid dots are for authentic sulfur ring mixture.   Conditions as in
Fig. 2.

hydrocarbon solutes of the same surface area is greatest in a purely organic eluent. It is strongly reduced by the presence of water in the eluent. Proton-free or almost proton-free solutes must be regarded as the most solvophobic solutes in pure methanol.

VII. THIOLS

A. Monothiols

n-Alkanethiols

Substitution of CH_2 by S in a methyl group leads to a thiol, which is the sulfur analog of an alcohol. Figure 14 shows chromatograms of n-alkanethiols and n-alkanols on an ODS column. To increase resolution, 30% H_2O was added to the eluent MeOH. In the same chromatographic system, alcohols (ROH) elute much faster than

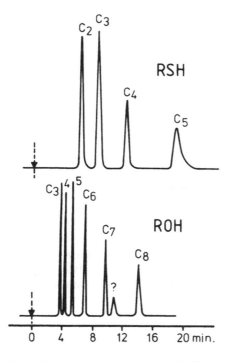

Fig. 14 Chromatograms of alkanethiols and alkanols. Column, Rad-PAK A; eluent 70% MeOH + 30% H_2O; flow rate, 1 ml/min. (From Ref. 35.)

thiols (RSH). The low retention of alcohols is due to their much stronger interaction with the eluent via hydrogen bridge formation. Figures 15 and 16 illustrate the dependence on eluent composition of parameters a and b, respectively, of the ln k' versus n_C relation. Both parameters increase with increasing eluent H_2O content, indicating strengthening of solvophobic forces. It should be noted that b(RSH) is slightly lower than b(alkane).

In Table 4, the I_K and ΔI_K values for several thiols and alcohols are given for a 70% MeOH + 30% H_2O eluent. Thiol formation causes an index decrease of 200 units on the average, replacement by OH a loss of 500 units. From the data it is evident that ΔI_K depends linearly on the chain length of R. In terms of ln k' parameters a and b for solute A that is a member of a homologous series [ln k'(A) = a(A) + b(A)n(A)] and the n-alkanes, the retention loss is expressed in Eq. (23):

$$\Delta I_K(A) = \frac{100}{b(\text{alk})} \{a(A) - a(\text{alk}) + n[b(a) - b(\text{alk})]\} \qquad (23)$$

That is, ΔI_K consists of a constant term which is characteristic for the type of substitution and a chain length-dependent term which reflects diminished retention of the methylene groups in the R chain of the substitution product.

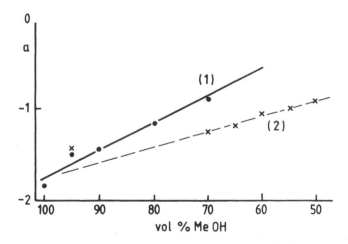

Fig. 15   Eluent composition dependence of intercept (a) in ln k' = a + bn for n-alkanes (1) and n-alkanethiols (2). Column, RadPAK A.

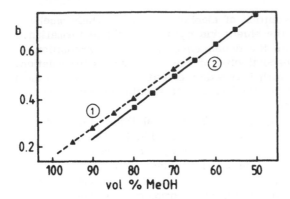

Fig. 16 Eluent composition dependence of slope (b) (methylene group contribution) in $\ln k' = a + bn$ for n-alkanes (1) and n-alkanethiols (2). Columns, RadPAK A.

Table 4 Retention Indices I_K and Retention Decrease ΔI_K of some Thiols and Alcohols[a]

Solute	I_K	ΔI_K
Et-SH	120	−180
Pr-SH	217	−183
Bu-SH	312	−187
Pent-SH	407	−193
Hex-SH	504	−196
Hept-SH	599	−201
Oct-SH	694	−206
Pr-OH	−110	−510
Bu-OH	−13	−513
Pent-OH	85	−515
Hex-OH	181	−519

[a]Eluent, methanol-water (80:30); column, 20 cm × 4 mm i.d. Nucleosil 5C$_{18}$; temperature, 25°C.
Source: Ref. 35.

The retention decrease of thiols under the conditions given in Table 4 is

$$\Delta I_K(RSH) = -170 - 4.4n_C \tag{24}$$

and that of alcohols is

$$\Delta I_K(ROH) = -501 - 2.9n_C \tag{25}$$

Apparently, the retention-lowering effect of alcohol formation is three times stronger than that of thiol formation. The ability of alcohols to form hydrogen bonds to both eluent components, MeOH and H_2O, is missing in the SH group of thiols. Dipole moments of thiols are too low to be responsible for $\Delta I_K = -170$. The retention-lowering effect of thia substitution is invariably observed if the solute molecule contains a sulfur atom in the immediate vicinity of alkyl protons. It has been postulated [45] that around such sites a local

$$CH_3-(CH_2)_{n-2}-CH_2-SH \tag{26}$$

polar center is formed, promoted by the influence of the surrounding highly polar eluent, which can be effectively solvated by the latter. The polar center formation can be thought of in terms of the electron donating property of alkyl groups, the ability of sulfur to accommodate an increased electron density, and the polarizing effect of methanol molecules. This could lead to the (at present entirely hypothetical) structure of the solvated group shown below:

Since the polar center and consequently also the solvation shell are located mainly at the $-CH_2-SH$ arrangement, we will call them "solvation patches." These solvation patches would permit an increased solute-solvent interaction and exclude the covered part of the solute molecular surface from solvophobic expulsion forces. Furthermore, the solvation patches might present an obstacle to attachment of the solute at the stationary bonded phase.

As seen from Eqs. (24) and (25), the retention-decreasing effect of thia substitution is not confined to the site of the local polar center. The retention contribution of the remaining CH_2 groups is also lowered by a small but clearly evident amount. The reason for this may be twofold. First, there may be a polarizing effect along the alkane chain which causes a slight increase in CH_2-eluent interactions. Second, it is highly probable that the orientation of eluent molecules, as found, for example, around alkanes, will be disturbed over a long range by the presence of the solvation patch. Since the ΔI_K values are strictly a linear function of n_C [Eq. (24)], the proposed effects seem to influence equally all carbon atoms of the chain.

Other Thiols

The retention-lowering effect of thia substitution is not restricted to chainlike alkanes. In 70% MeOH + 30% H_2O, cyclohexane has $I_K = 517$, methylcyclohexane $I_K = 600$, and cyclohexanethiol $I_K = 405$. If calculated from Eq. (24), $I_K = 404$ is predicted for cyclohexanethiol.

The retention differences between isomeric thiols are fairly small, as can be seen from Table 5. The normal thiol always has the highest retention. Since branching of the solute molecule lowers its

Table 5 Retention of Isomeric Thiols[a]

Solute	ln k'	I_K
n-C_3SH	0.8005	200
iso-C_3SH	0.6896	181
n-C_4SH	1.3678	298
sec-C_4SH	1.2622	280
iso-C_4SH	1.3155	289
$tert$-C_4SH	1.0814	248

[a]Column, 10 cm × 3 mm i.d. CP C_{18}; eluent, methanol-water (70:30).
Source: Ref. 35.

surface area, the retention decrease from n-C_3SH to *iso*-C_3SH or from n-C_4SH to *iso*-C_4SH and *tert*-C_4SH is not surprising. The lower retention of *sec*-C_4SH compared to n-C_4SH may be due to the slightly different alkyl environment that the SH group experiences.

When attached to a benzene ring, the SH group behaves differently. In 70% MeOH + 30% H_2O, benzene has I_K = 245, toluene I_K = 325, and thiophenol I_K = 241. Here, thia substitution causes the ΔI_K value to be only -84, which can be attributed mostly to the more acidic properties of the thiophenol SH proton. In the benzyl mercaptan (I_K = 275) derived from ethylbenzene (I_K = 405), the retention loss is still lower than in alkenethiols.

B. Dithiols

Data are available on $(1,\omega)$-alkanedithiols from C_2 to C_9, separated on an ODS column with 95% MeOH + 5% H_2O eluent. Figure 17 shows the $\ln k'$ values of alkanes, thiols, and dithiols in that system as a function of skeleton atom number. Evidently, substitution of both

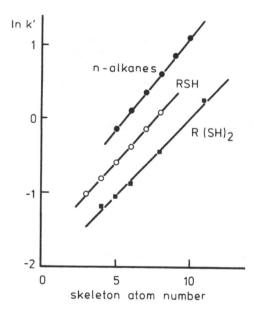

Fig. 17 Comparison of n-alkanes, n-alkanethiols, and $(1,\omega)$-alkanedithiols ($\ln k'$ versus skeleton atom number). Column, RadPAK 5μ C_{18}; eluent, 95% MeOH + 5% H_2O; flow rate, 1 ml/min. (From Ref. 35.)

of the terminal methyl by SH groups lowers retention twice as much
as monosubstitution does.

The corresponding I_K equations are

$$I_K[R(SH)_2] = -115 + 83.3n_C \tag{27}$$

$$I_K(RSH) = -38 + 88.5n_C \tag{28}$$

The index contribution of CH_2 in dithiols has decreased to 83.3 and
that in thiols to 88.5 units. The retention-lowering influence on the
alkyl chain is enhanced by the presence of a second SH group.

The value of 88.5 index units in thiols may be compared with 95.6
observed in 70% MeOH + 30% H_2O. The data indicate that in water-
rich eluents the methylene contributions approach the alkane value
of 100.

VIII. THIOETHERS

A. Monosulfides

Alkane Derivatives

Among dialkyl thioethers, which have the general structure R—S—R",
the symmetrical thioethers R_2S are a special case. Retention data
are available for several chromatographic RP systems [46].

In a 95% MeOH + 5% H_2O eluent on an ODS stationary phase, the
ln k' equation for the symmetrical R_2S is

$$\ln k' (R_2S) = -1.4689 + 0.2108n_C \qquad r = 0.9994 \tag{29}$$

The retention index follows Eq. (30):

$$I_K(R_2S) = -41.4 + 86.0n_C \tag{30}$$

For the nonsymmetrical methylthioalkanes MeSR'

$$\ln k' (MeSR') = -1.5009 + 0.2198n_C \qquad r = 0.9998 \tag{31}$$

and

$$I_K(MeSR') = -54.5 + 89.6n_C \tag{32}$$

were observed.

The retention of the thioethers may be compared with data for the
oxygen ethers. Since the latter exhibit low retention eluents with

higher water content are adequate. Some pertinent data are collected
in Table 6. For comparison, data for *n*-alkanes, alcohols, and thiols
are also included.

A plot of ΔI_K against skeleton atom number is shown in Fig. 18.
The smallest retention decrease is found for thiols, and the dependence of ΔI_K on chain length n_S is very weak in this water-rich
eluent. This is confirmed by the respective data in Table 6, which
show b = 0.6442 for alkanes and b = 0.6346 for thiols. The retention loss is greater if thia substitution leads to the thioether R_2S
instead of the thiol having the same n_C. The chain length dependence of R_2S is clearly more pronounced.

In the nonsymmetrical MeSR' the ΔI_K values are somewhat less
negative than those of R_2S and the chain length dependence is almost
the same. The retention loss is much stronger in ethers and alcohols
being about $\Delta I_K = -400$ for ROR and about -500 for ROH. While
the strongly negative ΔI_K of the oxygen compounds is certainly related to their ability to form hydrogen bonds, the retention loss in
the sulfides must be explained by the formation of local polar centers as in the case of thiols. The center in the symmetrical R_2S

$$CH_3-CH_2 \cdots \cdots CH_2-S-CH_2 \cdots \cdots \cdots CH_2-CH_3 \tag{33}$$

covers two to three skeleton atoms, and ΔI_K ranges between -240
and -300.

In MeSR' the center is thought to be somewhat hindered for expansion

$$CH_3-CH_2 \cdots \cdots \cdots CH_2-S-CH_3 \tag{34}$$

so that the retention loss is reduced.

From the temperature dependence of ln k', some sorption enthalpies,
ΔH, have been determined and are shown in Fig. 19 as a function of
$n_{skeleton}$ for a 70% MeOH + 30% H_2O eluent. The ΔH increase is
negative and linear with the chain length of R. In the range under
investigation, sorption enthalpies are all between -1 and -5 kcal/
mol. The $\Delta H(RSR)$ and $\Delta H(RSH)$ values are more negative than
$\Delta H(ROR)$ and $\Delta H(ROH)$.

The dependence of the ln k' values of thioethers on eluent composition is illustrated in Fig. 20, using *n*-butyl sulfide as an example.
From pure MeOH to 90% MeOH + 10% H_2O the relation is curved; for
higher water percentages it becomes linear. From the same figure
it is seen that in the whole range of eluent compositions the retention
index remains constant at $I_K = 623 \pm 7$ units.

Table 6 Coefficients of $\ln k' = a + bnC$ and $I_K = A + BnC$ for n-Alkanes and Some Monosubstitution Products[a]

Solutes	MeOH (%)	a	b	r[b]	A	B
n-Alkanes	60	−0.1995	0.5885	0.9999991	(0)	(100)
R_2S	60	−0.9338	0.5393	0.9991	−125.4	92.1
R_2O	60	−0.9547	0.5705	0.9997	−299.8	97.4
MeSR'	60	−9.9019	0.5550	0.997	−120.0	94.8
RSH	60	−0.7428	0.5740	0.99999	−92.8	98.0
ROH	60	−0.4478	0.5543	0.99998	−384.0	94.7
n-Alkanes	55	−0.0526	0.6442	0.99999	(0)	(100)
R_2S	55	−0.8767	0.5979	0.999	−127.9	92.8
R_2O	55	−1.9793	0.6323	0.9998	−299.1	98.2
MeSR'	55	−0.8335	0.6132	0.994	−121.2	95.2
RSH	55	−0.6484	0.6346	0.99997	−92.5	98.5
ROH	55	−2.3512	0.6024	0.999997	−356.8	93.5

[a]Eluents as indicated; column, 12.5 cm × 4 mm i.d. Nucleosil 5C18; temperature, 50°C.
[b]Correlation coefficient r for $\ln k'$ least-squares fit.

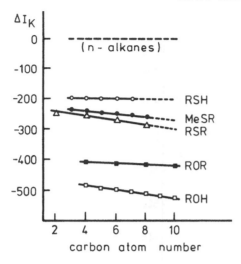

Fig. 18 Retention index decrease of thiols, alcohols, thioethers, and ethers (ΔI_K versus carbon atom number). Column, 12.5 cm × 3 mm Nucleosil 5μm C_{18}. Eluent, 55% MeOH + 45% H_2O; flow rate: 1 ml/min. (From Ref. 35.)

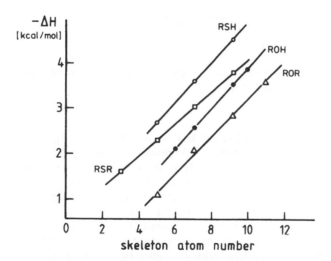

Fig. 19 Sorption enthalpies of some thiols, alcohols, thioethers, and ethers as a function of skeleton atom number. Column, 10 cm Nucleosil 5μm C_{18}; eluent, 70% MeOH + 30% H_2O; flow rate, 1 ml/min.

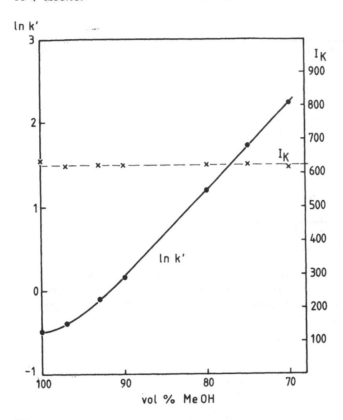

Fig. 20   Dibutyl thioether: ln k' and I_K versus eluent composition.
Column, 20 cm × 3 mm Nucleosil 5μm C_{18}.

The influence of eluent composition on the index increment $I_K(CH_2)$ of methylene groups in some substituted alkanes is shown in Table 7.   Although there is some data scatter, it is seen that the methylene group contribution increases when the eluent methanol content is lowered.   This seems to indicate that the solvation patches responsible for the low $I_K(CH_2)$ in water-lean eluents consist mainly or completely of MeOH molecules.   It can be assumed that the solvation patches are removed when the eluent MeOH content vanishes.   The thia-alkanes should then behave like alkanes, and $I_K(CH_2)$ should eventually reach the value 100.

In cycloalkane solutes, the thioether function lowers retention much as in n-alkyl sulfides.   With a 100% MeOH eluent, bicyclohexyl has I_K = 1182, only slightly below the n-dodecane value.   Dicyclo-hexyl thioether has I_K = 940 or ΔI_K = −360.   The I_K of the parent

Table 7 Influence of Thia Substitution and
Eluent Composition on Methylene Retention
Index Increment $I_K(CH_2)$[a]

Solute class	$I_K(CH_2)$ for MeOH			
	95%	70%	60%	55%
MeSR	89.6	93.4	94.8	95.2
RSH	88.5	97.2	98.0	98.5
R_2S	86	93.1	92.1	92.8
$R(SH)_2$	83.3	—	—	—

[a]System 10 cm × 8 mm RadPAK 5µm C_{18};
temperature, 25°C.

hydrocarbon, dicyclohexylmethane, is not available, but a good approximation is the mean of the I_K values of bicyclohexyl and 1,2,-di-
cyclohexylethane (I_K = 1350). Thus, the actual retention loss is
940 − 1265 = −325 units, as expected from Fig. 18 for n_C = 13.

Thia substitution in the cyclopentane ring (I_K = 431 for 70% MeOH +
30% H_2O eluent) leads to tetrahydrothiophene, which has I_K = 201.
The actual retention loss is 230 units, again well within the range
expected for a molecule of that size.

It also appears that the formation of local polar centers, buildup
of solvation patches, and, in turn, lowering of retention in thioethers
does not depend on a linear or cyclic structure of the alkyl groups.

Sulfur Atoms Between Aromatic Systems

For convenience, all the available data on thia compounds containing
aromatic groups are presented in Table 8. Data were obtained with
a 70% MeOH + 30% H_2O eluent.

Replacing the CH_2 group in ethylbenzene by sulfur leads to ΔI_K =
−79 units, which is about half the value expected for an aliphatic
C_8 thioether. When the CH_2 of diphenylmethane is thia-substituted,
the sulfur atom is bonded to two aromatic rings. The index change
now turns out to be positive ΔI_K = +30. The index difference between diphenyl disulfide and 1,2-diphenylethane is ΔI_K = +50. Comparable positive index changes are found in going from fluorene to
dibenzothiophene, or from dihydroanthracene to thianthrene.

Obviously, solvation does not take place at sulfur atoms between
benzene rings. Since there have been no detailed investigations so
far, only one or two comments can be made on this phenomenon.

Table 8 Retention of Aromatic Sulfur Compounds[a]

Solute	t_{ms}	ln k'	I_K	ΔI_K
Benzene	5.44	0.5048	245	−355
Toluene	7.09	0.9050	325	−375
Thiophenol	5.36	0.4801	241	−459
Ethylbenzene	9.55	1.3052	405	−395
Methylphenyl sulfide	7.11	0.9089	326	−474
Benzyl mercaptan	5.97	0.6512	275	−525
Biphenyl	14.42	1.8078	505	−695
Diphenylmethane	14.94	1.8491	513	−787
Diphenyl sulfide	17.02	1.9992	543	−757
Diphenylethane	22.33	2.3035	604	−796
Diphenyl disulfide	26.93	2.5083	644	−756
Fluorene	18.94	2.1202	567	−733
Dibenzothiophene	21.40	2.2564	594	−706
Anthracene	25.26	2.4387	630	−770
Dihydroanthracene	20.04	2.1834	580	−820
Thianthrene	23.34	2.3522	613	−787

[a]Eluent, methanol-water (70:30); column, 30 cm × 3 mm i.d. Micro-PAK CH10; temperature, 24°C.
Source: Ref. 35.

The most probable explanation is that the electron-withdrawing aromatic systems lower the electron density at the sulfur quite effectively, thus inhibiting the interaction with the eluent methanol. The observation that the methylene groups in diphenylmethane and diphenylethane show an extraordinarily low retention contribution points in the same direction, the only difference being that here electron withdrawal activates the methylene hydrogen to increase interaction with the eluent.

B. Multisulfides

Multisulfides [47] have the general formula R'−(S−R")n−R and contain at least two sulfur atoms, which are separated by at least one methylene group. There are also cyclic mutlisulfides. The ring is closed at R' and R in the above formula.

Retention data for some multisulfides are presented in Table 9. The 2,(ω-1)-dithia-alkanes can be regarded as a homologous series with the general formula $CH_3-S-(CH_2)_n-S-CH_3$. As a function of the number $n(CH_2)$ of methylene groups separating the sulfur atoms, I_K is given by

$$I_K = 73.6 + 69.4n(CH_2) \qquad r = 0.999 \qquad (35)$$

The methylene group index contribution is very low. Consequently, the retention decrease relative to the corresponding *n*-alkane has an unusually strong chain length dependence:

$$\Delta I_K = -326.4 + 30.6n(CH_2) \qquad (36)$$

A closer inspection of the data shows that the dependence described in Eq. (35) is not perfectly linear. If the dimethyl disulfide, with no methylene groups between the sulfur atoms, is included, the curve is strongly bent in the region n = 0 to n = 2. For Me_2S_2, an $I_K = 73.6$ is extropololated from Eq. (35). Experimentally, $I_K = 153$ was found.

The strong retention loss found in dithia-alkanes is attributed to the presence of two solvation patches [Eq. (37)]

$$CH_3-S-CH_2- \cdot \cdot \cdot \cdot \cdot -CH_2-S-CH_3 \qquad (37)$$

The nonlinearity of I_K for the small members of the series is thought to be due to some interference of solvation patches with each other.

The influence of the site of substitution can be studied nicely with the dithianonanes. In 2,8-dithianonane

$$CH_3-S-CH_2-CH_2-CH_2-CH_2-CH_2-S-CH_3 \qquad (I_K = 426)$$
$$(38)$$

the local polar centers are far apart and completely isolated. The retention loss $\Delta I_K = -474$ is exactly double the loss observed in methylheptyl sulfide.

In 3,7-dithianonane

$$CH_3-CH_2-S-CH_2-CH_2-CH_2-S-CH_2-CH_3 \qquad (I_K = 429) \quad (39)$$

Table 9   Retention Data for Multisulfides on ODS

Solute	MeOH (%)	ln k'	I_K	ΔI_K
2,5-DT-6	95	-0.7843	215	-385
2,6-DT-7	95	-0.6202	282	-418
2,7-DT-8	95	-0.4731	343	-457
3,6-DT-8	95	-0.4112	368	-432
2,8-DT-9	95	-0.2702	426	-474
3,7-DT-9	95	-0.2618	429	-471
4,8-DT-9	95	-1235	489	-414
2,9-DT-10	95	-0.1122	490	-510
3,8-DT-10	95	-0.0545	514	-486
4,7-DT-10	95	-0.0187	544	-456
4,9-DT-12	95	-0.3073	662	-538
2,5-8-TT-9	95	-0.5760	301	-599
1,3-DT-cyclopentane	70	0.0880	163	-337
1,3-DT-2-Me-cyclopentane	70	0.4299	231	-369
2,4-DT-1-Me-cyclopentane	70	0.4522	235	-365
2,5-DT-1,1-DMe-cyclopentane	70	0.7161	288	-412
2,5-DT-1,3-DMe-cyclopentane	70	0.8062	306	-394
1,3-Dithiane	70	0.0835	162	-438
1,4-Dithiane	70	0.2659	198	-402

Source: Data from Refs. 35 and 48.
Abbreviations: DT = dithia, TT = trithia, DMe = dimethyl.

the centers are still far enough apart not to interfere. The retention index remains almost unchanged.

In 4,6-dithianonane

$$CH_3-CH_2-CH_2-S-CH_2-S-CH_2-CH_2-CH_3 \qquad (I_K = 486)$$

$$(40)$$

where solvation patches can no longer fully develop, the retention index increases again.

In 4,5-dithianonane, which is actually a disulfide, this trend is continued:

$$CH_3-CH_2-CH_2-CH_2-S-S-CH_2-CH_2-CH_3 \qquad (I_K = 615)$$

$$(41)$$

The solvation patches can extend in only one direction along the chain. Now $\Delta I_K = -285$ corresponds to that of a thioether.

Basically the same observations are made with dithiadecanes, where the 2,9-isomer has $I_K = 490$ and the dibutyl disulfide has $I_K = 721$.

The 2,5,8-trithianonane

$$CH_3-S-CH_3-CH_2-S-CH_2-CH_2-S-CH_3 \qquad (I_K = 301) \qquad (42)$$

is a good example of the effectiveness of solvation in lowering retention. Since it is fully solvated, this molecule, which is the size of *n*-nonane, shows a retention corresponding to that of propane.

The retention effects in cyclic multisulfides are similar. The index of 1,3-dithiacyclopentane is lower than that of cylcopentane by 268 units. It is interesting to note that the addition of a methyl group, yielding 2,4-dithia-1-methylcyclopentane produces an index increase of only 68 units, whereas the I_K difference between methylcyclopentane and pentane is 97 units, which is very close to 100. The methyl group in the dithia compound seems to be partly involved in the solvation patch formation.

A more detailed study of thia- and selena-substituted cycloalkanes has been presented by Steudel and Strauss [48]. Some of their data have been converted to retention indices and are presented in Table 10. The numbers inside the rings are the I_K values and the numbers under the rings are the index differences between the respective solutes and the parent compound, cyclohexane. The

Table 10 Cyclohexane and Thiacyclohexanes[a]

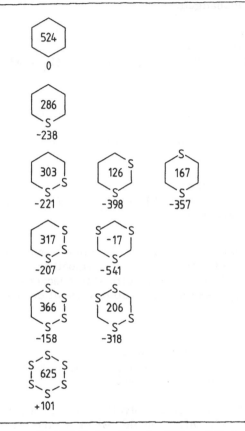

[a]Column: RadPAK A; eluent, 80% MeOH + 20% H$_2$O; flow rate, 1 ml/min.
Source: Data from Ref. 48.

introduction of the first sulfur atom is accompanied by $\Delta I_K = -238$. If a second sulfur is introduced vicinal to the first, the index is slightly increased again to 303. Placed in the 3 or 4 position, the second sulfur causes another retention loss to 126 or 167, respectively. The lowest retention is found for 1,3,5-trithian, with $I_K = -17$. Obviously, this molecule is completely solvated and shows a fairly strong interaction with the eluent (80% MeOH + 20% H$_2$O). The high eluent water content explains the relatively low I_K of cyclohexasulfur in this system.

IX. POLYSULFIDES

A. Aliphatic Polysulfides

Retention Phenomena

The solutes discussed in this section have the general formula
$R'—S_n—R''$ with $n \geq 2$. They contain at least two vicinal sulfur
atoms. It is generally assumed that polysulfide chains are not
branched, and the RPLC results also support nonbranched sulfur
chains. The longest sulfur chain identified by RPLC was found in
$nC_3—S_{21}—nC_3$ [49], but finding even longer chains seems to be
only a matter of patience in waiting for the elution of the next
longer chain.

Polysulfides are frequently formed in the oxidative degradation of
thiol solutions and other sulfur-containing materials. Organic solutions or suspensions of elemental sulfur yield a wide variety of polysulfides under the influence of light or heat. There are many places
where polysulfides are found.

Figure 21 shows the separation of diethyl polysulfides Et_2S_n from
$n = 2$ to $n = 12$ with 100% MeOH on ODS. The peak sequence is
strongly similar to the peak sequence of a homologous series. The
peak which is indicated as S_8 does not fit into the series and is
caused by cyclooctasulfur. This contamination originates from the
sample preparation process rather than decomposition on the column.
The single polysulfides can be isolated in high purity via preparative
RPLC. Samples composed of the pure components never contain
S_8 or any other sulfur ring.

In Fig. 22 the I_K values derived from Fig. 21 are plotted against
the sulfur chain length n_S as well as the skeleton atom numbers
$n_{skeleton} = n_S + 4$. Obviously, the diethyl polysulfides form a
homologous series. Up to the pentasulfide, the data points scatter
around the regression line slightly more than usual, the I_K values
of Et_2S_2 and Et_2S_4 being low and that of Et_2S_3 being high. This
effect is related to the alternating permanent dipole moment of polysulfides, which is high for the even- and low for the odd-n_S chains
[46]. In the longer sulfur chains, the alternations vanish, probably
because of the helical shape of sulfur chains.

The I_K equation for the above system is

$$I_K(Et_2S_n) = 49.1 + 136.2n_S \qquad r = 0.9999 \tag{43}$$

or, using the skeleton atom number,

$$I_K(Et_2S_n) = -495.8 + 136.2n_{skeleton} \tag{44}$$

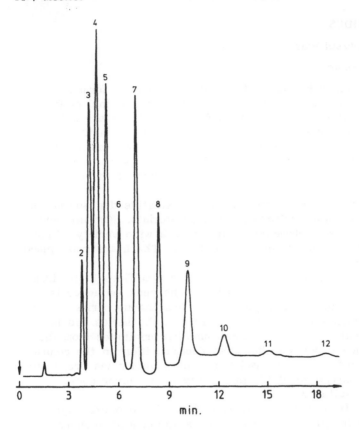

Fig. 21   Chromatogram of diethyl polysulfides Et_2S_n.   Sensitivity ×5 after the Et_2S_6 peak.   Conditions as in Fig. 2.

Each sulfur atom contributes 136.2 units to the retention index.   With small n_S, the I_K of the polysulfides is much lower than the I_K of the alkanes, due to the presence of two solvation patches.

$$CH_3-CH_2-S-S \cdot \cdot \cdot \cdot \cdot S-S-CH_2-CH_3 \qquad (45)$$

Since the I_K values of the polysulfides grows 1.36 times faster than those of the alkanes, there is an intersection between skeleton atom numbers 13 and 14, which in this case corresponds to $n_S = 9$ and $n_S = 10$.

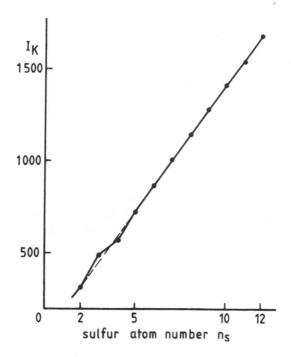

Fig. 22   Plot of I_K versus sulfur atom number n_S.   (Data from Fig. 21.)

The constant 49.1 in Eq. (43) must consist of the (positive) contribution from the ethyl group and the (negative) contribution from the solvation patches around the CH_2—S bond.   We can think of the index as being composed of "increments," which will be abbreviated δI_K.   In terms of increments, the I_K of diethylpolysulfides is

$$I_K(Et_2S_n) = 2\delta I_K(Et) + n_S \delta I_K(S) + 2\delta I_K(Et-S) \tag{46}$$

The last term is the bond increment just mentioned.   Since Eqs. (43) and (46) must be equal, we find, with an approximate value of 190 for $\delta I_K(Et)$

$$49.1 = 380 + 2\delta I_K(Et-S) \tag{47}$$

or

$$\delta I_K(Et-S) = -165$$

The influence of terminal alkyl residues on retention may be seen from Fig. 23, which shows the dicyclohexyl polysulfides prepared by γ-irradiation of a solution of S_8 in cyclohexane. Because of the larger terminal groups, the retention times are clearly longer than with the diethyl compounds. The I_K equation for dicyclohexyl polysulfides is

$$I_K(cycloC_6)_2S_n) = 872.7 + 129.7n_S \qquad r = 0.9997 \qquad (48)$$

Obviously, the index contribution of sulfur atoms in polysulfides is not constant, but depends on the size of the terminal alkyl or cycloalkyl group. This effect has been investigated with a larger set of dialkyl polysulfides on an ODS phase with 95% MeOH + 5% H_2O [45] and with pure MeOH. In Table 11 the indices found for 95% MeOH are given, together with calculated values. The latter were obtained

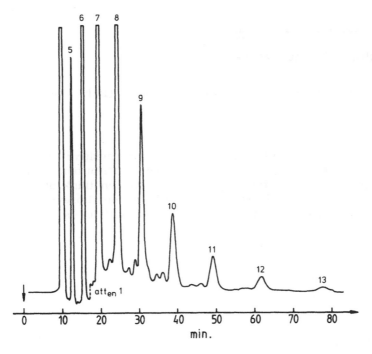

Fig. 23 Chromatogram of dicyclohexyl polysulfides. Attenuation from 4× to 1× as indicated. The figures at the peaks give the sulfur atom number. $(CHex)_2S_3$ to $(CHex)_2S_4$ and S_8 are not resolved. Small peaks are not identified.

Table 11   Observed and Calculated Retention
Indices with Relative Percent Deviation of
Polysulfides

Solute	I_K(obs)	I_K(calc)	Error (%)
Me_2S_2	153	187.4	−18.4
Et_2S_2	338	359.3	−5.9
Pr_2S_2	529	531.3	−0.4
Bu_2S_2	721	703.3	2.5
Pe_2S_2	899	875.3	2.7
Hx_2S_2	1077	1047.3	2.8
Me_2S_3	312	308.5	1.1
Et_2S_3	477	480.5	−0.7
Pr_2S_3	670	652.5	2.7
Bu_2S_3	848	824.5	2.9
Pe_2S_3	1025	996.5	2.9
Hx_2S_3	1206	1168.4	3.2
Me_2S_4	422	429.7	−1.8
Et_2S_4	582	601.7	−3.3
Pr_2S_4	763	773.7	−1.4
Bu_2S_4	935	945.7	−1.1
Pe_2S_4	1106	1117.6	−1.0
Hx_2S_4	1283	1289.6	−0.5
Me_2S_5	552	550.9	0.2
Et_2S_5	713	722.9	−1.4
Pr_2S_5	888	894.9	−0.8
Bu_2S_5	1056	1066.8	−1.0
Pe_2S_5	1225	1238.8	−1.1
Hx_2S_5	1400	1410.8	−0.8
Me_2S_6	679	672.1	1.0
Et_2S_6	839	844.0	−0.6
Pr_2S_6	1010	1016.0	−0.6

Table 11   (Continued)

Solute	I_K(obs)	I_K(Calc)	Error (%)
Bu_2S_6	1178	1188.0	−0.8
Pe_2S_6	1345	1360.0	−1.1
Hx_2S_6	1516	1532.0	−1.0
Me_2S_7	807	793.2	1.7
Et_2S_7	967	965.2	0.2
Pr_2S_7	1135	1137.2	−0.1
Bu_2S_7	1301	1309.2	−0.6
Pe_2S_7	1466	1481.2	−1.0
Hx_2S_7	1637	1654.1	−1.0
Me_2S_8	934	914.4	2.1
Pr_2S_8	1259	1258.4	0.0
Bu_2S_8	1425	1430.4	−0.4
Me_2S_9	1058	1035.6	2.2
Pr_2S_9	1385	1379.6	0.4
Me_2S_{10}	1182	1156.8	2.2

by simultaneous linear regression on n_C and n_S, using all data points. The experimentally determined indices for the 100% MeOH system are shown in Fig. 24. For the regression only data for $n_S \geqslant 4$ were used.   The disulfides and trisulfides exhibit noticeable deviations, which are caused by permanent dipole moment changes and non-linearities generally observable for the first members of homologous series.

Within the whole group of dialkyl polysulfides, two different types of homology can be defined. With constant n_S, the chain length of the terminal groups can be varied. We call this C-homology. On the other hand, S-homology is observed if the sulfur chain length is varied while the terminal groups are kept constant.

The I_K of the C-homologs can be expressed as

$$I_K = A_C + B_C n_C \tag{49}$$

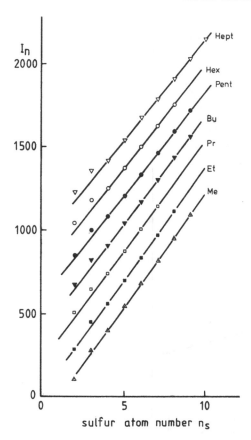

Fig. 24   Retention indices of 53 polysulfides from Me_2S_2 to $Hept_2S_{10}$ versus sulfur atom number.   Conditions as in Fig. 2.

where A_C and B_C are functions of sulfur atom number n_S.   The A_C and B_C found for the above system are shown in Fig. 25. A_C is a linear function of n_S:

$$A_C = a'_C + b'_C n_S \tag{50}$$

The slope of B_C of Eq. (49) is curved for $n_S = 2$ and 3 and can be extrapolated to a value of 100 for $n_S = 0$.   For the higher polysulfides, B_C can be reasonably well approximated by a linear function over n_S:

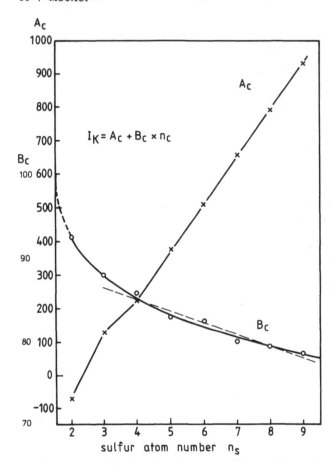

Fig. 25   Coefficients A_C and B_C for C-homology as a function of n_S. Data from Fig. 24.

$$B_C = a''_C + b''_C n_S \tag{51}$$

When looked at as S-homologs, the polysulfides yield

$$I_K = A_S + B_S n_S \tag{52}$$

The data are shown in Fig. 26. Both A_S and B_S depend linearly on carbon number n_C:

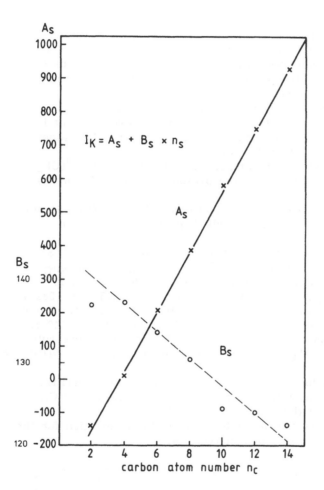

Fig. 26 Coefficients A_S and B_S for S-homology as a function of n_C. Data from Fig. 24.

$$A_S = a'_S + b'_S n_C \tag{53}$$

$$B_S = a''_S + b''_S n_C \tag{54}$$

It is clear that the retention index increment of sulfur atoms decreases with increasing terminal alkyl chain length. On the other hand, the methylene increment decreases with increasing sulfur chain length. On this basis, the parameters in Eqs. (49) to (54) can be given a clear interpretation.

A_C comprises the contribution from two solvation patches, $a'_C = 2\delta I(CH-S)$ and the contribution from the sulfur atoms present in this particular polysulfide, $b'_C n_S$, with $b'_C = \delta I^0(S)$. The superscript zero indicates nondisturbed atoms.

B_C contains the basic increment per CH_2 group, $a''_C = \delta I^0(CH_2)$, and the (negative) correction for the presence of n_S sulfur atoms, $b''_C n_S$.

A_S consists of the solvation patch increment $a'_S = 2\delta I(CH-S)$ and the contribution from the particular alkyl groups, $b'_S n_C$, with $b'_S = \delta I^0(CH_2)$.

B_S contains the basic increment of a sulfur atom, $a''_S = \delta I^0(S)$, and the correction for the presence of n_C methyl groups, $b''_S n_C$.

Equations (49) and (52) must yield the same index I_K. It is obvious that $a'_S = a'_C$. From the values found, -335.5 and -332.8, respectively, we take the average. Thus $\delta I(C-H) = 167.1$. Further, $b'_C = a''_S$. The average of 140.6 and 143.7 yields $\delta I^0(S) = 142.2$, the index contribution of a sulfur atom in a hypothetical pure sulfur chain. Finally, $b'_S = a''_C$. Both are found to be 91.0, so $\delta I^0(CH_2) = 91.0$.

These results hold for R_2S_n with $n_S > 4$. If the nonlinearities observed for di- and trisulfides were taken into account, $\delta I^0(CH_2)$ would eventually reach the expected value of 100.

The above increments can be combined to yield an equation for the retention index in terms of index increments:

$$I_K = 2\delta I(CH-S) + n_S[\delta I^0(S) + b''_S n_C] + n_C[\delta I^0(CH_2) + b''_C n_S] \tag{55}$$

For the system in use (100% MeOH), the numerical form is

$$I_K = 334.2 + n_S(142.2 - 0.75 n_C) + n_C(91.0 - 0.75 n_S) \tag{56}$$

This equation reproduces the experimental data with an accuracy of better than 1% relative standard deviation.

Figure 27 shows the relative percent deviation of ΔI_K of the polysulfides from the regression values as a function of n_S with an eluent

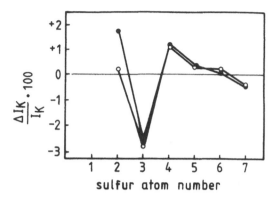

Fig. 27 Relative percentage deviation of I_K of polysulfides from regression values. Circles, *n*-propyl polysulfides; solid dots, average from all polysulfides. Column, RadPAK A; eluent, 95% MeOH + 5% H_2O; flow rate, 1 ml/min. (From Ref. 45.)

containing 95% MeOH + 5% H_2O. The diagram shows clearly that I_K is low for even n_S and high for odd n_S, as observed with the ethyl polysulfides in 100% MeOH. The influence on retention of permanent dipole moment changes from 1.52 (R—S—R) or 1.65 (R—S_3—R) debyes to 1.96 (R—S_2—R) or 2.16 (R—S_4—R) debyes is not very strong.

The effect of increasing thia substitution on the retention of a solute with a given skeleton number is demonstrated in Fig. 28. The I_K of tetradecane and its thia substitution products is plotted against the number of sulfur atoms present. The *n*-C_{14} has I_K = 1400 by definition. Substitution of a central CH_2 yields hexylheptyl thioether with I_K = 1055. When more CH_2 groups are replaced by sulfur, the index increases again. Between n_S = 11 and n_S = 12 the initial retention loss is overcompensated by excess retention introduced by the sulfur chains, and the index exceeds the alkane value of 1400. The I_K alternation between C_6—S—C_7 and C_4—S_4—C_6 caused by dipole moment changes is clearly visible.

It should be noted that the retention index change ΔI_K of dialkyl polysulfides (the monosulfide included) is most negative for the shortest sulfur chains (n_S = 1 or 2) becomes less negative for increasing chains, and reaches positive values for sufficiently long chains.

Nonsymmetric polysulfides R'—S_n—R'' can equally be separated. Figure 29 shows the chromatogram of a complex mixture of Me_2S_n (n_S = 2 to 10), $Pent_2S_n$ (n_S = 2 to 5), and Me—S_n-Pent (n_S = 2 to 7). The peaks corresponding to the respective homologous series

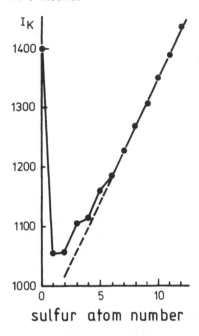

Fig. 28   Retention indices of tetradecane, hexylptyl sulfides, and polysulfides with skeleton atom number 14. Column, RadPAK 5μm C_{18}; eluent, 95% MeOH + 5% H_2O; flow rate, 1 ml/min. (From Ref. 35.)

can easily be identified. One may safely say that before RPLC was introduced the complete analysis of such a sample was not feasible.

Figure 30 shows the separation of the four isomeric butyl disulfides. Like the thiols, the normal compound has the highest retention and the tertiary, which is most branched, the lowest.

Eluent Composition Dependence

With increasing sulfur chain length the polysulfides become more similar to sulfur rings and their retention exceeds that of the reference *n*-alkane. It may also be anticipated that the main characteristic of sulfur retention—high I_K in pure MeOH which decreases on addition of H_2O to the eluent—has comparable effect on the group of polysulfides. As Fig. 31 shows, this is actually the case. For the *n*-butyl polysulfides from $n_S = 1$ to $n_S = 5$ the index is lowest in 100% MeOH and increasing with increasing eluent water content. The index of Bu_2S_6 is invariant to eluent composition. The higher polysulfides show an increasing similarity to sulfur rings; their index is highest in pure MeOH and decreases with decreasing eluent MeOH content.

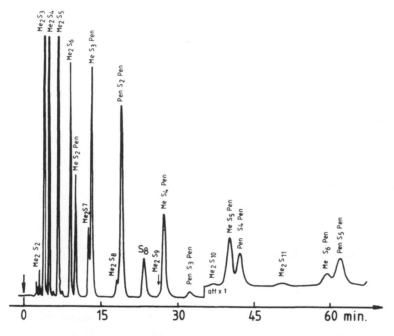

Fig. 29   Chromatogram of Me_2S_n (2–11), $Pent_2S_n$ (2–5), and $MePentS_n$ (2–6).   Column, RadPAK 5μm C_{18}; eluent, 85% MeOH + 15% H_2O; flow rate, 1.5 ml/min.

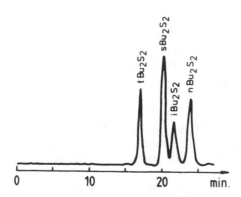

Fig. 30   Chromatogram of four isomeric butyl disulfides.   Column, RadPAK 5μm C_{18}; eluent, 85% MeOH + 15% H_2O; flow rate, 1 ml/min.

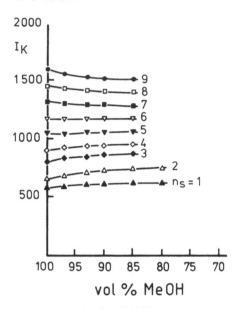

Fig. 31 Influence of eluent composition on retention indices of several di-n-butyl polysulfides. Column, RadPAK 5μm C$_{18}$.

The index change $\Delta I_K = I_K(Bu_2S_n) - 100(n_S + 8)$ of n-butyl polysulfides as function of n_S for various eluent compositions is shown in Fig. 32. The alternation of retention up to the pentasulfide is obvious. The strongest ΔI_K changes are seen in 100% MeOH. When going from the monosulfides to the disulfides, the ΔI_K change is negative between 100% and 95% MeOH and positive at lower eluent MeOH content. At 95% MeOH the change is zero. At Bu_2S_6, all curves intersect.

Similar observations have been made with the ethyl polysulfides and n-propyl polysulfides. Invariance of I_K to eluent composition is found between $-S_2-$ and $-S_3-$ for the ethyl and between $-S_3-$ and $-S_4-$ for the propyl compounds.

Temperature Dependence of Polysulfide Retention

The retention of all investigated polysulfides, except Me_2S_2 and Et_2S_2, is reduced when the column temperature is raised. As Fig. 33 shows, the corresponding ΔH get more negative with increasing solute molecular size.

Increasing eluent water content also shifts ΔH to more negative values. In 100% MeOH, the ΔH values of Et_2S_n are somewhat more negative than those of R_2S_2. The changes of I_K of $-S_n-$ chains

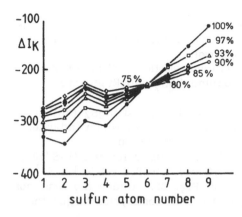

Fig. 32 Retention loss ΔI_K of some di-n-butyl polysulfides as a function of sulfur chain length and eluent composition, Column, RadPAK 5μ C_{18}.

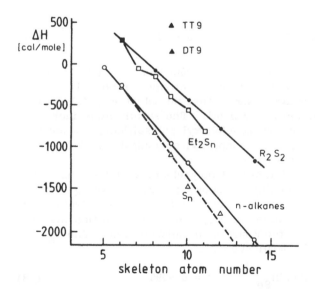

Fig. 33 Sorption enthalpies ΔH of several solute classes as a function of skeleton atom number. Column, 20 cm × 3 mm Nucleosil 5μ C_{18}; eluent, MeOH; flow rate, 1 ml/min; temperature, 25°C.

with dipole moment μ are parallelled by corresponding ΔH alterations. For equal skeleton atom numbers, the ΔH difference between the two solute groups is about 200 to 300 cal/mol.

In 80% MeOH, the ΔH difference is even smaller. Simultaneously, the I_K differences between corresponding R_2S_2 and Et_2S_n are reduced.

The sorption enthalpy in 100% MeOH of Me_2S_2 and Et_2S_2 (and the multisulfides 2,8-DT-9 and 2,5,8-TT-9) is positive, which indicates that in these cases entropy effects support retention.

If a formal phase volume ratio of 0.2 is assumed for the ODS/100% MeOH system, sorption entropies of $\Delta S = 3.3$ to 0.3 cal K^{-1} mol^{-1} are calculated for the solutes shown in Fig. 33. Under the same conditions, n-alkanes larger than C_8 and sulfur homocycles larger than S_{10} yield negative ΔS values.

In 80% MeOH + 20% H_2O the sorption entropies calculated for all solutes investigated are strongly negative. Although the ΔS data are poorly defined, it appears that sorption entropy is generally more negative the larger the solute molecular surface area.

B. Diphenyl Polysulfides and Related Compounds

The diphenyl polysulfides ϕ_2S_n form a perfect homologous series, as shown in trace b of Fig. 34. The retention indices can be expressed as

$$I_K(\phi_2S_n) = 345.2 + 125.6n_S \qquad r = 0.9998 \qquad (57)$$

In the same chromatographic system, benzene has $I_K = 150$. Consequently, the index increment for the heteronuclear bond is $\delta I_K(\phi - S) = +23$. This again shows that no solvation patches are formed around the benzene-sulfur bond. The sulfur index increment is close to the value found for dicyclohexyl polysulfides. Comparable observations were made with di-β-naphthyl polysulfides and $\delta I_K(S)$ is only slightly lower in this case.

Recently, the retention of diphenyl polyselenides, ϕ_2Se_n, has been described [56]. Trace d in Fig. 34 shows the respective I_K. The retention-determining backbones of polysulfides and polyselenides are respectively sulfur and selenium chains, proton-free entities whose retention is determined by their molecular surface area. The I_K equation in the ODS/100% MeOH system is

$$I_K(\phi_2Se_n) = 367.4 + 160.5n_{Se} \qquad r = 0.9996 \qquad (58)$$

Because of the larger surface area of selenium, the index increase $\Delta\Delta I_K$ per additional atom in the chain is higher than with sulfur. It

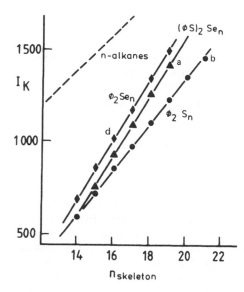

Fig. 34   Retention indices of diphenyl polysulfides, diphenyl polyselenides, bis-phenylthio polyselenides, and *n*-alkanes versus
skeleton atom number (one benzene ring counts as 6). Conditions as
in Fig. 33.

has been shown [56] that the respective index increases $\Delta\Delta I_K$ are
directly proportional to the corresponding surface area increases
$\Delta\Delta SA$:

$$\frac{\Delta\Delta I_K(Se)}{\Delta\Delta I_K(S)} = \frac{\Delta\Delta SA(Se)}{\Delta\Delta SA(S)} \tag{58}$$

From the constant A in Eq. (57) it can be seen that the benzene-
selenium bond increment is $\delta I_K(\phi{-}Se) = 34$, which is not very much
higher than the $\phi{-}S$ increment.

Trace a in Fig. 34 is derived from another compound class, the
bis-phenylthio polyselenides. It runs parallel to trace d, indicating
that a selenium chain is responsible for the I_K increase. The slightly
lower I_K of bis-phenylthio polyselenides is due to the presence of two
sulfurs instead of selenium and the slightly lower $\phi{-}S$ bond increment. The mutual influence of phenyl, sulfur, and selenium must be
negligible since the $\phi SSe_nS\phi$ curve can be extrapolated to the $\phi S_n\phi$
curve, where it intersects precisely at the I_K of diphenyl disulfide.

X. SUMMARY OF RETENTION EFFECTS
OBSERVED WITH SULFUR

In RP systems with a purely or mainly methanolic eluent, the retention contribution of sulfur atoms within solute molecules is higher than the retention caused by a methylene group. The retention difference between the solute molecular building units S and CH_2 depends on:

1. The respective eluent-accessible surface area contribution
2. The intramolecular environment
3. The eluent composition

The difference is greatest in pure MeOH, which seems to indicate that undisturbed sulfur aggregates like chains or rings interact with the eluent only via London forces, whereas alkyl residues exert some extra interaction which is thought to be mediated by the alkyl protons in the solute and the surrounding CH_3OH eluent. With sufficient water in the eluent, the extra interaction is suppressed by stronger solvophobic forces and the elution order is determined by the molecular surface area only. The increased interaction of solute alkyl parts with the eluent methanol seems to depend on the proton density per surface area rather than the absolute number of protons in the solute molecule.

If sulfur atoms are bonded to an alkyl group, the area around the CH_2—S fragment becomes solvated. The solvation patch enhances interaction with the eluent and, consequently, reduces retention. The degree to which retention is lowered appears to depend on the spatial extent of the solvation patch within the solute molecule.

Concomitant with solvation patch formation, a reduced retention contribution of methylene groups is observed. The reason for this reduction is not completely understood at present. It is probably to be found in orientational changes in the cavity-forming eluent molecules, induced by the solvation patch.

The combined effects mentioned above are illustrated in Fig. 35. The I_K for various solutes, all having $n_S + n_C = 9$, is shown as a function of the number of sulfur atoms present. The system was ODS/95% MeOH + 5% H_2O

By definition, n-C_9 has $I_K = 900$. The cyclononane I_K is slightly lower in this system. Due to the lower interaction with the eluent, cyclononasulfur has a high $I_K = 1260$. Substitution of the central methylene group of n-nonane produces a solvated local polar center and the index drops to 620. Substitution of more vicinal CH_2 groups leads to the polysulfides. The polar center is split and the halves move toward the ends of the chain. The increasing number of vicinal sulfur atoms again increases the index. Redistribution of sulfur atoms in 4,5-dithianonane to the 4 and 6 positions creates another polar

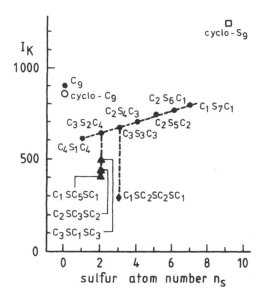

Fig. 35   Retention indices of various solutes having $n_{skeleton} = 9$ as a function of degree of thia substitution. Conditions as in Fig. 28. (Data from Ref. 45.)

center, and the index drops below 500. Shifting the sulfur atoms to the 2,8 and 3,7 positions further diminishes the unsolvated part of the solute molecules, and the index decreases to 415 and 430, respectively. Moving the sulfur atoms of dipropyl trisulfide from the 4,5,6 to the 2,5,8 positions creates a third polar center and the index falls to about 300.

   The results shown in Fig. 35 are useful in practice for the estimation of retention data for sulfur compounds. To recognize retention effects more precisely, one must take into account differences in solute molecular surface area. For example, the surface areas of all the nonanes are similar but not identical. In Table 12, the respective I_K and ΔI_K values of 11 "nonanes" have been normalized to the surface area of *n*-nonane, which is $SA(n\text{-}C_9) = 229.6$ Å^2. The system was ODS/100% MeOH. The I_K differences between C_8-SH, C_1-S-C_7, and C_4-S-C_4 are not dramatic, but they are clearly visible and reflect an expansion of a solvation patch along the alkyl chain. It can also be seen that in 100% MeOH, the I_K of a disulfide is slightly lower than that of the corresponding thioether. This indicates that in purely organic eluent some enlargement of the solvated molecular surface area overrides the additional retention conferred by the proton-free sulfur atoms.

Table 12 I_K and ΔI_K of Some
"Nonanes" Normalized to
$S\text{Å}(n\text{-}C_9) = 229.6 \text{ A}^2$

Compound	I_K	ΔI_K
$n\text{-}C_9$	900	0
$C_8\text{-}SH$	660	-240
$C_1\text{-}S\text{-}C_7$	650	-250
$C_1\text{-}S_2\text{-}C_6$	640	-260
$C_2\text{-}S_2\text{-}C_5$	626	-274
$C_4\text{-}S_1\text{-}C_4$	622	-278
$C_3\text{-}S_2\text{-}C_4$	617	-283
$2,8\text{-}DT\text{-}9$	417	-483
$2,5,8\text{-}TT\text{-}9$	293	-607
$C_2\text{-}S_5\text{-}C_2$	747	-153
$Cyclo\text{-}S_9$	1608	$+708$

The differently thia-substituted alkanes exhibit clearly different
behavior on changes of the eluent water content, because of their
different functionality. This is shown in Fig. 36 for some solutes,
again with $n_C + n_S = 9$. The strongest increase of retention with
eluent H_2O percentage is shown by n-nonane, because the addiitional
interaction with the methyl protons of MeOH is removed. The cyclononasulfur, lacking this interaction, has higher ln k' values at low
H_2O contents, but smaller slope. The intersection point is around
65% MeOH. Of the lower substituted solutes, the octanethiol has
highest retention and the steepest slope. This compound has the
longest alkyl chain and is, therefore, the one most comparable to
n-nonane in a chromatographic sense. The retention of the thioethers is clearly below that of the thiol and the retention increase
with increasing eluent H_2O content is weaker. This is a consequence
of a larger solvation patch, which makes the molecules somewhat more
hydrophilic. Within the eluent composition range shown, the disulfide
ln k' value is between the values for thiol and the thioether. It has
a larger polar center than the C_8SH, but exhibits a diminishing tendency of retention decrease, compared to the C_4SC_4. The C_4SC_4
and $C_3S_2C_4$ lines intersect at 95% MeOH.

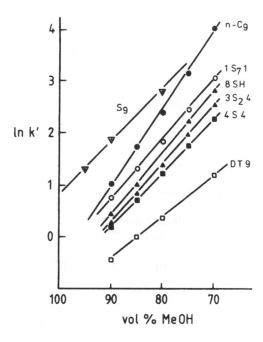

Fig. 36 Change of ln k' with eluent composition for several nonanes having varying degrees of thia substitution. Column, 10 cm × 3 mm Nucleosil 5μm C_{18}; flow rate 1 ml/min.

2,8-Dithianonane exemplifies the effect of two isolated polar centers without the retention-enhancing effect of vicinal sulfur atoms. The diminished slope of ln k' with eluent H_2O percentage shows that the hydrophilic character has increased.

REFERENCES

1. B. Meyer, *Sulfur, Energy and Environment*, Elsevier, New York, 1977.
2. L. Gmelin, *Handbuch Der Anorganischen Chemie* (Band Schwefel), 8th ed., Verlag Chemie, Weinhein, 1959.
3. J. H. Karchmer, ed., *The Analytical Chemistry of Sulfur and Its Compounds*, Wiley-Interscience, New York, 1972.
4. M. R. F. Ashworth, *The Determination of Sulphur-Containing Groups*, Vol. 3, Academic Press, London, 1977.
5. N. Kharasch, *Organic Sulfur Compounds*, Pergamon, London, 1961.

6. H. J. Moeckel, T. Freyholdt, and J. Weiss, Unpublished results.
7. H. J. Moeckel, T. Freyholdt, and H. Melzer, in *Koenigsteiner Chromatographietage 1982* (K. H. Franzen, ed.), Waters, Koenigstein, 1982.
8. H. J. Moeckel and T. Freyholdt, in *Koenigsteiner Chromatographietage 1980* (O. Kaiser, ed), Waters, Koenigstein, 1980.
9. K.-O. Hiller, Diplomarbeit, TU Berlin, 1975.
10. K.-O. Hiller, B. Masloch, and H. J. Moeckel, Z. Anal. Chem., *280*, 293 (1976).
11. R. Steudel, H. J. Maeusle, D. Rosenbauer, H. J. Moeckel, and T. Freyholdt, Angew. Chem. *93*, 403 (1981).
12. M. W. Prinzler, D. Pape, and M. Teppke, J. Chromatogr. *19*, 875 (1965).
13. J. H. Karchmer, in *Treatise on Analytical Chemistry of Inorganic and Organic Compounds*, Vol. 13 (I. M. Kolthoff and P. J. Elbing, eds.), Wiley, New York, 1966.
14. J. W. Vogh and J. E. Dooley, Anal. Chem. *47*, 816 (1975).
15. R. M. Cassidy, J. Chromatogr. *117*, 71 (1976).
16. J. K. Haken, M. S. Wainwright, and R. J. Smith, J. Chromatogr. *147*, 65 (1977).
17. J. K. Haken, M. S. Wainwright and R. J. Smith, J. Chromatogr. *147*, 65 (1978).
18. M. Gassiot, X. Guardino, J. Albaiges, G. Firpo, and Rodriguez-Vinals, J. Chromatogr. *147*, 540 (1978).
19. W. K. Al-Thamir, J. H. Purnell, C. A. Wellington, and R. J. Laub, J. Chromatogr. *173*, 338 (1979).
20. J. Sevcik, J. Chromatogr. *135*, 183 (1977).
21. E. Grobler, and G. Balisz, J. Chromatogr. Sci. *21*, 57 (1974).
22. M. S. Wainwright, J. K. Haken and D. Srisukh, J. Chromatogr. *179*, 160 (1979).
23. H. Engelhardt and G. Ahr, Chromatographia *14*, 227 (1981).
24. H. J. Moeckel and T. Freyholdt, Chromatographia *17*(4), 215 (1983).
25. G. E. Berendsen, P. J. Schoenmakers, L. De Galan, G. Vigh, Z. Varga-Puchony, and J. Inczedy, J. Liq. Chromatogr. *3*, 1669 (1980).
26. H. J. Moeckel and H. Melzer, in preparation.
27. E. Kovats, Helv. Chim. Acta, *41*, 1915 (1958).
28. D. C. Locke, J. Chromatogr. Sci. *12*, 433 (1974).
29. C. Horvath, W. Melander, and I. Molnar, J. Chromatogr. *125*, 129 (1979).
30. W. R. Melander and C. Horvath, in *High Performance Liquid Chromatography* (C. Horvath, ed.), Academic Press, New York, 1980, p. 113.
31. I. Halasz, Anal. Chem. *52*, 1393A (1981).

32. I. Halasz, Private communication.

33. H. J. Moeckel, Z. Anal. Chem. *318*, 116 (1984).

34. H. J. Moeckel, Z. Anal. Chem. *318*, 327 (1984).

35. H. J. Moeckel, J. Chromatogr. *317*, 589 (1984).

36. H. J. Moeckel and B. Masloch, Z. Anal. Chem. *281*, 379 (1976).

37. H. J. Moeckel, *Wiss. Erg. Ber. HMI,* HMI-319 (1979).

38. H. J. Moeckel and H. Melzer, in *Koenigsteiner Chromatographie-tage 1984,* (H. Aigner, ed.), Waters, Koenigstein, 1984.

39. R. Steudel and H. J. Maeusle, Angew. Chem. *51*, 165 (1979).

40. M. Schmidt, and W. Siebert, in *Comprehensive Inorganic Chemistry* (A. F. Trotman-Dickenson, ed), Vol. 2, p. 795, Pergamon, Oxford, 1973.

41. R. S. Pearlman, in *Physical Chemical Properties of Drugs* (S. H. Yalkowsky, A. A. Sinkula, and S. C. Valvani, eds.), Marcel Dekker, New York, 1980.

42. R. S. Pearlman, QCPE Bull. *1*, 15 (1981).

43. H. J. Moeckel and T. Freyholdt, Z. Anal. Chem. *368*, 401 (1981).

44. H. J. Moeckel, I. Hoefler, H. Melzer, J. Chromatogr. *388*, 267 (1987).

45. H. J. Moeckel, T. Freyholdt, J. Weiss, and I. Molnar, in *Practical Aspects of Modern HPLC* (I. Molnar, ed.), De Gruyter, Berlin, 1982.

46. H. J. Moeckel, Habilitationsschrift, TU Berlin, 1984.

47. M. J. Cardone, in *The Analytical Chemistry of Sulfur and Its Compounds* (J. H. Karchmer, ed.), Wiley-Interscience, New York, 1972.

48. R. Steudel and E.-M. Strauss, Z. Naturforsch. *388*, 719 (1983).

49. J. Weiss and H. J. Moeckel, Unpublished results.

50. A. Nahum and C. Horvath, J. Chromatogr. *203*, 53 (1981).

51. R. Steudel, R. Reinhard, and F. Schuster, Angew. Chem. *89*, 756 (1977).

52. R. Reinhard, R. Steudel, and F. Schuster, Angew. Chem. *90*, 55 (1978).

53. J. Steidel and R. Steudel, J. Chem. Soc. Chem. Commun. 1312 *(1982).*

54. R. Steudel, R. Strauss, and L. Koch, Angew, Chem. 97, 58 (1985).

55. A. T. Fojtik, F. Hoefler, H. Melzer, and H. J. Moeckel, in preparation.

56. H. J. Moeckel, F. Hoefler, and H. Melzer, Chromatographia *20*, 471 (1985).

2

The Application of Fleuric Devices to Gas Chromatographic Instrumentation

Raymond Annino   *The Foxboro Co., Foxboro, Massachusetts*

I. INTRODUCTION

This chapter will be concerned with the application of fleuric devices in gas chromatographic instrumentation. The use of these devices as sampling valves, detectors, logic gates, and, finally, as transducers in optical-to-pneumatic conversion elements of nonelectrical process control chromatographs will be discussed.

"Fleuric" is a word which has been coined to describe a fluid amplification device with no moving parts. Therefore, although this discussion deals exclusively with the pneumatic domain, it will exclude devices such as force balance amplifiers and diaphragm logic gates. which have been used in the design of nonelectric pneumatic chromatographs. Also excluded are the pressure and flow regulators used to control the flow of carrier gases. Likewise, although related, the flow switching method of Deans [1,2] for column selection and sampling will not be included in this discussion.

In each case, we will review briefly the principles governing the operation of the particular fleuric device which is germane to the application. However, it will not be possible, in the limited space provided, to discuss these principles in any great depth. Readers who are interested in pursuing the subject further are referred to the excellent texts by Kirschner [3] and Kirshner and Katz [4]. The original literature in this field is concentrated in the area of mechanical engineering, and most of it has been published as symposium proceedings of the American Society of Mechanical Engineers (ASME) and The Harry Diamond Laboratories (HDL) of the U.S. Army Material Command, which has been one of the most active laboratories in this field. Many of this laboratory's internal reports as well as the proceedings of the HDL-sponsored symposia can be obtained by writing directly to the laboratory. Finally, an excellent series of articles discussing the fundamentals of design for fleuric devices was written by Daryl Lethan and appeared in Machine Design through much of 1966 and 1967.

II. PRINCIPLES

In most fleuric devices, jets and the interaction of jets with each other and their environment are of primary importance in their operation. However, fluid mechanics, particularly concepts dealing with the flow of fluids as jets, is not an area of engineering which is familiar to most chromatographers.

Perhaps the easiest way to begin is with a physical picture. Consider a block of material which has the internal structure shown in Fig. 1. Fluid is forced to flow down channel A through a small nozzle to produce a "jet" exiting into channel B. Depending on the

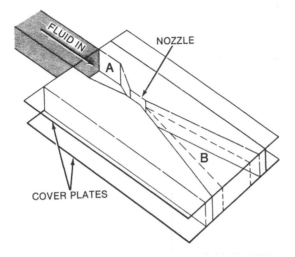

Fig. 1   Pictorial representation of a fleuric device package.

pressure available at the source feeding channel A, the density and
viscosity of the fluid that is present, and the dimensions and shape
of the nozzle, the jet profile will vary. If the emerging jet has a
low Reynolds number (Re), the jet stream will be laminar and will
tend to retain its shape. Higher Reynolds numbers produce turbulent jets, which will spread as they emerge and travel down the exit
chamber.*

In principle, a jet of sufficiently low Reynolds number can remain
laminar until all of its energy is dissipated in mixing with the surroundings. With the Reynolds numbers of jets used in actual practice,
however, the laminar stream will break into turbulence at some point
downstream of the nozzle. The region before turbulence, where the

---

*The Reynolds number is a dimensionless parameter used to charac-
terize flow. It is the ratio of the inertial to viscous forces operating in the fluid $Re = \rho VD/\eta$ where ρ is the density of the fluid,
kg/m^3; V its velocity, m/sec; η its fluid viscosity, kg/m-sec; and
D the tube diameter, m. Reynolds numbers below 2000 are characteristic of laminar flow. A transitional flow region exists between 2000
and 4000, and turbulent flow exists above this. In laminar flow
adjacent fluid elements move along parallel paths, and in turbulent
flow the flow paths are more or less randomly distributed in space.
The two flow patterns are illustrated in Fig. 2.

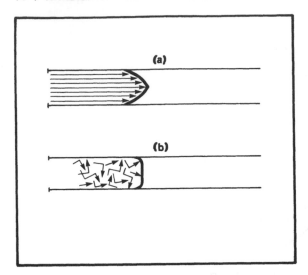

Fig. 2   Schematic representations of flow patterns associated with (a) laminar flow and (b) turbulent flow.

jet is becoming unstable, is called the transition flow region, to distinguish it from the zone of established and stable flow.

Jet flows of low Reynolds numbers will entrain less fluid than high Reynolds number jets. In general, the effect of disturbances on the velocity distribution of fluid within a jet depends on the Reynolds number; low number, laminar jets change more in profile for a given disturbance than do those with high Reynolds numbers. For example, a laminar jet can be forced into turbulence by an acoustic disturbance or a very small jet of fluid which is perpendicularly positioned.

The proper dimensions and geometry of a fleuric device are governed by all of the above considerations and the task for which it is being designed. However, in many cases the results reported by chromatographers were obtained with commercially available fleuric devices manufactured for other uses. Thus, it is quite possible that these results do not accurately reflect the optimum performance of fleurics in the particular application.

III. STREAM INTERACTION, BEAM DEFLECTION AMPLIFIERS

A. Bistable Operation—Turbulent Jet

The first type of beam deflection amplifier, which was actively studied during the early 1960s, contained a high-velocity turbulent

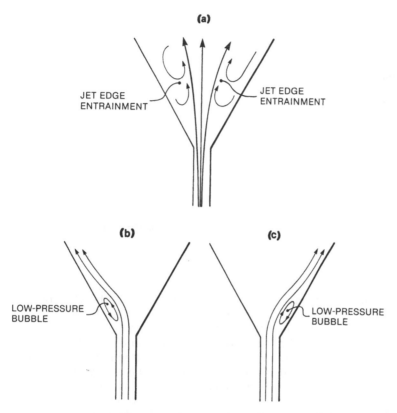

Fig. 3   (a) Schematic representation of an emerging turbulent jet, spreading as it travels down the cavity and entraining fluid at its edges.   This action starts a backflow of fluid to replace the entrained fluid.   (b) and (c)   Two stable positions of the turbulent jet.

jet. For the device shown in Fig. 1, the jet resembles a thin curtain of fluid whose two dimensions are fixed by the top and bottom plates. As shown by the dotted line, the jet expands as it leaves the nozzle and travels into chamber B. The high-velocity edges of this jet will entrain fluid. The backflow of fluid to replace that which was entrained will be reduced on the side of the jet closest to the wall. This inability to replace all of the entrained fluid will lead to a further decrease in pressure between the jet and the wall until finally the jet attaches itself to the wall.* These events are depicted in Fig. 3.

---

*This effect was first noted by Thomas Young in 1800. It was subsequently rediscovered by others, one of them being Henri Coanda, for whom it is now named.

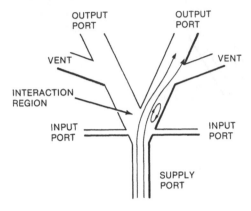

Fig. 4   Bistable stream-interaction amplifier configuration.

By adding control ports to the device and splitting the flow between two output ports, as shown schematically in Fig. 4, we can
construct a bistable stream-interaction amplifier. To switch the
stream from one wall to the other requires that fluid be added at
the control port on the side to which the jet is attached, at a rate
higher than the jet can entrain and remove it. This is usually accomplished by pulsing the control port with a positive pressure
pulse (usually 10 to 30% of supply), but other methods are also
effective depending on the condition in which the device is operated.
Switching times on the order of 1 msec are common.

The so-called Coanda effect thus leads to a digital device, and
applications of these devices have been typically in the logic/control
area since, by suitable control of internal geometry and external
connections, these fluidic amplifiers can be made to perform the
function of a number of the common logic gates. Some of these are
summarized in Fig. 5 together with the appropriate truth tables.
This type of amplifier became quite popular during the mid-1960s as
a logic component in various control systems. A number of them
are still available commercially from Corning Glass.*

---

*Electronic Material Department, Corning Glass Works, Corning,
NY 14830.

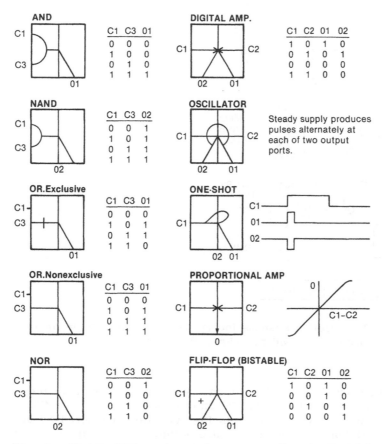

Fig. 5 Fluidic logic gates and corresponding truth tables.

IV. BISTABLE JET AMPLIFIERS IN GC INSTRUMENTATION

A. Turbulent Jet Amplifiers

The papers of Wade and Cram [5] and Cram and Chesler [6] are the first reports describing the use of these turbulent jet logic elements in gas chromatographic instrumentation. These authors demonstrated a number of sampling systems using these devices and then went on to describe a general class of fluidic sampling system including sample modulation, multiplexing, and sequential sampling.

A number of configurations for timing a sample injection were investigated. One example is shown in Fig. 6. It utilizes three

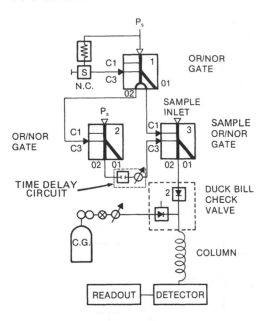

Fig. 6 Fluidic injection systems incorporating an adjustable pneumatic timed sample pulse width. (From Ref. 5.)

OR/NOR gates for timing and injection. With an OR/NOR gate the jet power is recovered by output O2 unless control port C1 or C3 is powered. In the system shown in Fig. 6, gate 1 (powered by air or carrier gas) normally has a pressure output at O2. This pressure is applied to the control ports of gates 2 and 3. Thus, the power jet of gate 2 is deflected into output port O1 and therefore no control pressure is available at C3 of the sample gate. However, since O2 of gate 1 is also connected to C1 of the sample injector gate, 3, the sample power jet is still deflected so that it is exiting O1 and not O2. When the normally closed switch, S, is activated, 10 to 30% of the supply pressure of gate 1 is applied to the control jet and the power jet is deflected to exit out of O1. The control pressure at C1 of the sample gate quickly drops to zero, as does the control pressure at C1 of gate 2. However, C3 of the sample gate is not immediately pressurized by the power now available at O2 of gate 2, since a time delay has been built into that line. Therefore, since both C1 and C3 of the sample gate are de-energized, the sample power jet returns to its normal position exiting O2 and thence into the column. After a time delay determined by the volume of the tank (capacitor) placed in the line between O2 of gate 2 and C3 of the sample gate, C3 is once again energized and the sample stream returns to exit O1. It should be noted here that it was found necessary to employ check valves in the circuit to avoid backflow of sample and obtain sharp sample profiles.

A simple but less flexible system is shown in Fig. 7, where a fluidic one-shot is used as the sample injector. When C1 is pressurized, the sample is switched to O1. The feedback path internal to this device, however, returns the jet to O2 after 10 msec, regardless of whether C1 is still pressurized or not (see Fig. 5 for the truth table of this device). These same workers demonstrated that acceptable injection peaks widths of 10 msec could be generated with 0.1% reproducibility. Thus, this simple device showed excellent promise as a sampling valve for high-speed chromatography where very narrow sample injection profiles are required.

A problem left unanswered was that of matching the impedance of the fluidic gate to the impedance of the column. The fluidic sample valve worked well only as long as the column head pressure was not too high. The difficulty exists because these devices were designed to operate at high flows. Therefore, to reduce their sensitivity to load, the designers created alternate flow passageways shown as VENTS in Fig. 4. Thus, when one tries to make fluid flow out O1 or O2 when these outputs are maintained at higher than atmospheric pressure, the fluid exits instead out of the vents.

The problem was solved quite simply (as everything is in hindsight) by Gaspar et al. [7]. These researchers reasoned that if

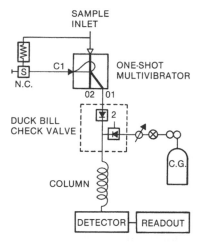

Fig. 7 Fluidic injection system with fixed sample pulse width. (From Ref. 5.)

one encapsulated the fluidic amplifier and raised the pressure of the capsule so that the vents now were at the elevated pressure, the device should once again be able to work as a sample valve against an elevated column pressure. They found this to be exactly the case and demonstrated that the fluidic gate could provide injection bands of a few milliseconds to tens of milliseconds at column pressures much higher than atmospheric. Moreover, with the now adjustable vent pressure matched to that of the column, the check valves which were required in the original circuit were no longer necessary. A simplified version of the circuit is shown in Fig. 8.

These workers were interested in fluidic samplers primarily as devices which could deliver the very sharp sample profiles so necessary for high-speed chromatography and were not interested in designing a totally nonelectric sampling system. Thus, the sample injection time was accomplished under computer control of electrically actuated solenoids. We will discuss in a later section a method for remotely controlling, via an optical fiber link, a totally fluidic version of this sampling system. Annino et al. [8] continued the work in Guiochon's laboratory on high-speed chromatography with short, narrow-bore capillary columns and also demonstrated the use of this sample switch in other applications such as correlation chromatography [9].

Annino and Leone [10] have published further on this system, illustrating the effect of various parameters on the operation of the device and showing that its repeatability makes it a good candidate for a sample valve for ensemble averaging (see Fig. 9). They

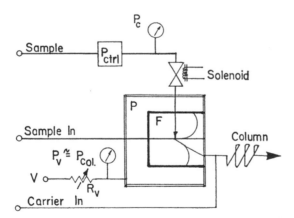

Fig. 8   Schematic representation of the fluidic sample system proposed by Gaspar et al. (From Ref. 10.)

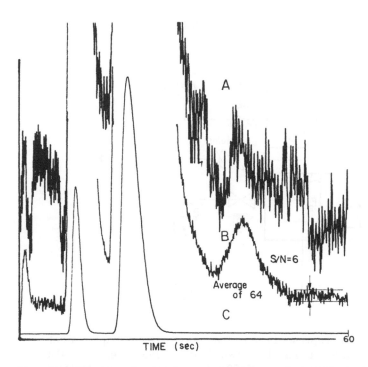

TIME (sec)

Fig. 9 Single pulse and ensemble chromatograms of air samples contaminated with methane. C = Single pulse chromatogram—unamplified. A = Same as C but amplified to see the methane contamination. B = Ensemble average of 64 chromatograms of the same sample. (From Ref. 10.)

also showed that these fluidic gates could be used quite effectively as parallel samplers with no significant cross contamination (see Fig. 10), even in the absence of check valves between the gates and column (see Fig. 11).

This fluidic sample injection system continues to be used in Guiochon's laboratory and by others in the study of high-speed chromatography [11,12].

All of the turbulent jet devices discussed thus far require flow rates of 2 to 5 liters/min (depending on the internal dimensions of the device) to maintain turbulent flow. This is certainly a disadvantage if sample is not plentiful. Also, the sample must be raised to above column head pressure for the proper flow to be maintained. In the next section, which deals with a laminar flow device, a solution to at least one of these problems (sample availability) will be presented.

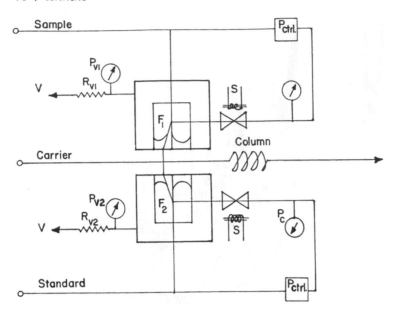

Fig. 10 Schematic representation of a gas chromatographic dual-sample standard injector based on two fluidic logic gates. F, OR/NOR logic gates, Corning No. 191453; P_{ctrl}, pressure controler to maintain P_C (the command pressure) at the proper value; V, vent; P_V pressure in the capsule; R_V, vent restrictor; S, solenoid. (From Ref. 10.)

B. Laminar Jet Amplifiers

In recent experiments R. Annino (unpublished work) has demonstrated that a laminar proportional jet amplifier (see Sec. IV.D) operated as a bistable device can be used as a sample injection switch in a manner similar to that described for the turbulent jet amplifier, albeit with the advantage of requiring much less sample for its operation.

The sample injection switch is constructed from planar 3.3 × 3.3 cm square laminates photochemically milled from 0.004-in. 455 stainless steel [13] with a 0.010-mil orifice size in the amplifier laminate. Each laminate has a particular functional purpose—amplifier, vent, resistor, capacitor, etc. Examples of some representative patterns are shown in Fig. 12, and a magnified view of the HDL-designed laminar proportional amplifier (LPA) is shown in Fig. 13. The individual laminates are arranged in a vertical stack, aligned via four bolt holes, and mounted to the same type of manifold described

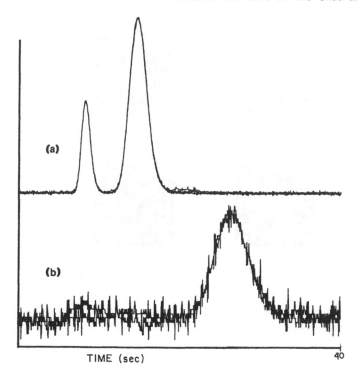

Fig. 11 Chromatograms of alternate injections of samples from two gates into the same column. (a) 100% air, (b) 4% methane in helium. Amplification: (a) 1, (b) ×50. (From Ref. 10.)

in reference 10. The particular laminates used in the fluidic switch are shown in order of their stacking in Fig. 14. More than one of each laminate is used, depending on the design requirements in terms of nozzle aspect ratio (ratio of width to depth), supply pressure, dropping resistors, etc. As indicated in the legend of Fig. 14, a total of 25 laminates were used to construct the sample switch. The standard amplifier hole format is that adopted by HDL and is designated as a C-format. Similar laminates are now commercially available in kit form.*

Sample flow can be reduced to as low as 20 ml/min with this version of a fluidic sampling valve. Preliminary results with this switch look quite promising. However, it has not yet been tested under the very high speed conditions in which the Coanda devices have

---

*Laminar Proportional Amplifier Fluidic Laboratory Starter Kit, TriTec, Inc., P.O. Box 56, 8925-11 McGaw Court, Columbia, MD 21045.

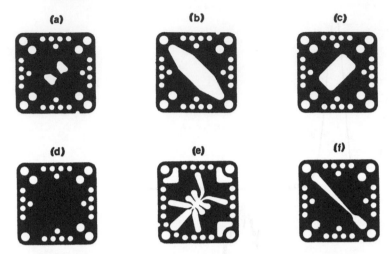

Fig. 12 Silhouettes of various laminates for the construction of fleuric amplifiers. (a) Exhaust (to collect vent flow from nozzle); (b) exhaust (lead vent flow to corner holes); (c) vent (to vent flow from nozzle); (d) gasket (separator plate); (e) LPA; (f) resistor (to reduce pressure between diametrically opposed holes).

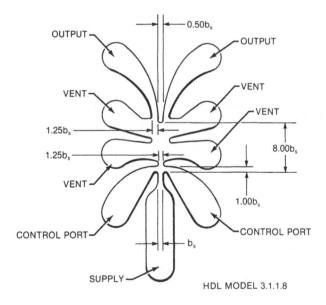

Fig. 13 Critical dimensions and silhouette of HDL Model 3.1.1.8 laminar proportional amplifier.

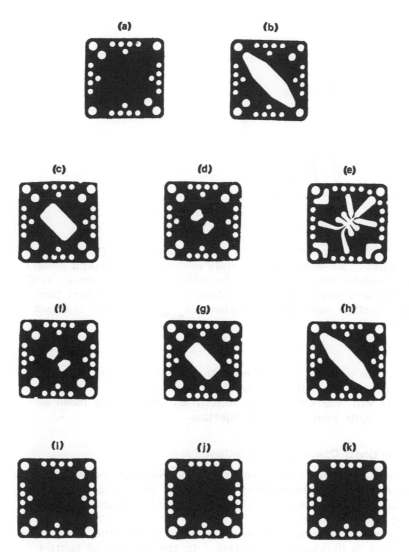

Fig. 14   Stacking order of the individual laminates used to make the LPA sample injector. A total of 25 laminates are used in the order: 1a, 5b, 2c, 1d, 5e, 1f, 2g, 5h, 1i, 1j, 1k.

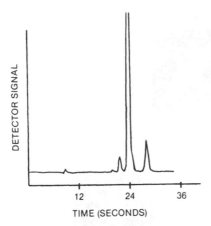

DETECTOR SIGNAL

12   24   36

TIME (SECONDS)

Fig. 15 Chromatogram of hexane isomers obtained with an LPA sample switch. Column, 7.3 m × 0.25 mm i.d. cross-linked methyl silicone (Analabs GB-1); 1.0 μm film thickness; temperature, 53°C; P_V = 8.59 psig; column flow, ∿4 ml/min helium; sample vent flow, 60 ml/min; helium vent flow, 60 ml/min; P_{ctrl} = 30 psig; injection width, 10 msec.

been found to perform well. Under the moderately fast chromatography conditions depicted by the chromatogram shown in Fig. 15, the results with the LPA sample switch, are identical to those obtained with split sample syringe injection.

C. Proportional Operation—Turbulent Flow

The bistable condition of the amplifiers described in the last section must be avoided for proportional operation of the device. This is most easily done by omitting walls in the vicinity of the power-jet nozzle as shown in Fig. 16.

In this type of device, the power jet flows across the interaction chamber and is divided by a splitter. In the absence of control flow, equal flows issue from each output port. When flow is introduced at a control port, it interacts with the power stream and deflects it toward the opposite port. The action is proportional to the pressure differential appearing at the control ports and amplification is obtained, since small pressure differentials at the control ports produce much larger differences at the output ports.

Although there are three fluid effects which can be employed in the deflection of jets—momentum exchange, pressure differential, and the Coanda effect—let us consider here a pure momentum exchange. If a power jet of cross-sectional area A has a density ρ and a

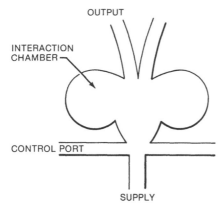

Fig. 16   Interior fluidic amplifier design for eliminating wall effects.

velocity v, the mass of fluid that emerges from the nozzle in time t is ρAvt and its momentum is $\rho Av^2 t$. Similarly, the fluid emerging from the control jet has momentum $\rho_c A_c v_c^2 t$, where the subscript c is used to distinguish the appropriate terms from those of the power jet.

Initially at right angles to the control jet, the power jet will be deflected an angle α, by the interaction of the two streams. Since momentum is conserved, the resulting deflection will be given by the vector addition of the two momentum vectors (see Fig. 17).

$$\tan \alpha = \frac{\rho_c A_c v_c^2}{\rho A v^2}$$

Thus, if the density of either of the two streams changes, so will the angle α and, consequently, the output pressure recovered at one of the output ports. This amplifier can, therefore, be used as a gas chromatographic detector. The carrier gas is teed at the tank to a variable restrictor and this line is used as the supply for the power jet. The exit flow from the column is connected to one of the control ports (an adjustable pressure may be required for the other control port to center the jet under normal operating conditions). When a component emerges from the column and into the control jet, the change in density of the control jet causes a proportional change in deflection of the power jet.

The use of such a device as a GC detector has been reported [19]. Unfortunately, this paper appears to be a preliminary communication and does not contain enough quantitative data to evaluate the

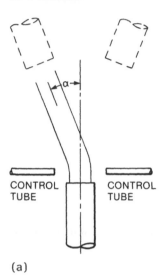

(a)

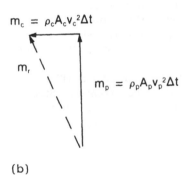

$$m_c = \rho_c A_c v_c^2 \Delta t$$

m_r

$$m_p = \rho_p A_p v_p^2 \Delta t$$

(b)

Fig. 17   Illustration of jet stream interaction and resultant deflection of the power jet.   (From D. L. Letham, *Machine Design*, June 23, 1966, courtesy of Machine Design.)

performance of the device.   Another report may have followed this one, but I have been unable to find it.

D.   Proportional Operation—Laminar Flow

Although the turbulent flow devices work well as bistable or monostable amplifiers, they are quite unstable and noisy when operated in the proportional mode.   Thus, in the late 1960s and early 1970s, the research focus shifted to investigating low Reynolds number laminar jet amplifiers.   These laminar proportional amplifiers (LPA) were found to be less noisy and much more stable than the turbulent jet variety. In addition, they were found to have a high threshold for detecting small differential pressures and a fairly large dynamic range. Their geometry is quite similar to that of their turbulent jet analog, except for the dimensions of the internal passageways. As mentioned, the laminar jet amplifier sample switch described in the last section was an LPA (described below) operated in the bistable mode.

In the beam-deflection type of device, commonly called a laminar proportional amplifier, the laminar jet can be deflected by lower-energy fluid entering from the control nozzles on either side of the power jet. As with the turbulent jet flow variety, the momentum flux of the power jet and the forces exerted on it by the two control jets determine the direction the fluid will assume after it leaves the interaction region. Typically very small differences in pressure at the control nozzles (fraction of an inch of water pressure) will produce large changes in the output pressure observed at the two output ports. The gain of the system is expressed as

$$G = \frac{P_A - P_B}{P_{c1} - P_{c2}}$$

Typically, this gain is on the order of 10 for a standard design LPA. However, the gain is reduced by the load into which the output is working. The effective gain is then determined by the ratio of the output to load resistance in the following manner:

$$G_b = \frac{G}{1 + R_0/R_1}$$

where G_b is the block loaded gain, R_0 the output resistance of the LPA, and R_1 the resistance of the load at the LPA output.

The relationshib between the control port and the output port pressures is shown in Fig. 18. In the region bounded by control pressures P_{c1} and P_{c2}, a proportional relationship exists between output and input. Thus, by staging, that is, using the output differential as the control port differential for another amplifier (as depicted in Fig. 19), the overall gain can be increased to quite large values. Typically, as with their electronic counterparts, the final gain is given by

$$G_F = G_1 \times G_2 \times G_3 \text{ etc.}$$

Theoretically, a three stage gain-of-10 amplifier would have an overall gain of 1000. An original signal pressure differential of 0.001 psi

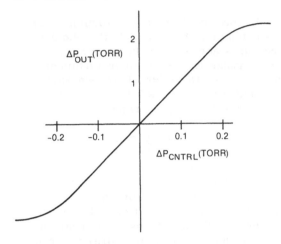

Fig. 18 Typical transfer characteristics of an LPA.

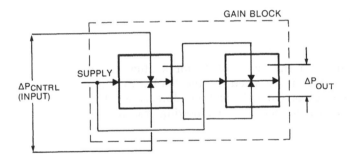

Fig. 19 Staging of LPAs.

would be amplified to 1 psi. However, as indicated previously, this gain is reduced by the load into which the output is working.

V. AN LPA-BASED OPTOPNEUMATIC INTERFACE

The necessity for designing sensors to be used in environments where explosive gases may be present persists as a design restraint for chromatographs to be used in process control applications. One solution which we have favored is to eliminate electricity as a source of power and substitute a more benign source such as pressurized gas or low-pressure steam. In line with this idea, a stand-alone nonelectrical gas chromatograph with its own on-board pneumatic computer has been commercially available for some years.* This unit, however, will analyze complex chromatograms for only two user-selected components and output a pressure signal which is proportional to the concentration of each of them.

The instrument would be much more versatile if

1. It could transmit its chromatogram to a remote location for computer-assisted analysis, thus becoming a multicomponent analyzer.
2. Its on-board pneumatic computer could be reprogrammed from a remote location and/or total control could be managed from the remote location.

A possible solution to this problem was described recently [15,16]. Of interest here is the optical-to-pneumatic conversion scheme which was used.

The problem was to design a transducer which would take a low-power optical signal ($\sim$0.7 mW) and convert it to a 3–15 psig signal containing sufficient power to move a mechanical sample valve from one position to another or to sequence a mechanical multiplexing unit.

The only way of converting optical to mechanical power is through one of its interactions with matter such as the photovoltaic or the photoacoustic effect. We chose the photoacoustic effect since it produces a conversion without traversing the electrical domain, which we were trying to avoid.

The basic concept, which is depicted in Fig. 20, involves modulating a 1-mW light source and transmitting this light via optical fiber so that it impinges on a blackened wall of the cavity which forms the pressure source for the control port of a staged LPA block. The periodic variations in temperature caused by the heat generated in the localized region of the light beam produce a periodic variation in the resultant pressure wave.

That this small pressure variation can be sufficiently amplified by a block of self-staged LPA amplifiers has been demonstrated in a preliminary HDL internal report by Gurney [17]. However, Gurney used a 20-mW He-Ne laser as his light source. With this much power, three stages of amplification, and a final rectifier stage to obtain his final DC output, he was able to obtain sufficient pressure to switch a sensitive diaphragm amplifier to supply the final driving power.

Laminar jet rectifiers are not yet a well-characterized fluidic device and have not performed well in our hands. We found in our research that a significant DC component existed at certain

---

*91-PCT Pneumatic Composition Transmitter manufactured by the Foxboro Company, Foxboro, MA.

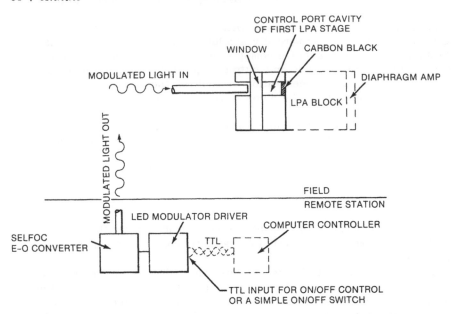

Fig. 20 Simplified system diagram of an optical-to-pneumatic conversion unit. Packaged LED/step index fiber/collimating lens from NSG America Inc., 136 Central Ave., Clark, NJ 07066.

frequencies even in the absence of a final rectifier stage. We elected to exploit this effect by modulating our light at a frequency which would maximize it (in our case this occurred at 625 Hz). Also, since we were interested in operating with low-power light sources (1 mW) it was necessary to use more stages of amplification to obtain the required power.

The LPAs were constructed with the C-format 3.3 × 3.3 cm stainless steel laminates described previously. In our version of the photofluidic interface (shown as an exploded view in Fig. 21), the primary photoacoustic stage is coupled [18,19] to seven other stages of amplification and finally to a diaphragm amplifier (shown in Fig. 22), which acts as the power stage and supplies the 15-psi power control signal needed to accomplish the various tasks.

At an optimum modulation frequency of 625 Hz, gating the modulated 1-mW light source fully "off" or "on" drives the output of the last stage of the eight-stage LPA block from a slightly negative pressure to 10 in. w.c. pressure. Coupling this output to a diaphragm amplifier whose switching point is close to 6 in. water column (w.c.) pressure ensures noise-free operation; i.e., no switching of the ceramic sample valve described below is observed in the presence of externally generated transient vibrations and acoustic noise.

Fig. 21   Exploded view of photofluidic interface.

The control signal is used to directly switch the sample valve, which is in the current commercial version of the pneumatic chromatograph (a 20-psi actuated ceramic-faced sample valve), or alternatively, as shown in Fig. 23, as the control port switching signal in a fluidic sampler. The chromatograms obtained with this version of the fluidic sampler are identical to those obtained with the solenoid-actuated system previously described. This implementation of the fluidic sample injector provides a very flexible system with an injection time and width which can be controlled from a remote location.

Finally, if one wishes to limit the control link to one fiber, the control signal can be used to sequence a mechanical multiplexing unit, which will allow the single control link to accomplish a multitude of tasks, such as reprogramming the on-board pneumatic computer.

A.   Laminar Proportional Amplifiers in Detection Schemes

Effective use of pneumatic resistances as the basis for chromatographic detection has been demonstrated by Novak and Janak [20,21] and Annino et al. [22,23,24]. The latter authors use a close-coupled

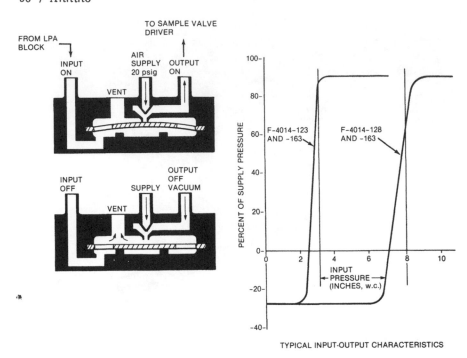

Fig. 22   Air Logic F-4014 series diaphragm amplifier connected to ceramic sample valve. Courtesy of Air Logic Division, 5102 Douglas Ave., Racine, WI 53402.

combination of linear and nonlinear resistances in a force balance amplifier circuit to obtain a signal which is quite insensitive to flow variations. Noval and Janak suggest a bridge circuit combination of resistances to accomplish the flow compensation. The GC signal is measured as an offset across the bridge similar to its TC Weatstone bridge analog. This differential pressure is quite small and requires a sensitive pressure sensor for its measurement.

Although not yet demonstrated as a GC detector, similar bridge circuits have been reported as effective binary gas concentration sensors [25], and an LPA used as a differential pressure sensor has been reported to be the most effective fluidic sensor for measuirng the small differential pressure across the bridge [26]. A schematic of the bridge circuit is shown in Fig. 24. With a few stages of amplification an LPA block was found by workers at HDL to produce 1 mm Hg/% CO_2 with an MDQ of 135 ppm CO_2 in air.

The same laboratory has reported [27] the extension of this work to the development of a sensor for hydrogen sulfide and hydrocarbon gases. However, detection thresholds are still greater than 100 ppm.

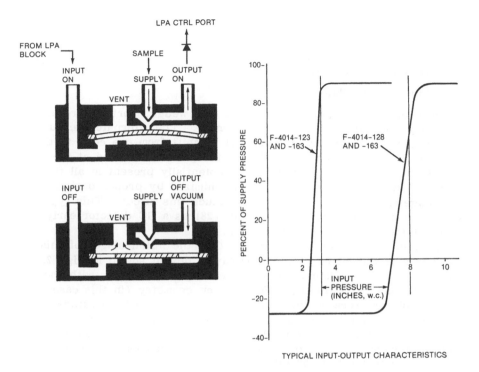

Fig. 23   Same as Fig. 22, connected to a fluidic sampler.

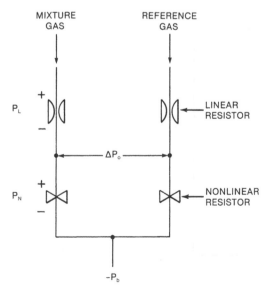

Fig. 24   Pneumatic bridge circuit. In this version the sample is pulled into the bridge by applying a vacuum. (From Technical Sheet FC-101 [April 1979], HDL, 2800 Powder Mill Road, Adelphi, MD 20783.)

VI. FLEURIC OSCILLATORS AS DETECTORS IN GC INSTRUMENTATION

A. Edgetone Oscillator

When a jet strikes the edge or vertex of a wedge, vortices are al-
ternately shed off each side of the vertex. The alternate pressure
fluctuations which occur cause the jet to shift from one edge to the
other with a consequent oscillation of the pressure sensed at the
outlets of the device. This effect is potentially present in all fleuric
devices (at the splitter) but may be minimized by proper design.
The edgetone oscillator was built to induce this effect. This type
of device was proposed by Testerman [28] as a GC detector early
in the history of gas chromatography.

The fleuric oscillator which was used in the initial stages of this
work was of the jet edge resonator cavity design shown in Fig. 25.
In this type of oscillator, the jet edge frequency couples with the
cavity resonance frequency. For a given geometry (in this case
cavity volumes were on the order of 1.5 µl), if the temperature is
maintained constant, the output frequency becomes a function of the
properties of the gas flowing through the device. The basic operating frequencies of these oscillators with hydrogen as carrier gas
were found to be 30 to 50 kHz.

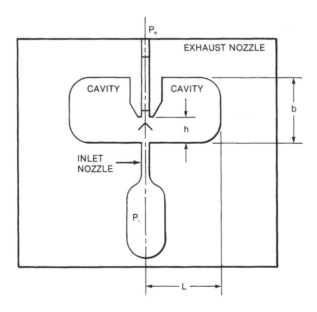

Fig. 25   Edgetone oscillator. (From Ref. 29.)

Two oscillators were used in the detector system. One was connected through the appropriate resistor to receive pure hydrogen and generate a reference frequency, while the other was connected to the exit end of the column. The two oscillator frequencies were mixed, and the resultant beat frequency was measured. This detector was found to have a very large dynamic range, a sensitivity of 1200/1, and a time constant of 10 msec.

A similar device for the measurement of carbon dioxide in breath has been reported [29], as has its incorporation in a respiratory and anesthetic gas analyzer [30].

B. Feedback Oscillators

It is quite obvious that if one were to feed back a portion of the output to the control ports of a beam deflection amplifier in such a manner as to provide positive feedback, oscillation of the jet between the two output ports would take place. Such a device is shown in Fig. 26. This detector was also examined by Testerman and his students and found to give comparable results to the edge-tone oscillators at a decreased cost of manufacture [31]. The device was patented as a fluid oscillator suitable for use as a gas chromatographic detector [32].

Similarly, this type of oscillator has been investigated by others as a detector for certain gases in hazardous areas [33].

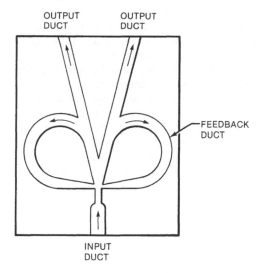

Fig. 26   Feedback oscillator.

VII. CONCLUSION

One of the problems which accompanied the 1960 announcement at HDL of the invention of a family of "no-moving-part" devices was that it so fired the imagination of a large body of individuals that, in their enthusiasm, they oversold the technology. A great many organizations started research programs in the field. Few are left today.

And yet, in the 25 years that have elapsed, a few dedicated research groups have added immensely to our understanding of the operation and design of these devices, and we are in a much better position today to predict their behavior under a variety of operating conditions than we were then.

It is hoped that the preceding discussion of fleuric devices in GC instrumentation will stimulate other researchers to consider this technology when confronted with certain design restraints, but I certainly do not wish to imply that it will provide the solution to all our problems. Complex calculations and signal handling are best done with readily available microprocessors. However, the degree of process hardening necessary to ensure reliable operation of such units when situated in harsh environments may argue for the split architecture proposed by Annino and Caffert [16].

As with all ideas and concepts, there is an appropriate place where fleuric devices can be utilized as the best solution to a particular problem. A case in point is the development at Grumman Aircraft [34] of a fleuric-based flight control system as a backup to the electronic-based flight control system normally in use. Although it is recognized that a fleuric-based flight control system will never be competitive with electronic controls in applications requiring high performance, in those situations where modest performance is acceptable (just get the aircraft home) and high reliability is required (especially immunity to environments of vibration, shock, and electromagnetic interference), fluidics may be the only answer.

Similarly, our interest has been stimulated not only by the basic simplicity and nonwearing qualities of fleuric devices but also by the constant need to design process analyzer units that will not blow up regardless of the environment in which they are placed. Designing an electrically powered unit to "intrinsically safe" standards is certainly one approach, but perhaps (in my opinion) the best and most elegant solution may be to eliminate the electricity altogether.

REFERENCES

1. D. R. Deans, Chromatographia *1*, 18 (1968).
2. D. R. Deans, J. Chromatogr. *289*, 43 (1983).
3. J. M. Kirshner, ed., *Fluid Amplifiers*, McGraw-Hill, New York, 1966.

4. J. M. Kirshner and S. Katz, *Design Theory of Fluidic Components*, Academic Press, New York, 1975.
5. R. L. Wade and S. P. Cram, Anal. Chem. *44*, 131 (1974).
6. S. P. Cram and S. N. Chesler, J. Chromatogr. *99*, 267 (1974).
7. G. Gaspar, P. Arpino, and G. Guiochon, J. Chromatogr. Sci. *15*, 256 (1977).
8. G. Gaspar, R. Annino, C. Vidal-Madjar, and G. Guiochon, Anal. Chem. *50*, 1512 (1978).
9. R. Annino, M.-F. Gonnord, and G. Guiochon, Anal. Chem. *51*, 379 (1979).
10. R. Annino and J. Leone, J. Chromatogr. Sci. *20*, 19 (1982).
11. G. Gaspar, C. Vidal-Madjar, and G. Guiochon, Chromatographia *15*, 125 (1982).
12. C. P. M. Schutjes, C. A. Craemer, C. Vidal-Madjar, and G. Guiochon, J. Chromatogr. *279*, 269 (1983).
13. M. F. Eycon, Jr., and D. J. Schaffer, Report HDL-CR-82-288-1, April 27, 1982.
14. J. Trompler and Frau E. Rokosinyi-Hollos, Chromatographia *2*, 256 (1969).
15. R. Annino and C. Caffert, A Non-Electrical Gas Chromatograph Utilizing Direct Optical to Pneumatic Conversion and Pneumatic to Optical Conversion for Transmission and Control, presented at the Pittsburg Conference, Atlantic City, N.J., March 1986.
16. R. Annino, C. Caffert and E. L. Lewis, Anal. Chem. *58*, 2516 (1986).
17. J. O. Gurney, Jr., Photo-Fluidic Interface, Harry Diamond Laboratories, Adelphi, Md., April 1982.
18. F. M. Manion, Fleurics: 33., Design and Staging of Laminar Proportional Amplifiers, AD-751, 181, Harry Diamond Laboratories, Washington, D.C., Sept. 1972.
19. T. Drzewiecki, Fleurics 42: Some Commonly Used Laminar Fluidic Gain Blocks, Report HDL-TM-82-10, Harry Diamond Laboratories, Aldelphi, Md. (1982).
20. J. Novak and J. Janak, Pneumatic Detector of Gas Chromatographic Fractions, U.S. Patent 3,354,696, Nov. 28, 1967.
21. J. Novak and J. Janak, J. Chromatogr. *138*, 1 (1977).
22. R. Annino, J. Curren, Jr., R. Kalinoski, E, Karas, R. Linquist, and R. Prescott, J. Chromatogr. *126*, 301 (1976).
23. R. Annino and R. Voyksner, J. Chromatogr. *142*, 131 (1977).
24. E. Karas, R. Annino, R. Kalinoski, "Pneumatic Detector for Chromatographic Analyzer," U.S. Patents 4,033,171; 4,095,455; and 4,006,624.
25. F. Villarroel, Analog Fleuric Gas Concentration Sensor, U.S. Patent 3,771,348, Nov. 13, 1973.
26. Fluidic Gas-Concentration Sensor, HDL Technical Sheet FC-101, April 1979.
27. G. E. Stevens and M. F. Funke, Development of a Fluidic Gas Concentration Sensor, Report HDL-CR-80-103, June 16, 1980.

28. M. K. Testerman, Ultrasonic Whistles as Gas Chromatographic Detectors, in *Gas Chromatography*, N. Brenner, J. Callen, and M. Weiss, eds.), Academic Press, New York, 1962, pp. 183–188.
29. F. Villarroel and J. W. Joyce, Fleuric Carbon Dioxide Concentration Sensor, in *Proceedings of the ASME Winter Annual Meeting*, Nov. 29–Dec. 3, 1970.
30. J. M. Calkins, C. K. Waterson, R. J. Saunders, and V. J. Samoy, Ann. Biomed. Eng. *10*, 83 (1982).
31. M. K. Testerman, Research/Development, p. 24 (May 1968).
32. M. K. Testerman and P. C. McLeod, Jr., Fluid Oscillator Analyzer and Method, U.S. Patent 3,273,377, Sept. 20, 1966.
33. E. A. Farber, Fluidic-Oscillator Gas Analyzer, work done for Kennedy Space Center, NASA Tech. Briefs, Summer 1978.
34. R. L. Benoit and J. B. Leonard, Mech. Eng. p. 31 (Nov. 1983).

3

High-Performance Hydrophobic Interaction Chromatography of Proteins

Yoshio Kato   *Toyo Soda Manufacturing Co., Ltd., Yamaguchi, Japan*

I. INTRODUCTION

Hydrophobic interaction chromatography was introduced in 1957 under
the name salting-out chromatography as a technique for separating
substances, according to their hydrophobicities, by gradient elution
with a decreasing salt concentration, which reduces the solubility of
substances such as ammonium and sodium sulfates [1]. The first
applications were to small molecules such as alcohols, polyglycols,
and amines [1–3]. A few years later it was applied to proteins
[4,5]. Serum proteins were purified successfully. Since then, the
technique has been used for the separation of many kinds of proteins. In addition, high-performance hydrophobic interaction chromatography with microparticulate supports was introduced in 1983
and made it possible to separate proteins rapidly with high resolution. It is now one of the most important techniques for protein
separation. In this chapter supports developed recently for high-
performance hydrophobic interaction chromatography are briefly
described first. Then the effects of various chromatographic conditions are discussed in detail. Finally, some applications to protein
purification and analysis are introduced.

II. SUPPORTS

Initially, cross-linked dextrans and agaroses were used to separate
proteins [4–6]. Agarose derivatives carrying both hydrophobic
and charged groups were also examined extensively [7–14]. However, conventional hydrophobic interaction chromatography has been
carried out mainly on uncharged alkyl and aryl derivatives of agarose
[15,16].

The first supports for high-performance hydrophobic interaction
chromatography were reported in 1983 [17]. They were silica-based
supports, prepared by covering a silica surface with hydrophilic
compounds and then incorporating butyl or phenyl groups on them,
as illustrated in Fig. 1. Since then various silica-based supports of
a similar type have been developed [18–24]. These silica-based
supports are rigid, 5 to 10 μm in particle diameter, 250 to 300 Å in
mean pore diameter, and weakly hydrophobic. Owing to the rigidity,
they can be operated under high pressure and hence rapid separations are possible. The small particle sizes of 5 to 10 μm provide
high resolution. The pore diameters of 250 to 300 Å are large
enough for proteins to penetrate into the pores, which facilitates
high protein adsorption capacity and high resolution. The hydrophobicities of the supports are adjusted so that proteins are adsorbed
on them in the presence of 1 to 2 M antichaotropic ions such as sulfate and phosphate, whereas the proteins are not adsorbed on them
in the absence of the salts. Consequently, proteins are first adsorbed
on the supports and then desorbed during chromatography by

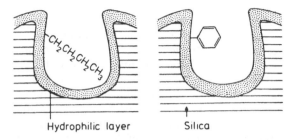

Fig. 1   Schematic structures of silica-based supports prepared first
for high-performance hydrophobic interaction chromatography.

gradient elution with a decreasing salt concentration, just as in conventional hydrophobic interaction chromatography.   This elution
condition is so mild that proteins are usually recovered in their
native state.

Hydrophilic resin-based supports were also developed [25−27].
The basic properties of the resin-based and silica-based supports
are similar except for chemical stability.   Resin-based supports are
chemically very stable, and it is possible to operate them in the pH
range 2 to 12.   A more extreme pH is also acceptable for washing
or regeneration.   This is a great advantage of resin-based supports
over silica-based ones.   Use of silica-based supports is generally
limited to pH values around 8.

Several silica-based and hydrophilic resin-based supports are
now commercially available.   Accordingly, it is now easy to carry
out high-performance hydrophobic interaction chromatography.

III. CHROMATOGRAPHIC CONDITIONS

In high-performance hydrophobic interaction chromatography, proteins are usually separated on columns of 5 to 25 cm × 4.6 to 7.5 mm
inner diameter (i.d.) with 30- to 60-min linear gradients of ammonium
sulfate from 1.5−2.5 M to 0 in common buffers, e.g., 0.1 M phosphate buffer (pH 7.0), at flow rates of 0.5 to 1.0 ml/min at ambient
temperature.   However, the results obtained depend considerably on
the operational variables.   Consequently, it is important to under-
stand the effects of operational variables and select them properly
in order to obtain optimum separations.

A. Type of Support Ligand

The effects of the type of ligand of support have been studied by
many workers [17−22,24,27−30] and it has been found that retention

and selectivity depend substantially on the type of ligand. Proteins are generally more retained on the supports with more hydrophobic ligands. If the ligand is too hydrophobic, it is difficult to elute proteins in native states from the column. In contrast, if the ligand is very hydrophilic, very high concentrations of salt are required to retain proteins. Therefore, the ligand should be moderately hydrophobic. In some cases, however, hydrophilic ligands are preferable. Unstable proteins are sometimes eluted as broad peaks due to partial denaturation when they are separated on supports with moderately hydrophobic ligands. Also, there is a possibility of low recovery for rather hydrophobic proteins. Supports with hydrophilic ligands can provide satisfactory results even for unstable and rather hydrophobic proteins [27, 29].

Selectivity varies more or less with the type of ligand. Accordingly, it may be possible to improve separations by changing the type of ligand.

B. Column Length

Slightly higher resolution is attainable with longer columns in high-performance hydrophobic interaction chromatography of proteins, because the peaks of proteins become narrower with longer columns. When proteins were separated on one, two, and three 75 × 7.5 mm i.d. columns, the two-column system resulted in 10 to 20% narrower peaks than the one-column system, as shown in Fig. 2, and the three-column system yielded 5 to 10% narrower peaks than the two-column system [3]. The separation time increased only slightly with the longer columns.

C. Type of Salt

Retention of proteins depends significantly on the type of salt in hydrophobic interaction chromatography [18, 19, 21, 22, 30–34]. Protein retention is usually increased by using a salt with a greater salting-out ability. When ammonium sulfate, sodium sulfate, and potassium phosphate were compared, sodium sulfate was most effective in retaining proteins and ammonium sulfate and potassium phosphate were comparable, as shown in Table 1 [31]. Therefore, sodium sulfate may be more suitable than ammonium sulfate, which has commonly been used in conventional hydrophobic interaction chromatography. However, because the solubility of sodium sulfate in water is limited (to 1.5 M at 25°C) and depends substantially on the temperature, sodium sulfate is not appropriate when high concentrations of salts are required. In such cases, ammonium sulfate should be better. In contrast, ammonium sulfate is not suitable at alkaline pH because pH control is difficult owing to loss of ammonia. Even at pH 8 the

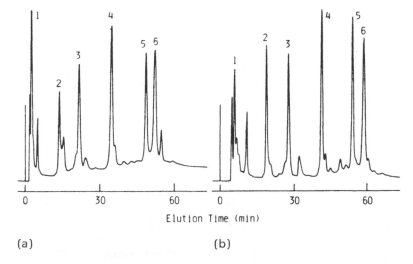

Elution Time (min)

(a)   (b)

Fig. 2   Chromatograms of a protein mixture obtained on column
systems consisting of one (a) and two (b) TSKgel Phenyl-5PW
columns (Toyo Soda, Tokyo, Japan) of 75 × 7.5 mm i.d. at a flow
rate of 1 ml/min with a 60-min linear gradient of ammonium sulfate
from 1.7 M to 0 in 0.1 M phosphate buffer (pH 7.0).   Samples:
(1) cytochrome c, (2) myoglobin, (3) ribonuclease, (4) lysozyme,
(5) α-chymotrypsinogen, (6) α-chymotrypsin.   (From Ref. 31.)

stability is not high.   Sodium citrate is more effective than sodium
sulfate for some proteins [35].   The principal parameter that determines the effect of the type of salt on protein retention is the molal
surface tension increment of the salt, provided there are no specific
interactions between proteins, support, and salt [32].   Use of a salt
of greater molal surface tension increment results in increased retention of proteins.
   Selectivity also depends on the type of salt [22,33].   Therefore,
it may be possible to improve separations by changing the type of
salt.   Figure 3 shows chraomatograms of commercial urease obtained
with gradient elutions of decreasing ammonium sulfate and sodium
sulfate concentration.   Components a and b are separated better with
ammonium sulfate than with sodium sulfate, while components c and d
are separated better with sodium sulfate than with ammonium sulfate.

D.   Initial Salt Concentration

The initial salt concentration has a significant effect on the retention
of proteins [18,20,28,31–33].   Increasing the initial salt concentration

Table 1   Effect of Type of Salt on Elution Volumes of Proteins[a]

| | Elution volume (ml) | | |
Protein	Ammonium sulfate	Sodium sulfate	Potassium phosphate
Cytochrome c	2.70	5.35	2.55
Myoglobin	8.07	15.18	7.50
Ribonuclease	13.50	18.72	13.41
Lysozyme	20.10	24.99	20.70
α-Chymotrypsinogen	27.18	29.58	26.70
α-Chymotrypsin	30.06	32.28	29.49

[a]Proteins were separated on a TSKgel Phenyl-5PW column of 75 × 7.5 mm i.d. at a flow rate of 1 ml/min with a 30-min linear gradient of ammonium sulfate, sodium sulfate, or potassium phosphate decreasing from 1.5 M to 0 in 0.1 M phosphate buffer (pH 7.0).
Source: Ref. 31.

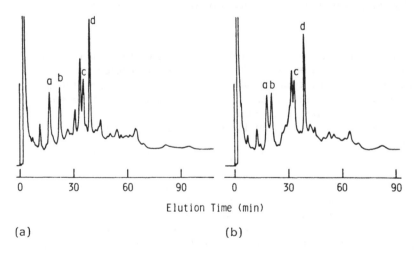

(a)   (b)

Fig. 3   Chromatograms of commercial urease obrained on a TSKgel Phenyl-5PW column of 75 × 7.5 mm i.d. at a flow rate of 1 ml/min with a 60-min linear gradient of ammonium sulfate from 1.5 M to 0 (a) or sodium sulfate from 1.1 M to 0 (b) in 0.1 M phosphate buffer (pH 7.0). (From Ref. 33.)

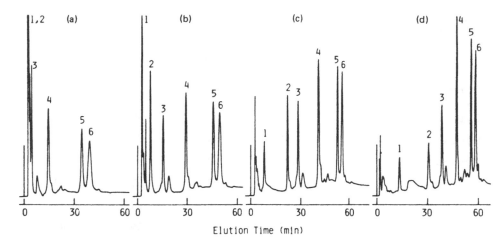

Fig. 4   Chromatograms of a protein mixture obtained on a TSKgel Phenyl-5PW column of 75 × 7.5 mm i.d. at a flow rate of 1 ml/min with a 60-min linear gradient of ammonium sulfate from 1.0 M (a), 1.5 M (b), 2.0 M (c), and 2.5 M (d) to 0 in 0.1 M phosphate buffer (pH 7.0).   Samples as in Fig. 2.   (From Ref. 31.)

generally leads to an increase in protein retention, as illustrated in Fig. 4.   However, the effect is different for individual proteins. The initial salt concentration greatly affects the retention of proteins that elute early but has little effect on the retention of proteins that are tightly retained.

Resolution also varies with the initial salt concentration.   However, there is no definite trend (refer to Fig. 4), and the optimum initial salt concentration for high resolution depends on the sample.   When proteins are separated at different initial salt concentrations but with a constant gradient of decreasing salt concentration, only a parallel shift of the whole chromatogram occurs except for components that elute very early.

Initial salt concentration also affects the recovery of proteins in some cases.   Some proteins are recovered in low yields at initial salt concentrations that correspond to the concentrations that cause intermediate retardation of the proteins; see Table 2.   At initial salt concentrations at which the proteins are not retained or are retained strongly, recoveries are very high.

E.   Eluent pH

Retention of proteins varies more or less with eluent pH [19,22,30, 31,34,35].   This is probably due to changes in the ionization states

Table 2 Recovery of Proteins at Various Initial Ammonium Sulfate Concentrations[a]

Protein	Recovery (%)			
	0.5 M	1.0 M	1.8 M	2.5 M
Cytochrome c	99	100	96	78
Myoglobin	98	95	66	70
Ribonuclease	101	99	92	98
Lysozyme	99	96	97	100
Ovalbumin	97	83	88	86
α-Chymotrypsinogen	93	92	89	101
Trypsin inhibitor	98	72	89	94
α-Chymotrypsin	97	70	78	90

[a]Proteins (each 0.5 mg) were adsorbed on a TSKgel Phenyl-5PW column of 75 × 7.5 mm i.d. in 0.1 M phosphate buffer containing 0.5, 1.0, 1.8, or 2.5 M ammonium sulfate (pH 7.0) and desorbed with a 60-min linear gradient of ammonium sulfate decreasing to 0 at a flow rate of 1 ml/min.
Source: Ref. 31.

of acidic, basic, and terminal amino acids with pH. When the amino acids become uncharged, the proteins should become more hydrophobic and are retained more. When the amino acids become charged, the opposite trend should appear. However, there is no obvious relationship between retention and isoelectric point of individual proteins.

It is often possible to change the selectivity for some components in a sample by manipulating the eluent pH. Accordingly, the eluent pH is another parameter that can be utilized to obtain optimum separations.

F. Slope of Salt Concentration

Resolution is affected by the slope of the salt concentration [31,36]. In general, higher resolution is attained by decreasing slopes of salt concentrations, as shown in Fig. 5. However, this effect is insignificant when the slope is gentle. Also, the effect tends to be less pronounced at higher flow rates. The separation time becomes longer and the dilution of proteins increases almost in inverse

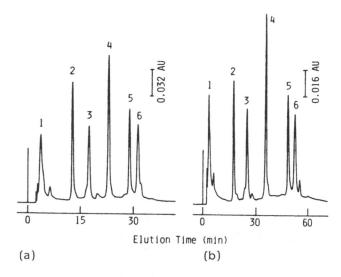

Fig. 5 Chromatograms of a protein mixture obtained on a TSKgel Phenyl-5PW column of 75 × 7.5 mm i.d. at a flow rate of 1 ml/min with a 30-min (a) or 60-min (b) linear gradient of ammonium sulfate from 1.8 M to 0 in 0.1 M phosphate buffer (pH 7.0). Samples as in Fig. 2. (From Ref. 31.)

proportion to the slope of the salt concentration. In addition, the recovery of proteins sometimes depends on the salt concentration slope. Some proteins tend to be recovered in slightly low yield in the case of gentle slopes.

G. Addition of Organic Solvents, Chaotropic Agents, and Surfactants in Eluents

Addition of organic solvents to eluents in low concentration generally promotes desorption of proteins [31]. This effect depends on the type of the organic solvent. When various alcohols were compared, isopropanol was most effective. Retention of proteins was reduced by 10 to 20% on addition of 5% isopropanol in the final buffer (see Table 3). The retention decreases continuously with increasing concentration of organic solvent up to a certain level. However, the addition of more organic solvent seems to promote adsorption rather than desorption. Chaotropic agents such as urea and guanidine hydrochloride have effects similar to those of organic solvents.

Selectivity often varies to some extent when organic solvents and chaotropic agents are added or their concentrations are changed [31].

Table 3 Effect of Addition of Organic Solvent in the Eluent on the Elution Volumes of Proteins[a]

Protein	Elution volume (ml)					
	Buffer	Methanol	Ethanol	Isopropanol	n-Butanol	Acetonitrile
Cytochrome c	7.95	8.16	8.10	5.73	8.25	8.43
Myoglobin	15.54	15.15	14.55	11.85	—[b]	15.15
Ribonuclease	19.80	18.90	18.15	14.70	18.06	18.84
Lysozyme	24.69	23.10	21.90	18.18	22.35	22.50
α-Chymotrypsinogen	29.67	28.80	27.90	25.14	26.40	28.50
α-Chymotrypsin	32.13	30.15	27.90	25.14	28.95	28.50

[a]Proteins were separated on a TSKgel Phenyl-5PW column of 75×7.5 mm i.d. at a flow rate of 1 ml/min with a 30-min linear gradient from 0.1 M phosphate buffer containing 2 M ammonium sulfate (pH 7.0) to 0.1 M phosphate buffer (pH 7.0) containing 5% organic solvent.
[b]Myoglobin was not eluted.
Source: Ref. 31.

Addition of surfactants, e.g., 3-[(3-cholamidopropyl)dimethylammonio]-1-propane sulfonate (CHAPS), at concentrations below the critical micelle concentration also alters the selectivity [37]. Therefore, it is expected that separation can be improved by the addition of organic solvents, chaotropic agents or surfactants.

Addition of small quantities of an organic solvent or chaotropic agent to the eluent does not seem to cause denaturation of proteins. Ferredoxin-NADP$^+$ reductace activity was recovered in more than 90% yield when it was separated with a gradient elution from 0.1 M phosphate buffer containing 1.5 M ammonium sulfate (pH 7.0) to 0.1 M phosphate buffer (pH 7.0) containing 5% isopropanol, 2M urea, or 1 M quanidine hydrochloride [31]. Furthermore, it is likely that the addition of small quantities of organic solvents can reduce the possibility of protein denaturation in some cases. It was found that addition of 5 to 10% methanol to the eluent could greatly improve the peak shape when proteins were eluted as broad peaks [38]. The broad peaks probably resulted from partial denaturation of the proteins during separation, and the addition of methanol was effective in eliminating or reducing the denaturation.

H. Flow Rate

The flow rate affects resolution, separation time, and dilution of the sample [31,34,36]. In separations at constant gradient time, slightly higher resolution is achieved at lower flow rates. This flow rate dependence of the resolution is almost negligible at short gradient times and becomes pronounced as the gradient time increases. The dilution of proteins during separation decreases almost in proportion to the flow rate, while the separation time increases only slightly with decreasing flow rate, as seen from Fig. 6. The resolution increases slightly with decreasing flow rate for separations at constant gradient volume. In this case, however, the separation time becomes longer in almost inverse proportion to the flow rate, as seen from Fig. 7. Dilution of the proteins decreases with flow rate. The recovery of some proteins tends to decrease slightly at higher flow rates [31].

I. Temperature

Temperature has an effect on retention because the strength of hydrophobic interactions increases in general with temperature. As expected, proteins elute later as the temperature increases [21,29,38]. However, the influence of temperature on hydrophobic interactions is not the same for all proteins and the retention of individual proteins varies differently.

In hydrophobic interaction chromatography, proteins are usually eluted from the column in their native state. However, conformational

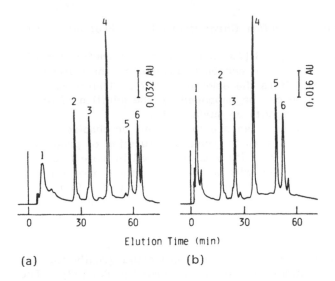

(a)   (b)

Fig. 6   Chromatograms of a protein mixture obtained on a TSKgel
Phenyl-5PW column of 75 × 7.5 mm i.d. at a flow rate of 0.5 ml/min
(a) or 1 ml/min (b) with a 60-min linear gradient of ammonium sulfate
from 1.8 M to 0 in 0.1 M phosphate buffer (pH 7.0).   Samples as in
Fig. 2.   (From Ref. 31.)

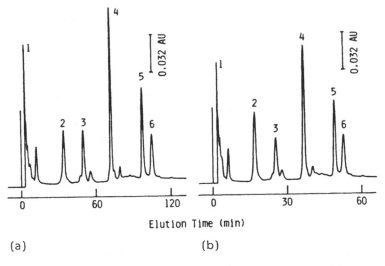

(a)   (b)

Fig. 7   Chromatograms of a protein mixture obtained on a TSKgel
Phenyl-5PW column of 75 × 7.5 mm i.d. at a flow rate of 0.5 ml/min
(a) or 1 ml/min (b) with a linear gradient of ammonium sulfate from
1.8 M to 0 in 0.1 M phosphate buffer (pH 7.0).   The gradient volume
was 60 ml in each case.   Samples as in Fig. 2.

changes may be induced in some cases depending on the stability of the protein, the hydrophobicity of the support, and the temperature. If conformational changes occur, proteins elute later as broad peaks. Because conformational changes tend to occur at higher temperatures, separations at subambient temperatures should be preferable for unstable proteins [29,38].

J. Sample Loading

Resolution usually decreases as the sample loading increases above a certain limit, although it is almost constant up to the limit [19,25, 36]. Accordingly, sample loading should be kept lower than the limit to obtain the highest resolution. However, the decrease in resolution above the limit is not so serious, and hence more of the sample can be applied if a slight decrease in resolution is acceptable. Figure 8 shows chromatograms of a crude sample of phosphoglucose isomerase obtained on a column of 15 × 2.15 cm i.d. with various sample loadings. Up to 100 mg the sample could be applied without loss of separation efficiency. Slightly broader peaks were observed with sample loadings above 100 mg. The limit of sample loading that permits the highest resolution depends on the column size, support, sample, and so forth. The limits of sample loading on commercial columns seem to be 10 to 50 mg/cm^2 of column sectional area for crude samples and 0.1 to 1 mg/cm^2 for pure samples.

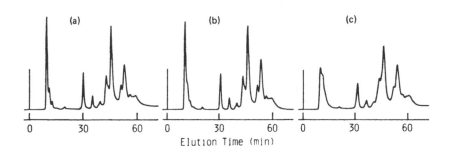

Fig. 8 Chromatograms of phosphoglucose isomerase crude sample obtained on a TSKgel Phenyl-5PW column of 15 × 2.15 cm i.d. at a flow rate of 4 ml/min with a 60-min linear gradient of ammonium sulfate from 1.5 M to 0 in 0.1 M phosphate buffer (pH 7.0) with sample loadings of 50 (a), 100 (b), and 200 mg (c). (From Ref. 36.)

IV. APPLICATIONS

The usefulness of high-performance hydrophobic interaction chromatography has been demonstrated for purifications and analyses of various proteins. Some examples are given here.

The rat liver enzymes carbamoyl-phosphate synthetase I and ornithine transcarbamylase from crude mitochondrial extracts were purified in 20 min with 3- to 15-fold increases in specific activity [39]. Recoveries of enzymatic activity were more than 95%.

High-performance hydrophobic interaction chromatography was used in the final step of aromatic L-amino acid decarboxylase purification from human pheochromocytoma. Specific activity was increased 12-fold in the 51% yield. The purified L-amino acid decarboxylase showed a single band of sodium dodecyl sulfate-polyacrylamide gel electrophoresis [40].

Figure 9 shows an example of enzyme purification on a semipreparative column. A 200-mg sample of commercial crude lipoxidase was separated. The peak corresponding to lipoxidase between the two vertical lines was fractionated. Recovery of enzymatic activity in the fraction was 86%. The degree of purification based on the specific activity was 5.6-fold. The results of purity tests of the fraction by high-performance liquid chromatography (HPLC) are shown in Fig. 10. One major peak and several very small peaks are seen in each chromatogram of the fraction. Because the major peaks showed enzymatic activity (except in the case of reversed-phase chromatography), they must correspond to lipoxidase. This means that

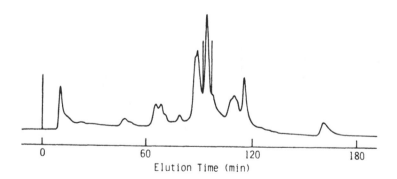

Elution Time (min)

Fig. 9 Purification of lipoxidase on a TSKgel Phenyl-5PW column of 15 × 2.15 cm i.d.. A crude sample of 200 mg was separated at a flow rate of 4 ml/min with a 120-min linear gradient of ammonium sulfate from 1.5 M to 0 in 0.1 M phosphate buffer (pH 7.0). (From Ref. 36.)

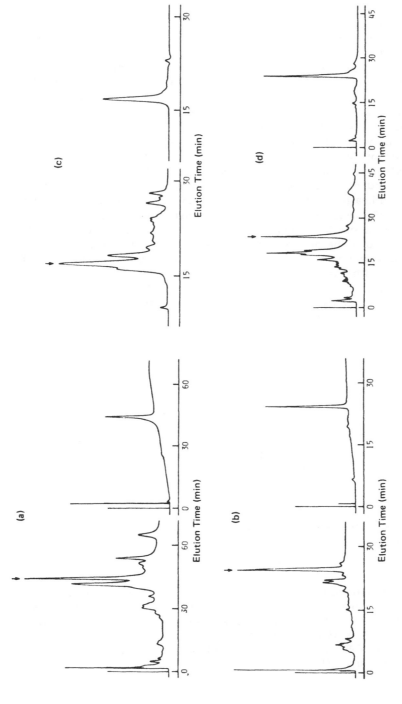

Fig. 10 Chromatograms of a crude lipoxidase sample (left) and the fraction in Fig. 9 (right) obtained by hydrophobic interaction chromatography (a), reversed-phase chromatography (b), gel filtration (c), and ion exchange chromatography (d). Peaks indicated by arrows in the chromatograms of the crude sample correspond to the main peaks of the fraction. (From Ref. 36.)

lipoxidase of high purity was obtained from commercial lipoxidase containing large amounts of impurities. Phosphoglucose isomerase and lactate dehydrogenase were also purified successfully on the same column [36].

Figure 11 shows another example of large-scale separation. The separation of 16 ml of human serum (containing about 1.2 g proteins) was performed on a 20 × 5.5 cm i.d. column. In preliminary experiments with various sample loadings, almost identical patterns were observed for up to 16 ml of human serum. The albumin peak was collected between the two vertical lines. Recovery of albumin was determined by immunodiffusion and found to be 92% in the fraction. The purity was examined by immunoelectrophoresis (Fig. 12). Only a single band corresponding to albumin is seen in the pattern of the fraction, indicating that very pure albumin was obtained in one step of high-performance hydrophobic interaction chromatography. In a purification of crude α-amylase on the same column, up to 1.5 g of the sample could be applied without loss of separation efficiency. The separation was completed within 1 hr, including regeneration of the column with the initial eluent, and α-amylase of high purity was obtained with 90% yield of enzymatic activity [27].

The analysis of estrogen receptors labeled with [^{125}I]iodoestradiol-17β was examined and it was found that high-performance hydrophobic

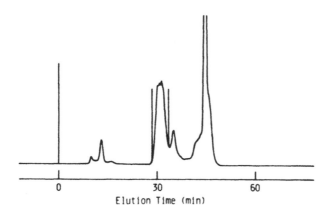

Elution Time (min)

Fig. 11   Purification of human serum albumin on a TSKgel Ether-5PW column (Toyo Soda) of 20 × 5.5 cm i.d.. Human serum (16 ml) containing about 1.2 g of proteins was separated at a flow rate of 40 ml/min with a 36-min linear gradient of ammonium sulfate from 1.7 M to 0.68 M, followed by step gradient to zero in 0.1 M phosphate buffer (pH 7.0). (From Ref. 27.)

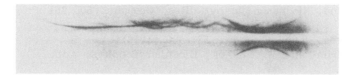

Fig. 12   Immunoelectrophoresis patterns of the albumin fraction in Fig. 11 (bottom) and the original human serum (top). (From Ref. 27.)

interaction chromatography may be useful for isolating isoforms of steroid hormone receptors [41].

High-performance hydrophobic interaction chromatography was applied successfully to the quantitative determination of intracellular ferritin-associated radioactive iron. A single ferritin-containing radioactive peak was separated from the bulk of the non-ferritin-bound iron peak. Total recovery of radioactivity was 98% [42].

V. CONCLUSION

High-performance hydrophobic interaction chromatography is now one of the most important techniques for separating proteins. It separates proteins according to their hydrophobic properties. Reversed-phase chromatography is another approach that has been used for the same purpose; however, since proteins are separated in buffers containing large quantities of organic solvents at low pH, there is always a risk of denaturation, and also proteins are often recovered in low yield. In contrast, because of its mild eluting conditions, hydrophobic interaction chromatography can be performed with almost quantitative recovery of both protein mass and enzymatic activity, and hence it is more versatile than reversed-phase chromatography.

REFERENCES

1. R. N. Sargent and W. Rieman, J. Phys. Chem. *61*, 354 (1957).
2. R. N. Sargent and W. Rieman, Anal. Chim. Acta *17*, 408 (1957).
3. R. N. Sargent and W. Rieman, Anal. Chim. Acta *18*, 197 (1958).
4. R. N. Sargent and W. Rieman, Anal. Chim. Acta *30*, 101 (1964).
5. J. Porath, Nature (London) *196*, 47 (1962).
6. P. Swanljung, Anal. Biochem. *43*, 382 (1971).
7. R. J. Yon, Biochem. J. *126*, 765 (1972).
8. Z. Er-el, Y. Zaidenzaig, and S. Shaltiel, Biochem. Biophys. Res. Commun. *49*, 383 (1972).

9. B. H. J. Hofstee, Biochem. Biophys. Res. Commun. *50*, 751 (1973).

10. B. H. J. Hofstee, Anal. Biochem. *52*, 430 (1973).

11. R. Jost, T. Miron, and M. Wilchek, Biochim. Biophys. Acta *362*, 75 (1974).

12. B. H. J. Hofstee and N. F. Otillio, J. Chromatogr. *159*, 57 (1978).

13. G. Halperin, M. T. Finkelstein, and S. Shaltiel, J. Chromatogr. *317*, 103 (1984).

14. S. Shaltiel, Methods Enzymol. *104*, 69 (1984).

15. J. Porath, L. Sundberg, N. Fornstedt, and I. Olsson, Nature (London), *245*, 465 (1973).

16. S. Hjertén, J. Rosengren, and S. Påhlman, J. Chromatogr. *101*, 281 (1974).

17. Y. Kato, T. Kitamura, and T. Hashimoto, J. Chromatogr. *266*, 49 (1983).

18. D. L. Gooding, M. N. Schmuck, and K. M. Gooding, J. Chromatogr. *296*, 107 (1984).

19. J. L. Fausnaugh, L. A. Kennedy, and F. E. Regnier, J. Chromatogr. *317*, 141 (1984).

20. J. L. Fausnaugh, E. Pfannkoch, S. Gupta, and F. E. Regnier, Anal. Biochem. *137*, 464 (1984).

21. N. T. Miller, B. Feibush, and B. L. Karger, J. Chromatogr. *316*, 519 (1984).

22. J.-P. Chang, Z. E. Rassi, and Cs. Horváth, J. Chromatogr. *319*, 396 (1985).

23. L. A. Kennedy, W. Kopaciewicz, and F. E. Regnier, J. Chromatogr. *359*, 73 (1986).

24. A. J. Alpert, J. Chromatogr. *359*, 85 (1986).

25. Y. Kato, T. Kitamura, and T. Hashimoto, J. Chromatogr. *292*, 418 (1984).

26. S. Hjertén, K. Yao, Z.-Q. Liu, D. Yang, and B.-L. Wu, *354*, 203 (1986).

27. Y. Kato, T. Kitamura, and T. Hashimoto, J. Chromatogr. *360*, 260 (1986).

28. M. N. Schmuck, K. M. Gooding, and D. L. Gooding, LC Magazine *3*, 814 (1985).

29. S.-L. Wu, K. Benedek, and B. L. Karger, J. Chromatogr. *359*, 3 (1986).

30. D. L. Gooding, M. N. Schmuck, M. P. Nowlan, and K. M. Gooding, J. Chromatogr. *359*, 331 (1986).

31. Y. Kato, T. Kitamura, and T. Hashimoto, J. Chromatogr. *298*, 407 (1984).

32. W. R. Melander, D. Corradini, and Cs. Horváth, J. Chromatogr. *317*, 67 (1984).

33. T. Kitamura and Y. Kato, J. High Resolut. Chromatogr. Chromatogr. Commun. *8*, 306 (1985).

34. S. Hjertén, K. Yao, K.-O. Eriksson, and B. Johansson, J. Chromatogr. *359*, 99 (1986).
35. J. L. Fausnaugh and F. E. Regnier, J. Chromatogr. *359*, 131 (1986).
36. Y. Kato, T. Kitamura, and T. Hashimoto, J. Chromatogr. *333*, 202 (1985).
37. D. B. Wetlaufer and M. R. Koenigbauer, J. Chromatogr. *359*, 55 (1986).
38. S. C. Goheen and S. C. Engelhorn, J. Chromatogr. *317*, 55 (1984).
39. N. T. Miller, B. Feibush, K. Corina, S. P. Lee, and B. L. Karger, Anal. Biochem. *148*, 510 (1985).
40. H. Ichinose, K. Kojima, A. Togari, Y. Kato, S. Parvez, H. Parvez, and T. Nagatsu, Anal. Biochem. *150*, 408 (1985).
41. S. M. Hyder, R. D. Wiehle, D. W. Brandt, and J. L. Wittliff, J. Chromatogr. *327*, 237 (1985).
42. D. Josic, E. Mattia, G. Ashwell, and J. V. Renswoude, Anal. Biochem. *152*, 42 (1986).

4

HPLC for Therapeutic Drug Monitoring and Determination of Toxicity

Ian D. Watson   *Royal Infirmary, Glasgow, Scotland*

I. INTRODUCTION

A. Need for Therapeutic Drug Monitoring and Clinical Toxicological Assays

The indications for measurement of serum* drug concentrations are that there be (a) no readily quantifiable pharmacological manifestation, (b) adequate demonstration that pharmacological effects, whether

---

*Although serum is referred to throughout the text, the term plasma may readily be substituted. For the overwhelming majority of drugs it does not matter which fluid is assayed.

therapeutic or toxicological, are related to serum drug concentration, and (c) agreement that knowledge of the serum drug concentration will affect therapeutic decision-making in the individual. If such criteria are applied to the range of drugs currently available only a small proportion are worthy of routine analysis for the purposes of therapeutic drug monitoring (TDM) and clinical toxicology (CT), although many drugs are needlessly measured purportedly for these purposes. The discussion in this chapter will be confined to drugs for which there is at least some supporting evidence that assays are necessary to monitor therapy and diagnose and follow intoxication.

In TDM there must be a so-called therapeutic range for the drug of interest; the kinetics in the population of interest should be reasonably well understood and thus dosage individualization should be intended and practiced. In clinical toxicology a knowledge of serum concentration should be diagnostically helpful and allow initiation of appropriate therapy.

B. Assays for Therapeutic Drug Monitoring and Clinical Toxicology

As drug concentrations in CT are higher than those in TDM for a given drug, it is possible to use less sensitive techniques for such analyses. In the past colorimetry and spectrophotometry were used, e.g., for salicylates and barbiturates, but these suffered from inaccuracy and imprecision. TDM requires more sensitive techniques such as fluorimetry but suffers from similar limitations.

Gas-liquid chromatography (GLC) lent itself to provision of the quantitative and qualitative information required in CT and TDM; however, GLC requires a reasonable amount of time and skill and generally remained the prerogative of specialists. More recently, immunological assays have been introduced, greatly simplifying assays and improving turnaround times. Although radioimmunoassay has been used, particularly where there is a need for sensitivity, homogenous enzyme immunoassay and fluorescence polarization immunoassay are rapid, simple to operate, accurate, and precise, particularly for TDM [1—3]. However, such immunoassays require the use of manufacturers' reagents and can be costly.

From the early 1970s liquid chromatography (LC) has been shown to be cheaper than immunoassay for anticonvulsants [4]. High-performance liquid chromatography (HPLC) was applied to the measurement of drugs, and as LC systems have become more widely available an increasing number of methods related to the determination of drugs for TDM have been published. LC was an attractive method of analysis as it did not have the limitation of volatility of GLC or the frequent need for derivitization, yet it gave quantitative and qualitative information. As awareness of the pharmacokinetics of monitored drugs improved, it became increasingly obvious that for a significant number of drugs measurement of pharmacologically active metabolites was necessary.

As reversed-phase LC became more popular [5], it had the effect of allowing the utilization of much less complex sample preparation techniques. Allied to these advantages was the flexibility of an LC system for method development to the extent that today LC is widely used for endogenous and exogenous metabolites [6,7] and nearly every clinical laboratory has access to an LC—frequently for drug analysis.

It is the purpose of this chapter to consider the place of LC in therapeutic drug monitoring and clinical toxicology.

C. Sample Preparation

Drug analysis in biological fluids is hampered by the complexity of the matrix: drug metabolites, endogenous metabolites, proteins, lipids, etc. can all contribute to the imprecision and inaccuracy of the analysis.

Sample preparation and analytical strategy in TDM and CT have been considered [8–14]. The need for correct sampling procedures has also been emphasized [15].

Liquid-Liquid Partitioning

Historically, this is the usual method of analyte isolation and due to its great flexibility it has been widely and successfully used. Liquid-liquid partitioning is the distribution of the compound of interest between two immiscible liquids; the degree of distribution can be expressed as a partition coefficient:

$$P = \frac{O_c}{A_c} \tag{1}$$

where P is the partition coefficient for a liquid-liquid system, O_c is the concentration in the organic phase, and A_c the concentration in the aqueous phase.

There are similar expressions for the phase volume ratio (V) and the fraction of analyte in the organic phase (O_f). By appropriate substitution it can be shown that

$$O_f = \frac{PV}{PV + 1} \tag{2}$$

Partition between the phases can be affected by a number of variables to improve recovery and minimize interfering extractants. These factors include variation in ionic strength, solvent strength, solvent selectivity, and pH control. A pH is usually chosen to maintain ion suppression (i.e., pH $>$ 2) from the pK_a of the analyte (equivalent to $<$1% analyte ionized). Back-extraction is frequently used and it is not uncommon even today to find multiple extraction methods; the latter should be avoided if possible to improve analytical throughput provided assay performance is not degraded.

Although liquid-liquid partitioning is flexible, there are several disadvantages to the technique. Safety problems from flammability, explosion, toxicity, or carcinogenicity are significant factors, losses due to solvent-mediated decomposition of the analyte or adsorption to glassware may cause difficulty, as may the concentration of solvent impurities during the commonly used evaporation stage, and emulsion formation is often a source of difficulty. Alternative techniques are gaining favor.

Liquid-Solid Extraction

The wide variety of liquid-solid extraction options have recently been surveyed [13]; those commonly used may be subdivided into ion-exchange resins, silica adsorbents (including reversed-phase materials), and hydrophobic polymers (e.g., XAD-2, which is mainly reliant on van der Waals forces). It is not proposed to discuss these in great detail, but useful guides have been published [8,13, 14,16].

Briefly, liquid-solid extraction may be considered as a distribution ratio between the liquid and solid phases:

$$D = \frac{V_s A_r}{V_b A_s} \qquad (3)$$

where D is the coefficient of distribution, V_s the volume of solution, V_b the resin bed volume, A_r the amount of solute on resin, and A_s the amount of solute in solution. Thus varying D determines the degree of affinity for the solid phase; large values of D favor retention whereas small values favor elution.

Prior to the availability of small commercial columns, liquid-solid extraction was not an attractive option due to column irreproducibility and time required to ensure that the resin was free from impurities. However, it lends itself to batch analysis, an important consideration in TDM, and to automation (either off-line, as with Vac-Elute, Prep, or Sep-Pak cartridges, or on-line as with AASP*). Such developments have significant implications for sample throughput, assay imprecision, and cost, as has been discussed [13,14].

---

*Trademarks: Vac-Elut (Analytichem), Prep (Du Pont), Sep-Pak (Waters), AASP (Analytichem).

Dilutional Methods

Protein is the major problem in biological sample handling. In extraction procedures it remains in the aqueous phase. However, reversed-phase LC, which is compatible with aqueous solutions, is widely used in TDM and CT and a number of protein removal procedures have been developed. Since these involve dilution of the sample they are not readily suitable for trace analysis.

Protein Precipitation: A number of water-miscible organic solvents have been used as protein precipitants. Acetonitrile is the most popular [17], although acetone has been claimed to have some merit [18]. However, incomplete protein precipitation shortens column life and can lead to poor assay performance due to late-eluting peaks [19]. The use of inorganic protein precipitants may give rise to similar problems and lead to poor analyte recovery [20]. One method of improving recoveries in serum, blood, or tissue is to use the proteolytic enzyme subtilisin Carlsberg, which greatly simplifies analysis, particularly for toxicological purposes, where matrix problems, particularly with tissue, can cause incomplete recovery [21, 22].

Ultrafiltration: Ultrafiltration may be used for drugs that are not extensively protein-bound or that have their protein binding minimized by pretreatment. This has been the basis of assays for total drug in serum [23, 24] and is an ideal method for measuring free (i.e., unbound) drug levels routinely [25, 26]. In the author's experience such approaches are less satisfactory when applied to basic drugs because of variable recoveries caused by membrane binding of the analyte. Microultracentrifugation has been proposed for the determination of free drug levels; it has been applied to phenytoin analysis [27] but should be suitable for the analysis of free basic drugs.

Direct Injection

Direct injection of serum onto an LC column is rarely practiced as protein precipitation causes blockages and rapidly degrades column selectivity. Injection onto a preconcentration column has been used, and Roth and Beschke [28] describe a fully automated assay with the flexibility required to allow the assay of a number of drugs. Use of column switching in sample cleanup has been reviewed [29]. Micellar chromatography, which allows direct injection of serum into the liquid chromatography shows promise [30–32] (Fig. 1).

Internal Standardization and Calibration

Calibrants should be prepared in a matrix not too dissimilar to that of the material to be tested. It is generally accepted that there will be sample-to-sample variation in extraction and that the use of

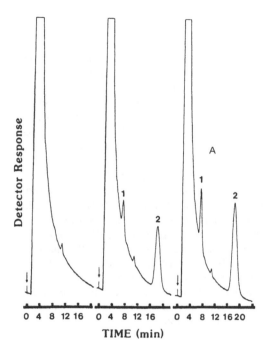

Fig. 1 Micellar chromatography, A: chromatogram of serum contain-ing 400 µg/liter propranolol (1) and 3 mg/liter quinidine (2). Chroma-tographic conditions: column 10-µm µBondapak C_{18}; mobile phase, 0.03 M SDS with 10% propranolol; flow rate, 1.0 ml/min; fluorescence detection, excitation wavelength 215 nm, emission cutoff 300 nm. (From Ref. 30.)

an internal standard allows compensation for such vagaries; commonly such a standard is an analog of the analyte, although it has been claimed that any compound will be satisfactory provided it is repro-ducibly ionized at the extraction pH [19,33].

However, as has been noted [34], the use of an internal standard does not guarantee compensation for analytical variables, as it is equally possible that the internal standard will decompensate, i.e., that it will not have parallel physicochemical behavior in all conditions of routine practice. The fallacies and pitfalls of using internal standards as analytical compensators have been thoroughly considered [35], which supports the earlier view and gives criteria for the choice of an internal standard. It has been suggested [19] that the k' of the internal standard should be within 30% of the k' of the analyte.

II. ANTICONVULSANTS

Anticonvulsants were among the first drugs for which therapeutic ranges were established [36,37], although serum drug concentration should be used as an adjunct to clinical judgment and not as a guide to achieving a concentration in a "therapeutic" range [38]. Current opinion favors one drug for epilepsy in the individual patient [39]; a range of anticonvulsants are used in therapy, reflecting the different clinical requirements. The most frequently used anticonvulsants, their therapeutic ranges, and their pharmacologically significant metabolites are listed in Table 1.

The chromatographic analysis of anticonvulsants has been reviewed [40–42]. Often the assays are designed to allow simultaneous analysis of a number of anticonvulsants and their metabolites. This is advantageous if (a) the anticonvulsant has a pharmacologically significant metabolite, (b) a patient is prescribed more than one anticonvulsant, or (c) the population of epileptics served by the laboratory receives a number of anticonvulsants although each individual may only receive one. Most recent papers emphasize multiple analysis, usually of hydantoin, suximide, barbiturate, and tricyclic anticonvulsants (Fig. 2); simultaneous determination of benzodiazepine anticonvulsants and valproate is the expection.

Table 1   Commonly Used Anticonvulstants and Their Metabolites

Drug	Therapeutic range, mg/liter (μmol/liter)	Metabolite(s)
Phenytoin	10–20 (40–80)	Hydroxyphenytoin
Phenobarbitone	20–40 (80–160)	Hydroxyphenobarbitone
Primidone	8–12 (32–48)	Phenobarbitone Phenylethylmalonamide
Carbamazepine	4–10 (17–42)	10,11-Epoxycarbamazepine CGP 10,000
Ethsuximide	40–100 (285–710)	Desethylsuximide
Methsuximide	Rarely used	N-Desmethylsuximide
Valproate	50–100 (350–700)	
Clonazepam	0.01–0.09 (0.04–0.29)	

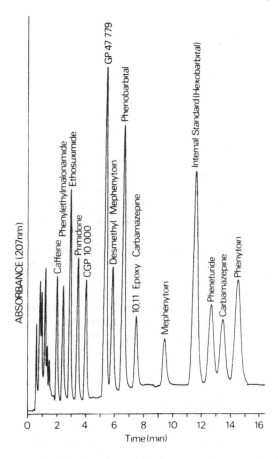

Fig. 2   Chromatographic separation of anticonvulsants at therapeutic concentrations added to drug-free serum and extracted as a patient sample.   (From Ref. 58.)

A. Assays for Individual Drugs

Clonazepam

Clonazepam is used primarily in petit mal seizures and motor seizures in children, although it may be used for a variety of indications. The therapeutic concentrations of clonazepam are considerably less than for other anticonvulsants with a relatively wide therapeutic range.   The indications for monitoring clonazepam are few, as toxicity (ataxia, drowsiness) may be detected clinically.

A normal-phase LC assay has been described [43] but carbamazepine is not fully separated from clonazepam. A number of reversed-phase procedures have been described [44–49].

Liquid-liquid extraction with hexane-ethyl acetate [47,48] or chloroform [44] has been used; however, liquid-solid extraction yields coefficients of variation (CVs) of less than 5% [46,49] at the midpoint of the therapeutic range.

Acetonitrile-acetate buffer mixtures have most frequently been used as mobile phases [44,46–48] on C_{18}, C_8, and C_6 packings operating at ambient temperatures, although acetonitrile and phosphate buffer at 50°C have been used [49]. Clonazepam has been monitored at 306 nm [44,48] and 313 nm [46]. As might be anticipated, other benzodiazepines are a potential source of interference; chlordiazepoxide interfered with some assays [44,48]. This interference was avoided by Taylor et al. [46], which may be attributable to the inclusion of methanol in the mobile phase.

Determination of the metabolites may be performed [47], but a different set of chromatographic conditions must be used. In addition, a carbamazepine metabolite interferes with determination of 7-amino-clonazepam and a further extraction step is necessary.

As noted by Burke and Thenot [42], it is surprising that electrochemical detection has not been used, although technically this might not be entirely appropriate for a routine therapeutic monitoring assay.

Valproate

The indications for valproate determination are few, but nevertheless it is frequently measured. Since it has weak ultraviolet (UV) absorption and can be readily determined by GLC, there are few LC procedures. These usually utilize derivatization to produce an adequate chromophore. In addition to the procedures noted by Burke and Thenot [42], other procedures have been reported using phenacyl bromide [50–52], ODS columns, and acetonitrile: buffer or water eluants, acetonitrile precipitation for sample preparation with detection at 254 or 245 nm, and nonanoic acid [50,52] or cyclohexane carboxylic acid [51] as internal standard.

Valproate has been determined using UV detection as part of an anticonvulsant analytical profile [53]. This is discussed in more detail below. Given the poor pharmacological correlation with serum valproate level and the ease with which valproate may be analyzed with GC, there seems to be no advantage in using LC assays.

Other Anticonvulsants

Few papers now report assays for single anticonvulsant drugs. Most report simultaneous analysis of a number of compounds, although the metabolism of anticonvulsants (for review see [54]) is of some interest [55] and the accuracy of the analysis can be affected by sampling [56]. Liquid chromatographic assays for individual anticonvulsants have been reviewed [41,42].

B. Simultaneous Determination of Anticonvulsants

Economically and technically, it is attractive to have assays that allow determination of any drug in a "panel" of drugs by a single procedure; such an approach is more complicated if metabolites also require quantitation. Sample profiling has been developed for determining anticonvulsants, particularly those with similar physicochemical properties, thus allowing common sample preparation, chromatographic, and detection steps.

Sample Preparation

Anticonvulsant assay by LC does not lack sensitivity, provided adequate sample is available; therefore the aim of sample preparation is to remove potential interferences and to isolate the analytes from proteins. Liquid-liquid and liquid-solid extraction and protein precipitation have all been used.

Liquid-Liquid Extraction: Despite the time-consuming nature of liquid-liquid extraction it continues to be used, usually for the anticonvulsants, phenytoin, phenobarbitone, carbamazepine, and primidone. Solvents used include ethyl acetate [57–60], diethyl ether [61–63], a chlorhydrocarbon solvent [64], or a chlorhydrocarbon solvent with either an alcohol [65] or diethyl ether [66]; benzene has been used [67] but cannot be recommended on safety grounds. Recoveries of carbamazepine, phenytoin, and phenobarbitone are usually better than 90% [58,60,65]; that for primidone should be better than 85% but ethsuximide recovery tends to be of the order of 75% [58,60]. Wad [58] noted that evaporation should be performed at 37°C to avoid degradation of ethsuximide; with such a limitation diethyl ether might be a better extraction solvent than ethyl acetate. There was also a high level of imprecision associated with the estimation of this analyte. Subsequently, Wad [68] suggested a solvent demixing procedure, avoiding an evaporation step, which improved the precision for ethsuximide without deficit to carbamazepine, phenobarbitone, and phenytoin, although the precision and accuracy for primidone worsened. There was no precision information on the numerous metabolites previously examined.

Other authors have used solvent demixing for sample preparation [53,69]. Alric et al. [69] were able to show statistically that between-extract variance was negligible. Unfortunately, they did not evaluate assay accuracy. It has been claimed that solvent demixing yields heavily contaminated extracts, but Kushida and Ishizaki [53], monitoring at 210 nm, found no accuracy problems with an ammonium acetate-acetonitrile technique. Solvent demixing, although often a dilutional sample preparation procedure, can be used to give an element of concentration [53].

Deproteinization procedures are inevitably dilutional. Acetonitrile has been used most commonly as a precipitant [70–74] but depends entirely on the chromatographic step for selectivity; in addition, column blocking is a common problem. The use of acetonitrile and propan-2-ol (50:50) has been claimed to give a more complete protein precipitant than acetonitrile [75], as has acetone [18].

Perchloric acid has been used [76] but rapidly eluting interferences are a notable limitation and the low pH is not conducive to long column life.

Liquid-Solid Extraction: Manual liquid-solid extraction has been used for ethsuximide, phenytoin, phenobarbitone, primidone, and carbamazepine [77]. Recoveries were better than 90% with precisions better than 5%. Although the Technicon Evaporation to Dryness Module has been used for automated liquid-liquid extraction and determination of anticonvulsants [78], liquid-solid extraction has the greatest potential for automation. Anticonvulsants are a major work load component in therapeutic drug monitoring programs and automation has been applied. XAD-type materials have been used off-line [79,80] on a Du Pont Prep or C_{18} cartridges on a Vac Elut (Analytichem International) [77] or AASP for column switching with direct injection. Automated loading of serum samples for anticonvulsant analysis onto an extraction column with injection of a cleaned-up sample via column switching [81–83] is of considerable interest. This work follows from the publications of Roth and colleagues [28, 84,85].

With automated on-line sample loading there is a requirement for additional equipment—pump, switching valves, etc. There are significant differences between packing materials from different manufacturers and even between batches from the same manufacturer [81–83]. Lichrosorb RP-18 (10 μm) (Merck) was favored by Kuhnz and Nau [82] and Nucleosil-3 C_{18} (30 μm) was used by Juergens [81], whereas Nazareth et al. [83] prepared their own C_{18} material by using Lichrosorb Si 60 (30 μm). Back-flushing has been most frequently used [81–83], although Juergens [81,86] has used fore-flushing with a very short precolumn (0.5 cm). Recoveries for the drugs and metabolites investigated by Juergens [81] were better than 95% for primidone, phenobarbitone, phenytoin, carbamazepine, ethsuximide, carbamazepine 10,11-epoxide, phenylethylmalonamide, and N-desmethylethsuximide. Similar results were found by other workers [82,83]. The procedures of Kuhnz and Nau [82] and Juergens [81, 86] have been validated for sera from patients. It seems probable that automation of anticonvulsant analysis with either in-house configurations or commercial equipment will expand as it is a cost-effective

Table 2 Reversed-Phase Chromatographic Methods for Anticonvulsants[a,b,c]

Packing	Column dimensions (nm)	Wavelength (nm)	Mobile phase	Drugs in order of elution (run time in min)	Reference
RP8 (7 μm)	100 × 4.6	204, 214 200	A: 0.01 M KH_2PO_4 + KOH + H_2O; B: 17.5% ACN, pH 5.3, at 75°C (gradient)	PR, PEMA, ET, PB NDS, CBE, PT, CBZ NI, CL, IS (MPS) (MP) (25 min)	91
Techsil 5C8 (5 μm)	250 × 5	204	MeOH:H_2O (50:50)	ET, PR, PB, PT CBZ, IS (HX) (12 min)	66
Fast LC8 (5 μm)	150 × 4.6	214	0.0025 M NaH_2PO_4 in 45% CH_3OH + 65 μl/liter TEA	PR, PB, PT, CBZ (12 min)	83
Lichrosorb RP8 (10 μm)	250 × 4	207 or 204	CH_3CN:H_2O gradient at 35°C	PEMA, ET, PR, CGP 10,000 GP47,779, DMP, PB, CBE, MPT, IS(HX,) PTU, CBZ, PT (21.5 min)	58, 68
μBondapak C_{18} (10 μm)	300 × 3.9	195	CH_3CN:CH_3OH:0.04 M PO_4 buffer, pH 6, at 30°C (17:28:55)	ET, PR, PB, NDS, IS (TB), PT, CBZ (20 min)	18
RPC 18 (5 μm)	100 × 4.6	210 or 195	CH_3CN:CH_3OH0.02 M PO_4 buffer, pH 3.7 at 50°C (13.5:35: 31.5)	ET, PR, PB, IS (CP), PT, CBZ (2.5 min)	77

Table 2 (Continued)

Packing	Column dimensions (mm)	Wavelength (nm)	Mobile phase	Drugs in order of elution (run time in min)	Reference
Zorbax C_{18} (7 μm)	100 × 4.6	250	CH_3CN:H_2O:0.5 M H_3PO_4 (36:64:0.01)	ET, PB, PT, CBZ, IS (MP) (2.5 min)	63
μBondapak C_{18} (10 μm)	300 × 3.9	208	CH_3CN:THF:CH_3OH (70:22:8)	PR, ET, PB, CBZ, PT, NI, CL, DI (10 min)	92
NovaPak C_{18} (5 μm)	100 × 8	200	Acetone:CH_3OH:CH_3CN:0.01 M PO_4 buffer, pH 7.95 (10:21:8:61)	PEMA, ET, PR, PB, HPT, CBE, CGP 10,000, NDS IS (TB), CBZ, PT, (9 min)	74
Shandon ODS 2 (5 μm)	250 × 4.6	205	A: PO_4 buffer, pH 4:CH_3CN (90:10); B: PO_4 buffer, pH 4:CH_3CN (40:60), gradient at 70°C	PEMA, PR, CBD, PB, NDS, CBE, PT, CBZ (11 min)	81
Spherisorb ODS 2 (5 μm)	250 × 4.6	200	CH_3CN:H_2O (21:80) at 60°C	PEMA, ET, PR, CBD, PB, CBE, PT, CBZ (25 min)	82
μBondapak C_{18} (10 μm)	250 × 4.6	195 or 254	PO_4 buffer pH 4.4: CH_3OH:CH_3CN (130:42:28)	PR, PB, IS (HX), PT, CBZ (16 min)	65
Sherisorb 5 ODS (5 μm)	250 × 4.6	195	CH_3CN:CH_3OH:PO_4, buffer, pH 4 (21:24:55)	ET, PB, PT, CBZ, IS (TPH) (7 min)	64

Column (size)	Dimensions	Wavelength (nm)	Mobile phase	Drugs	Ref.
Rad-Pak C$_{18}$ (10 μm)	100 × 8	200 and 254	CH$_3$OH:H$_2$O (45:55)	ET, PR, PB, CBE, MS, MPB, PT, CBZ (14 min)	60
Spherisorb 5 ODS (5 μm)	150 × 4.6	254	H$_2$O:CH$_3$OH:CH$_3$CN (62:22:14)	CBD, PB, HCB, IS, (ACB), CBE, IS, (HMCA), PT, CBZ (15 min)	59
Lichrosorb RP18 (5 μm)	250 × 4	210	CH$_3$OH:THF:0.05 M PO$_4$ buffer, pH 5.9, at 50°C (44:1:55)	PR, PB, IS (MPR), VP, PT, CBZ (10 min)	53
RPC18 (5 μm)	150 × 4	210	CH$_3$OH:CH$_3$CN:0.1 M KH$_2$PO$_4$, pH 3.7 (24:16:1)	PR, PB, PT, CBZ (9 min)	26

[a]Drugs and metabolites: CBD, carbamazepine 10,11-dihydro-10,11-diol; CBE, carbamazepine-10,11-epoxide; CGP 10,000, metabolite of CBZ and oxcarbazepine; CL, clonazepam; CBZ, carbamazepine; DI, diazepam; DMP, desmethylphenytoin; ET, ethsuximide; GP47,779, metabolite of oxcarbazepine; HCB, 10-hydroxy-10,11-dihydrocarbamazepine; HPT, hydroxyphenytoin; MPB, methylphenobarbitone; MPT, Mephenytoin; MS, methsuximide; NDS, N-desmethylsuximide; NI, nitrazepam; PB, phenobarbitone; PEMA, phenylethylmalonamide; PR, primidone, PT, phenytoin; PTU, pheneturide; VP, valproate

[b]Other abbreviations: TEA, triethylamine; CH$_3$OH, methanol; CH$_3$CN, acetonitrile; THF, tetrahydrofuran, PO$_4$, phosphate; H$_2$O, water.

[c]Internal standards (IS): ACB, allylcyclopentenylbarbiturate; CP, cyclopal, HMCA, hydroxymethylcarbamoylacridane; HX, hexobarbitone; MP, methylphenylphenylhydantoin; MPR, methylprimidone; MPS, methylpropylsuximide; TB, tolybarbitone; TPH, tolyphenylhydantoin.

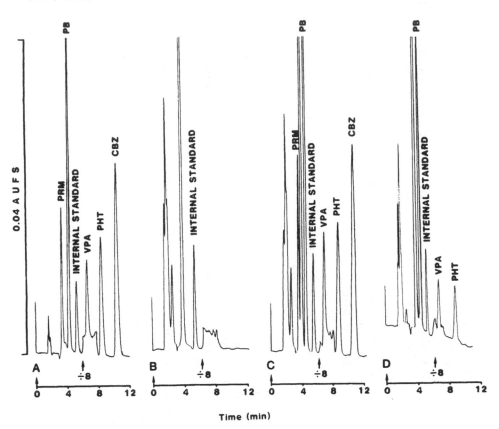

Fig. 3 Chromatograms of (A) standard mixture of sodium valproate
(VPA) (1 μg), phenobarbitone (PB) (0.2 μg), phenytoin (PHT) (0.1
μg), carbamazepine (CBZ) (0.1 μg), primidone (PRM) (0.1 μg), and
internal standard (0.06 μg); (B) blank plasma; (C) 100 μl of plasma
spiked with VPA (10 μg), PB (2 μg), PHT (1 μg), CBZ (1 μg), and
PRM (1 μg); and (D) plasma from an epileptic patient on VPA, PB,
and PHT. (From Ref. 53.)

alternative to the expensive immunoassay procedures currently used
for anticonvulsants in most laboratories performing therapeutic drug
monitoring.

Chromatography

Normal-phase chromatography was used in a few early reports [87–
90] and it has been suggested that this mode could be useful [41].
When compared to a reversed phase, a normal phase cyanopropyl
column had better selectivity [75].

Use of either C_8 or C_{18} packings predominates. Some recent publications are noted in Table 2. Most authors use a mobile phase composed of methanol, acetonitrile, and a phosphate buffer, resulting in
comparable elution sequences; some have attempted to produce very
rapid methods [63,77,83] but the potential for interference in such
systems is high. Acetone and tetrahydrofuran resulted in rapid
chromatography and reversal of carbamazepine and phenytoin retention [74,92]. The latter solvent allowed determination of benzodiazepines within 10 min [92] and was also used in the only procedure
so far reported that allows the simultaneous determination of valproate
with the other anticonvulsants [53]. Ion-pairing agents have been
infrequently applied [57,83].

Most detection was performed at wavelengths below 215 nm. Dual-
wavelength monitoring has been used by some authors and at high
valproate can be quantified sensitivity (Fig. 3). Internal standards,
most commonly baributrates, were used in most analyses, although
they were omitted from some automated procedures.

Automated assays involving gradient elution were used for the most
sophisticated chromatograms, and for broad-spectrum laboratories
with a large work load this would be the preferred option. For
other laboratories the wealth of reproducible chromatographic information makes isocratic reversed-phase LC an appropriate choice.

III. ANTIMICROBIAL AGENTS

Although there may be a defined upper limit of toxicity for antimicrobial agents, the effective concentration is defined by the minimum
inhibitory concentration of the target organism; usually this is gen-
eralized to "susceptible organisms" but may be higher (or lower) in
some instances. Typically, analyses are performed using microbio-
logical assays [93], although immunoassays are now in use for the
aminoglycoside drugs.

The need for antimicrobial agent monitoring has been reviewed
[94,95], as have assay procedures [96,97], particularly liquid chromatography [98–100].

A. Aminoglycosides

The aminoglycosides must be monitored due to their association with nephrotoxicity and ototoxicity.

Chromatography

Reversed-phase chromatography has been used, usually with C_{18}, acetonitrile phosphate buffers [101–103], Tris [104], and chelating agents such as EDTA [105,106]. Methanol with buffer or in conjunction with acetonitrile gave satisfactory chromatography [106–108], as did ion-pairing with sulfonates [103,109–111].

Detection

Because of the poor UV absorption of aminoglycosides, most procedures employ derivatization. Amikacin, gentamicin, tobramycin, and other aminoglycosides have been derivatized with o-phthaldehyde [104–108,110–112] to allow fluorescence detection, although other reagents have been used [113,114]. Barends and co-workers have produced chromophores by using 1-fluoro-2,4-dinitrobenzene [115–118], as have other workers [119]. Trinitrobenzenesulfonate has been used by Kabra and others [101,102,120]. Detection was at 365 and 340 nm, respectively. Recently, streptomycin has been determined without derivatization with monitoring at 195 nm [103]. Precolumn derivatization predominates; not only is it technically easier but the production of chromophores requires incubation at temperatures of 70 to 80°C for 10 to 45 min. Fluorophores produced with o-phthalaldehyde can be made rapidly and are suitable for postcolumn reaction systems [110,111].

Sample Preparation

Aminoglycosides are not well extracted from serum by organic solvent or precipitation procedures [112]. The trend has been to use liquid-solid extraction, usually with C_{18} [101–103,108]. Sephadex has also been used [118].

Essers [108] described an automated system with direct injection of serum onto a precolumn, in situ o-phthaldehyde derivatization, and subsequent fluorimetric detection. The system is applicable to wide range of aminoglycosides with minor modifications of the methanol-sodium acetate composition of the mobile phase.

Isomers

There has been a great deal of interest in the fact that there are four components in commercial gentamicin (gentamicin C_1, C_{1a}, C_2, and C_{2a}) [108,120–123] (Fig. 4), particularly as it has been suggested that C_1 may be less toxic [124] and that C_2 may cause

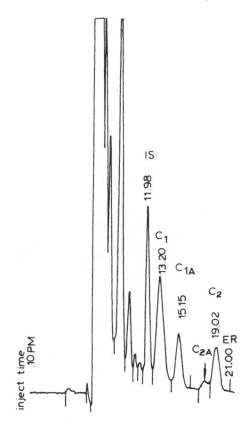

Fig. 4   Chromatogram of a separation of gentamicin from a serum sample containing 3.7 mg/liter gentamicin and 2.5 g/liter tobramycin (internal standard).   (From Ref. 108.)

nephrotoxicity [120]. It may be that therapeutic monitoring of gentamicin should be directed to component analysis, for which the only current satisfactory method is HPLC.

B. Antimalarials

The antimalarial group of drugs includes quinine, chloroquinine and its analogs, and pyrimethamine. Monitoring is important because of toxicity, e.g., chloroquine retinopathy, and as a compliance check. The analysis of antimalarials has been summarized [125] and only recent LC assays are considered below.

Interest in quinine as an antimalarial has been revised due to the emergence of chloroquine-resistant strains of *Plasmodium falciparum*.

One assay used acetonitrile precipitation with reversed-phase ion-pair chromatography and fluorescence detection [126] but was relatively imprecise (CV > 7%), although, no internal standard was used. In addition, quinidine interfered. A further potential interferent, dihydroquinine, was not considered.

Normal-phase ion-pair chromatography for chlorquine analysis of blood spots on filter paper was used following solvent extractions with both fluorescence and UV detection [127]. Reversed-phase chromatography is usually performed with an acetonitrile phosphate buffer eluant [128,129]. A similar procedure with UV detection at 340 nm has been reported for the analysis of chloroquine and amodiaquine and their metabolites and yielded good precision despite absolute recoveries of 75% [130].

Many authors have used ion-pairing for the analysis of antimalarials [see 125]. For primaquine, octanesulfonic acid was used [131]; a recent assay for the same analyte used electrochemical detection [132].

Pyrimethamine and dapsone are used in conjunction; therefore simultaneous assy for both is pertinent, especially for the latter due to the methemoglobinemia that results from its use. In addition, acetylator status can be determined by measuring dapsone and its acetylated metabolite following a test dose [133,134]. Recent procedures have tended to use reversed-phase chromatography with pentanesulfonic acid as the ion pair [135,136], although quinine, another antimalarial, is used as internal standard. The same separation can be achieved by gradient chromatography [137], but the run time was over 30 min as opposed to 12 min with an isocratic reversed-phase ion-pair procedure [136].

C. Antifungals

Antifungal analysis has been reviewed [125,138].

Amphotericin B

This drug is extremely nephrotoxic and is used only in severe systemic fungal infections. The first LC assay lacked an internal standard [139]. Two subsequent procedures using protein precipitation for sample preparation corrected this deficit [140,141], but low recoveries (60%) were obtained. Complete recovery has been claimed with phenyl extraction columns [142]. A sensitivity of 5 µg/liter was achieved with reversed-phase chromatography and UV monitoring at 386 nm, a 20-fold improvement over previous methods [140]. A further method has recently been published [143].

Imidazoles

These are important as oral antifungals. Ketoconazole can cause liver damage, although other imidazoles have not been implicated.

Liquid chromatographic assays have been reported for econazole, miconazole [144–146], and ketoconazole [147–152]. Typically, reversed-phase chromatography with UV detection at around 230 nm is used. It has been reported that acetonitrile precipitation in conjunction with diethylamine as a mobile-phase ion pair gives the simplicity and specificity desired [152].

Griseofulvin: Several liquid chromatographic procedures for griseofulvin have been reported [153–156].

Fluorocytosine

A number of assays for 5-fluorocytosine have been reported [157–162], but as 5-fluorocytosine is converted by fungal cells to 5-fluorouracil, measurement of the latter would appear more germane. 5-Fluorouracil measurement is discussed under antineoplastics. 5-Fluorocytosine measurement is helpful in patients with renal failure.

D. β-Lactams

Penicillin

Penicillins act by inhibiting bacterial cell wall formation and consequently are effective only against rapidly dividing organisms. Gram-positive bacteria are usually susceptible, but this extends to many Gram-negative organisms with the broad-spectrum penicillins. Bacterial resistance arises from β-lactamase production, which cleaves the β-lactam ring. In addition, antimicrobial activity may be lost on oral dosing by penicillins susceptible to gastric acid.

Sampling and Sample Preparation: Penicillins must be stored at −80°C to avoid degradation. Plasma pH is acceptable but urine should be bufferred to a neutral pH.

Diluted urine may be injected directly into an LC [e.g., 163], but determination in plasma is more difficult as penicillins are degraded in acid or base. Liquid-liquid extraction may be preferred over simple precipitation methods. Nonetheless, early procedures used either perchloric [164–166] or trichloroacetic [167–168] acid. Acetonitrile or methanol precipitation is now favored and is appropriate given the high concentrations of penicillins in serum [169–173]. Precision is acceptable, but liquid-liquid extraction may be preferred either to reduce imprecision or to increase assay sensitivity. Halohydrocarbons tend to be preferred [166,169,174–176]. Protein precipitation prior to extraction has been used in procedures in which mercuric mercaptides of imidazole penicillanic acids are detected in the UV at around 310 nm [165] and for fluorescence detection of derivatives [168]. Ion-pair extraction resulted in low recoveries [177].

Because of the acid lability of penicillin, solvent extraction requires some care as possible residual acidity may cause losses; this may be circumvented by performing an ether extraction and evaporating the solvent over a neutral buffer [24].

Liquid-solid extraction with C_{18} has been used [178,179], although a large late-eluting peak was seen in the procedure of Lee and Brooks [178]. Ion-pair extraction on octadecyl columns gave good accuracy and precision with recoveries of around 80% [180].

Chromatography: Reversed-phase chromatography is universally used. Most authors employ C_{18} or C_8, although phenyl-bonded phases have been used [176], with eluants of methanol or acetonitrile with phosphate or occasionally acetate. UV detection is adequate for most purposes, and although sensitivity may be improved with fluorescence detection of derivatives, such sensitivity is rarely necessary routinely. Cationic surfactants have occasionally be used [180,181].

The choice of internal standard varies from another penicillin, e.g., carbenicillin in a ticarcillin assay [182], chloramphenicol [183], a cephalosporin [176], and thienyl butyric acid [24], to compounds that might be found in the sample, such as benzoic acid (from foodstuffs) [180] or hydroxyphenytoin (a major metabolite of phenytoin) [174]. The analyses involving derivatization did not use internal standards [165,168], although external standardization of a C_{18} extraction assay gave acceptable precision [178].

Simultaneous assay of more than one β-lactam is possible [164,166, 167,170,177,181]. Annesley et al. [176] used a solvent gradient.

Diastereomers of penicillins and cephalosporins can be separated in biological fluids with methanol phosphate eluants [24,182,184].

The liquid chromatographic analysis of penicillins has been reviewed [97,185].

Cephalosporins

Cephalosporins are structurally related to penicillins, having a cephem ring instead of a penam ring. The indications for monitoring, mechanism of action, and causes of resistance are parallel to those of penicillins.

There is a vast number of cephalosporins, and they are usually classified according to their time of arrival on the marketplace, first generation, second generation, etc. As a generalization, the "younger" a cephalosporin, the wider its spectrum of activity.

The differences in structure of the different substituents on the cephem ring can be quite minor, and hence in some cases different cephalosporins can be assayed with the same chromatographic system. Cephalosporin analyses were summarized by Rouan [97], who noted that cephalosporin assays fall into two groups, single and multiple. Single assays for cephalosporins continue to appear [186—188] but are equally applicable to other cephalosporins [186,188]. The more versatile procedures include an ion-pairing agent, usually tetrabutyl-ammonium, in the mobile phase [e.g., 189]. Such mobile phases are also suitable for the analysis of epimers, e.g., of cefotetan [190].

An assay that is applicable to a wide range of cephalosporins with only minor modifications of the mobile phase [191] represents an approach to class analysis that is particularly appropriate for this group of drugs. UV detection is the norm; a wavelength of 254 nm is frequently used and the commonest form of sample preparation is organic solvent precipitation. Such simple assay requirements meet the needs of laboratories monitoring therapy.

β-Lactamase Inhibitors

The clinical relevance of measuring β-lactamase inhibitors is not known, but assays using derivatives suitable for pharmacokinetic studies have been published for clavulanic acid [24,163,192,193] and sulbactam [194].

E. Folate Inhibitors

Sulfonamides inhibit bacterial incorporation of p-aminobenzoic acid into dihydrofolate and possibly bacterial dihydrofolate reductase conversion of dihydrofolate to tetrahydrofolate [195], a reaction that is inhibited by trimethoprim. Originally it was believed that there was synergy between the sequential inhibitors of bacterial folate metabolism and sulfonamide/trimethoprim were prescribed together, although trimethoprim is now available alone, as are sulfonamides. The usual combination is trimethoprim with sulfamethoxazole, sulfamoxole, or sulfadiazine, usually in a ratio of 1:5.

Sulfamethoxazole and other sulfonamides used in combination with trimethoprim are usually analyzed simultaneously. A reversed-phase procedure allowing simultaneous determination of sulfamethoxazole, N^4-acetylsulfamethoxazole, and trimethoprim in serum and urine with simple dilution and protein precipitation was described [196,197]. In this assay trimethoprim was poorly resolved from early-eluting material.

Gochin et al. [198] reported a technique which gave excellent resolution, but caffeine, widely found in drinks, coeluted with sulfamethoxazole. A procedure utilizing dual-wavelength detection required 15 min per chromatogram [199]; electrochemical detection has also been used [200].

Reversed-phase ion-pair chromatography has been used for trimethoprim analysis in serum and urine [201] and also saliva and sputum [202], but sulfonamides did not chromatograph in the conditions used.

The above procedures used reversed-phase chromatography [203]. Normal-phase chromatography for the separation of trimethoprim and sulfadiazine and its N^4-acetyl metabolite has been used [203], and the procedure may be applied for the assay of trimethoprim and sulfa-methoxazole and its N^4-acetyl metabolite. The analysis is simple, precise, and sensitive; however, no information has been given on its accuracy. Another procedure required a large serum sample (2 ml) [204].

F. Other Drugs

The analysis of antitubercular drugs has been reviewed [205], as have other antimicrobial agents not considered above [97,125]. Only reports subsequent to these reviews are noted below.

Acyclovir

A procedure employing trifluoroacetic acid as protein precipitant and reversed-phase ion-pair chromatography with PRP-1 as stationary phase was reported [206]; unfortunately, its performance was not characterized. It should be compared to the reversed-phase ion-pair assay by Land and Bye [207].

Chloramphenicol

An acetonitrile solvent demixing sample preparation with a methanol:aceto-nitrile:phosphate buffer eluant for chloramphenicol determination [208] appears to offer no advantage over other liquid chromatographic procedures.

Ciprofloxacin and Flumequine

Reversed-phase procedures for quinoline antibiotics have been reported, using methanol phosphate buffer with either tetrahydrofuran [209] or acetonitrile [210] and fluorescence detection for ciproflaxacin or strong anion exchange with UV detection for flumequine [211].

Isoniazid

Ion pairing is commonly used and was employed in a recent novel assay involving ultrafiltration and propionyl derivative formation followed by chromatography; with such a procedure no internal standard was considered necessary [212].

Vancomycin

The procedure of Uhl and Anhalt [213] was used to validate a fluorescence polarization assay [214,215] but, following "improvement," still suffered from potential interference by theophylline, paracetamol (acetaminophen), and salicylate, three very common drugs [216].

A new LC assay claims no interference from these compounds [217]. Another procedure has also recently been published [218].

IV. ANTINEOPLASTICS

Anticancer drugs are, by their nature, highly toxic and thus monitoring might appear warranted; however, relationships between serum concentration, toxicity, and efficacy are not necessarily established. Antineoplastic assay techniques have been reviewed [219,220].

A. Alkylating Agents

Alkylating agents act by irreversible chemical reaction with cellular constituents.

Chlorambucil and Bisulfan

Several methods have been developed for estimation of chlorambucil [221–224], in one of which serum may be injected directly on-column [224], but none has been reported for bisulfan.

Liquid-solid extraction is most useful where concentrations exceed 5 mg/liter [225–227], particularly if fluorescence detection is used [225]; this assay did not incorporate an internal standard. Ion pairing with dodecyl sulfonate resulted in more efficient chromatography [226,227].

Cyclophosphamide

There has been a report of cyclophosphamide determination in serum [228] by a relatively simple liquid-solid extraction procedure followed by liquid-liquid extraction with UV detection at 195 nm. It would be apposite if an assay were to be developed for the pharmacologically active metabolites, acrolein, and phosphoramide mustard.

B. Antibiotics

Cytotoxic antibiotics act on DNA. The liquid chromatography of this class of cytotoxics has been reviewed [219].

Mitomycin and Bleomycin

There are a number of liquid chromatographic procedures for mitomycin [229–236], usually using liquid-solid extraction [232–234,236] with detection at around 360 nm and sensitivities of 5 µg/liter or better.

Bleomycin, a mixture of structurally closely related compounds, is most frequently determined by radioimmunoassay (RIA). An LC procedure has been published [237] but is chromatographically inefficient and lacks sensitivity.

Anthraquinones

The most extensively analyzed cytotoxic antibiotics are the anthraquinones, typically doxorubicin and daunorubicin [238—249], although these procedures and others may be used for the new anthraquinones such as 4'-deoxydoxorubicin [242,248,249] (Fig. 5). The relationship between serum concentration of the anthracyclines and/or their metabolites and toxicity and therapeutic effect is as yet not proved.

Liquid-solid extraction with reversed-phase material tends to be favored [239,240,245,247] although liquid-liquid extraction can be used [243,246]. Complete recovery may not be achieved [243], but this may be improved by using an ammonium formate buffer instead of the phosphate buffers commonly employed [249]. Automation with "loop" columns, either phenyl [244] or C_8 [242], can be used for anthracycline analogs and their metabolites. Direct injection of plasma into an ODS column preequilibrated with protein was effective [241] but required a stepped gradient. Internal standards, when used, are analogs of the analyte.

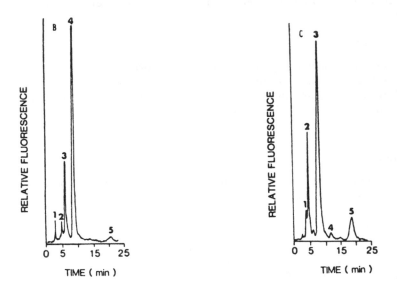

Fig. 5   Chromatography of anthracyclines.   (B) Serum extract 20 min after 30 mg/m^2 4'-doxorubicin. Peaks: serum peaks (1), 4'-deoxyrubicinol (2), 4'-deoxydoxorubicin (3), daunorubicin (internal standard) (4), 4'-deoxyrubicinol aglycone (5).   (C) serum extract 20 min after 40 mg/m^2 doxorubicin. Peaks: doxorubicinol (1), doxorubicin (2), daunorubicin (internal standard) (3) doxorubicinol aglycone (4), doxorubicin aglycone (5).   (From Ref. 248.)

Reversed-phase chromatography is used in recent procedures; acetonitrile:phosphate, acetate, or formate eluants are common, with acetonitrile being preferred to alcohols [250, 251]. The effect of the addition of surfactants to eluants for anthracycline analysis has attracted interest [246, 250]. Addition of anionic surfactants was found to increase retention [250] although not strictly according to expectation [246]. Cationic surfactants reduced retention [245, 246]. The nonionic surfactant Brij-35 was found to be a useful modifier, resulting in higher efficiencies [246] but at the concentrations used micellar chromatography was probably occurring [31].

Monitoring the native fluorescence of the anthracyclines gives sensitivities of 2 μg/liter or better, typically with excitation at 480 nm and emission at 580 nm.

C. Antimetabolites

Antimetabolite antineoplastics act by inhibiting a vital metabolic route, thus diminishing the pool of metabolic precursors or resulting in lethal synthesis.

Purine and Pyrimidine Antagonists

Purine analogs are extensively metabolized and pyrimidine antagonists are rapidly metabolized; thus the relationship between serum concentration of the drug and/or its metabolites is not clear. Present LC assays for 6-mercpatopurines have used a reversed phase with methanolic eluants [252, 253] or gradient anion exchange chromatography [254] with photometric detection at 313, 342, or 331 nm.

Reversed-phase LC is also favored for determination of the pyrimidine antagonists cytosine arabinoside and its principal metabolite uracil arabinoside. Ultrafiltration/precipitation sample preparation gave a sensitivity of 2 μg/liter, but the accuracy of the uracil arabinoside assay suffered from variable interference from endogeneous compounds [255]; a number of endogenous peaks were detected with a chromatography time of 30 min. Others did not obtain satisfactory results with such sample preparation and resorted to direct injection of plasma with chromatography times of less than 10 min. No interference was noted but sensitivity was only 50 μg/liter [256].

5-Fluorouracil

5-Fluorouracil is used as a cytotoxic agent and is the active metabolite of 5-fluorocytosine used as an antimycotic. Several methods are available, predominantly reversed-phase with neutral or acid eluants [257–259], although cation exchange [260] and anion exchange materials have been used [261]. C_{18} and a porous polymer packing (PRP-1) have been compared [258], and a cetrimide-containing eluant on the latter column was deemed superior.

Detection in the UV at 254 or 280 nm is most frequently used. Fluorescence detection of the 4-bromethyl-7-methoxycoumarin derivative was shown to be exquisitely sensitive [259] although the accuracy was not investigated.

Methotrexate

Unlike other antineoplastics, methotrexate serum concentrations have been correlated with toxicity [262] and folinic acid rescue can be modified if required [263]; these findings were an impetus to the introduction of effective assays, which have been reviewed [220,264]. 7-Hydroxymethotrexate, deoxyaminopteroic acid (DAMPA), and glutamate metabolites have all been implicated in toxicity; these can be efficiently quantitated only by liquid chromatography.

Liquid-liquid extraction and protein precipitation techniques are adequate for methotrexate measurement at high dose but not for measurement at the low levels (40 µg/liter) at which clinical decisions are made 72 hr postdose. High recoveries were obtained with liquid-solid extraction with either reversed-phase [265,266] or ion-exchange [267–269] materials; recovery of methotrexate was approximately 95%, 7-hydroxymethotrexate 75%, and DAMPA 90% [268].

As with most procedures reversed-phase chromatography is predominantly used. Ion pairing with hexanesulfonic acid allowed the potential for simultaneous assay of methoxtrexate, 7-hydroxymethotrexate, DAMPA, and the glutamate metabolites (Fig. 6) and the putative metabolite 7-hydroxy-DAMPA [268,270].

Detection is performed either at methotrexate's maximum of 307 nm or at 313 nm with a detection limit of the order of 10 µg/liter [268], which is similar to that achieved with on-line enrichment [271], electrochemical detection [272], and fluorescence following derivatization [273].

D. cis-Platinum

cis-diamminedichloroplatimum is of proven efficacy in the treatment of germinal neoplasms of the testes and ovaries and in head and neck cancers [274]. Liquid chromatography is the preferred mode of assay for this compound; anion exchange chromatography has been used [275,276], but it has claimed that solvent-generated anion exchangers are chromatographically more acceptable [220].

Ultrafiltration is the preferred method of removing serum proteins, although this may be followed by liquid-liquid extraction to enhance cleanup [277]. Detection was originally performed by atomic absorption [275,276], giving a sensitivity of 40 µg/liter, but 7 ml of serum was required to obtain an adequate volume of ultrafiltrate, although smaller volumes may be used [277]. Photometric detection of platinum complexes has resulted in assays with a sensitivity down to 2.5

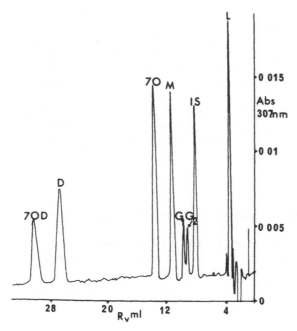

Fig. 6   Chromatogram of methotrexate, its metabolites, and leucovorin: leucovorin (L), L-triptophyl-L-glutamic acid (internal standard), triglutamyl methotrexate (G2), diglutamyl methotrexate (G2),
methotrexate (M), 7-hydroxymethotrexate (7O), deoxyaminopteroic acid
(D), and 7-hydroxydeoxyaminopteroic acid (7OD). (From Ref. 268.)

μg/liter, using diethyldiethiocarbamate complexation with either nickel
[278] or palladium [277] for an internal standard. The specificity
of precolumn derivatization is dubious and a postcolumn procedure
with sodium bisulfite derivatization and potassium dichromate activa-
tion has been reported, although the sensitivity is only 40 μg/liter
[279]. Electrochemical detection, either reductive or oxidative, may
be used; the latter is most sensitive, with a detection limit of around
3 ng on-column [280]. Some procedures have been found to be applicable to other platinum cytotoxic agents [278,280].

E.   Podophyllotoxins

Semisynthetic podophyllotoxins have been used in treatment of a wide
variety of neoplasms.
   Liquid-liquid extraction with halohydrocarbon solvents [281,282] or
ethyl acetate [283], and liquid-solid extraction [284,285] yield acceptable recoveries.

Reversed-phase chromatography has been universally utilized with a number of authors preferring phenyl materials [281–283]. Although earlier assays used photometry [286–288] or fluorimetry [284, 286,289], electrochemical detection is now preferred because of its greater sensitivity, which is of the order of 2 to 5 µg/liter. The potential metabolites of etoposide and temiposide can be determined [281–283].

A PRP-1 preconcentration column in a fully automated system for samples pretreated with the proteolytic enzyme subtilisin Carlsberg improved recovery, with an analysis rate of five samples per hour [284] and a sensitivity of 10 µg/liter.

F. Other Cytotoxics

Aminoglutethimide used in the palliative treatment of breast cancer has been determined by LC [290–292].

The S-adenosylmethionine decarboxylase inhibitor methyl GAG can be measured by reversed-phase ion-pair chromatography following protein precipitation with perchloric acid [293,294].

Reversed-phase chromatography of BCNU and methyl CCNU with photometric detection has been reported [295–297].

V. ANALGESICS

Although salicylate can be used to treat rheumatoid arthritis, measurement of analgesics for therapeutic monitoring purposes is performed only infrequently. However, some analgesics can be lethal in overdose and are especially dangerous as patients may be asymptomatic on presentation. There is a need to measure salicylate and acetaminophen (paracetamol) serum concentrations as an aid to diagnosis and prognosis [298,299].

The determination of salicylate and acetaminophen in serum has been reviewed [300]. The assay of these and other analgesics has been briefly reviewed [301].

A. Salicylate

As toxic concentrations are of the order of grams per liter, assays with simple dilutional sample preparation are adequate [301–306]. In one early assay serum was injected directly on-column [307].

Liquid-liquid extraction is still used [308–310], usually for pharmacokinetic studies. Avoidance of hydrolysis of acetylsalicyclic acid requires stringent sample handling, with fluoride inhibition, temperature at 0°C, and rapid transfer [308,311,312].

Reversed-phase chromatography with acidic eluants and UV detection are the norm [302–311,313–316]. Fluorescence may be used but is no more sensitive than UV [304].

Because of the clinical significance of the results, it is important that assays be accurate. Theophylline interference [313] and other interferences limit some procedures [302,303,317], although inaccurate colorimetric assays are used routinely much more frequently than accurate assays such as liquid chromatography.

A number of assays determine acetylsalicyclic acid [303–306,308, 309,311,314,315] and also the metabolites of salicylate, including salicyluric acid, gentisic acid, and gentisuric acid [302–306,308,310, 314].

If high sensitivity should be required, assays have been described with sensitivities of 0.1 mg/liter or better [302–306,308–310,314].

B. Acetaminophen (Paracetamol)

Unlike salicylate, acetaminophen assay has not, until recently, been simple. In photometric assays, inaccuracy is a persistent problem; this can be circumvented by using liquid chromatography [318,319]. Acetaminophen analysis has been reviewed [300,319,320].

There are many liquid chromatographic assays [313,321–327], most of which are reversed-phase procedures. Exceptions are an early ion-exchange procedure [321,328], which excessive analysis time made unsuitable for routine use, and a normal-phase procedure with a chloroform:tetrahydrofuran mobile phase [326], which was rapid and had a sensitivity of 1 mg/liter, but this mode does not lend itself to simple preparation. Reversed-phase assays have utilized protein precipitation sample preparation to simplify analysis [324,325,329], although a surprising number of procedures retain cumbersome liquid-liquid sample preparation.

Reversed-phase chromatography with acidic eluants is favored with formic or acetic acid [313,322–339]. Determination of all clinically significant acetaminophen metabolites, i.e., glucuronide, sulfate, and mercaptide in plasma, can best be performed by the method of Adriaenssens and co-workers [324,325], in which propan-2-ol is used to enhance selectivity. Ion pairing has been used only once, in an assay for urinary metabolites [323].

Sensitivities of 1 mg/liter or better can be achieved readily with photometric detection [324,329,336,337,339]. Although electrochemical detection has been applied [339–341], such sensitivity is not required in routine practice.

Early acetaminophen assays suffered from significant interference from theophylline [328] or other drugs [327]. A recent assay used theophylline as an internal standard [337]. The procedure described by

Buchanan et al. [325] epitomizes a rapid, simple, sensitive accurate assay for acetaminophen.

C. Simultaneous Assay for Acetaminophen and Salicylate

Concurrent poisoning with acetaminophen and salicylate is not uncommon and several procedures allow simultaneous determination [313, 317,342]. Theophylline coelutes with salicylate in one assay [313], a problem circumvented in the assay of Kinberger and Holman [317].

D. Ibuprofen

Ibuprofen is becoming available without prescription in some European countries and in overdose causes hypotension and coma, but definitive evidence to confirm its high therapeutic index is still required [343]. Recent reversed-phase assays with acidic eluants and monitoring at 229, 220, or 196 nm have been reported [344–348]. One procedure with a cyanopropyl column and octane sulfonic acid as an ion-pairing agent [349] allowed simultaneous determination of codeine.

Earlier procedures [350–356] have been criticized for requiring samples of 0.5 ml or greater; "micro" procedures use acetonitrile precipitation [347], ethyl acetate extraction [344], and C_{18} liquid-solid extraction [346].

VI. ANTIASTHMATICS

Although there is a large range of drugs for the treatment of asthma and chronic obstructive airway disease, encompassing steroids, mast cell stabilizers, xanthines, β_2 agonists, and anticholinergics, xanthines are the only type for which monitoring is of proven value [357]. Of these, theophylline is the archetype.

A. Theophylline

Theophylline analysis has been reviewed [357–360], as has the liquid chromatography of this and other antiasthmatics [361].

The literature on liquid chromatography of theophylline is large and many methods have not been adequately validated, a not infrequent complaint [360–362]. Only well-characterized procedures will be considered below (see also reviews cited above).

It is well known that a major source of inaccuracy in the LC of theophylline is coelution with the caffeine metabolite 1,7-dimethylxanthine (paraxanthine) [360,363–372].

Only 1 of 20 laboratories using reversed-phase LC used assays capable of discriminating theophylline from 1,7-dimethylxanthine [360],

and methods suffering from such interference continue to be published [373]. Inaccurate assays employ binary eluants comprising
aqueous acid and organic modifier; in addition, a number of authors
have noted interference from antibiotics, principally β-lactams [379,
380], which may be used to control exacerbation of bronchitis in
patients in whom theophylline may be used.

Eluants containing tetrahydrofuran resolve 1,7-dimethylxanthine
from theophylline sufficiently to enable accurate measurement of both
[364,370]; separation may also be effected by using buffers with a
pH of 3.6 to 5.2 [367,368], and a combination of these results in
complete resolution [371,372]. The potential hazard of tetrahydrofuran make it an unattractive option. Ion-pairing with a cationic
surfactant gives complete resolution [366,381,382]. Normal-phase
chromatography has not been reported to suffer from difficulties
with accuracy [383,384].

Liquid-liquid extraction with chloroform-isopropanol [383–385] or
ethyl acetate [367] has been employed, but the utility of reversed-
phase chromatography may be best met by using protein precipitation
with solvents [363,381] or zinc sulfate [369,370]; the use of plasma
ultrafiltrates has been advocated [23]. Direct injection of serum onto a precolumn with subsequent reversed-phase ion-pair chromatography has been described [382]. A number of theophylline analogs
have been used as internal standards: 8-chlorotheophylline [e.g.,
369], β-hydroxyethyltheophylline [e.g., 381], ethyltheophylline [384],
and 3-isobutyl-1-methylxanthine [383].

Rowe et al. [360] recommended β-hydroxyethyltheophylline as an
internal standard for reversed-phase, reversed-phase ion-pair, and
normal-phase LC on the basis of k'. 8-Chloroethyltheophylline can
be used in reversed-phase LC provided its pH-dependent k' is
recognized.

VII. CARDIOVASCULAR DRUGS

The number of cardioactive drugs is large and increasing and there
is a concomitant rise in the number of analytical procedures, mainly
for the purpose of obtaining pharmacokinetic data. The necessity for
measurement of therapeutic drug concentrations for drugs that give
pharmacodynamically measurable responses is still a subject of debate.
Cardiovascular drugs may be considered as encompassing the drug
classes, cardiac glycosides (e.g., digoxin), antiarrhythmic drugs
(quinidine, flecainide, lidocaine, etc.), β-blockers (propranolol,
nadolol, etc.), vasodilators (e.g., organic nitrates, calcium antagonists), antihypertensive agents (e.g., debrisoquine, hydralazine,
captopril), and diuretics (e.g., frusemide). Discussion here will be
confined to drugs for which there is an indication of a need for routine
monitoring and those drugs for which liquid chromatography can

uniquely provide the data required for pharmacokinetically accurate information. A recent review comprehensively covers the assay of cardiovascular drugs [386].

A. Cardiac Glycosides

Digoxin is the most commonly used cardiac glycoside, with a quoted therapeutic range of 0.8 to 2.0 µg/liter; there is, however, considerable debate as to the validity of this [387,388]. Among a number of factors affecting response to digoxin, such as age, renal function, hypokalemia, and hypercalemia, is the possibility of analytical inaccuracy due to interference from digoxin metabolites in the radioimmunoassay [389]. An assay that employed LC prior to radioimmunoassay gave a ratio of 0.8 as compared to direct immunoassay [390], which was attributed to metabolites and has been confirmed by others using LC-radioimmunoassay procedures [391,392]. In addition, unidentified endogenous compounds have been found to have digoxin-like immunoreactivity, particularly in neonates [391,393–398], hepatic failure [399,400], renal failure [397,401], hepatorenal failure [402], and during the third trimester of pregnancy [403,404].

To obtain accurate digoxin concentrations, extraction into dichloromethane [390] or onto C_{18} [390] or C_8 [391] reversed-phase columns with an acetonitrile-phosphate buffer mix or 12% aqueous propan-2-ol has been used. The collected fractions were then determined by radioimmunoassay (Fig. 7). Recently, the endogenous digoxin immunoreactive substance has been isolated [405], but it awaits characterization.

There is increased sensitivity to digoxin in neonates and in renal failure, and current therapeutic monitoring practice requires reevaluation in light of the potential inaccuracy of the immunoassay procedures used; such reevaluations may result in a redefined therapeutic range. Certainly, reference and pediatric laboratories offering a monitoring service should seriously consider adopting an LC-immunoassay procedure for use with the patient groups known to suffer from the methodological inaccuracy. A fluorimetric procedure has been reported, but it requires a large sample of serum [406] and is not applicable to neonatal sampling.

B. Separation of Isomeric Cardioactive Drugs

Beta-Blockers

Beta-blockers are used in the treatment of hypertension and ischemic heart disease; estimation of serum concentrations is not routinely performed except in specialist centers.

Many β-blockers have chiral centers, which can result in significant pharmacodynamic [407,408] and pharmacokinetic differences [409].

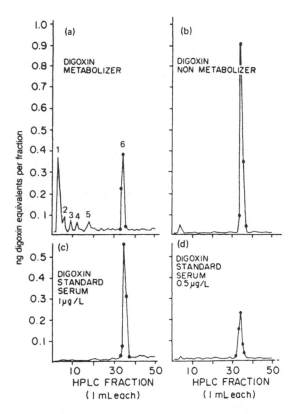

Fig. 7   HPLC/RIA measurement of "true" serum digoxin on a C_8 reversed-phase column, eluted with aqueous isopropanol (200 ml/liter) at 1 ml/min. Each 1-ml fraction was assayed for digoxin by RIA and all the values under each peak summed. (a) Patient with extensive digoxin metabolism: RIA digoxin, 1.9 μg/liter; LC/RIA digoxin, 0.7 μg/liter. Peaks: polar metabolites (1 and 2), digoxigenin (3), mono-digitoxoside (4), bisdigitoxoside (5), dogoxin (6). (b) Patient with little digoxin metabolism: RIA digoxin, 1.5 μg/liter; LC/RIA digoxin, 1.5 μg/liter. (c) and (d) Digoxin standards. (From Ref. 391.)

Propanolol sulfate esters can be resolved by using chiral derivatives, 2,3,4,6-tetra-O-acetyl-β-D-glucopyranosyl isothiocyanate [410] (Fig. 8) or (+)-1-phenylethyl isocyanate, to derivatize propranolol and its 4-hydroxy enantiomeric metabolites [411] in preference to N-trifluoro-acetyl(-)prolyl chloride originally used to derivatize propranolol [412], which tended to racemize on storage [413]. The foregoing were determined in serum.

Sedman and Gal [414] reported the use of 2,3,4-tri-O-acetyl-α-D-glucopyranosyl isothiocyanate and 2,3,4,6-tetra-O-acetyl-β-D-glucopyranosyl isothiocyanate for a number of β-blockers. Ion pairing with (+)-10-camphorsulfonate has been used [415]. A chiral packing consisting of γ-aminopropyl modified with (R)-N-(3,5-dinitrobenzoyl) phenylglycine was used to separate the cyclic 2-oxazolidone derivatives of enantiomeric propranolol [416] (Fig. 9). Alpha-1 acid glycoprotein columns might be useful [417].

Antiarrhythmics

A chiral α_1-acid glycoprotein column has been used for the separation of the R and S isomers of disopyramide [417–419]. Mexiletine

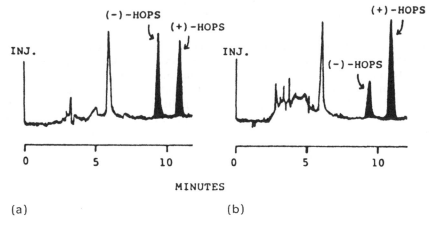

Fig. 8 Separation of isomers: reversed-phase HPLC of 2,3,4,6-tetra-O-acetyl-β-D-glucopyranosyl isothiocyanate diasteroisomer derivatives of (a) synthetic 4'-hydroxypropranolol sulfate (HOPS) (250 ng of each isomer) and (b) HOPS isolated from human plasma [(-)HOPS, 124 μg/liter; (+) HOPS, 373 μg/liter]. (From Ref. 410.)

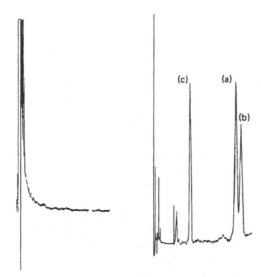

Fig. 9 Separation of isomers: chromatogram of whole blood extract from (1) subject before administration of racemic propranolol and (2) subject 2.5 hr after administration of an 80-mg dose of recemic propranolol. Peaks: *l*-propranolol oxazolidone (a), *d*-propranolol oxazolidone (b), and *dl*-pronethalol oxazolidone (c). (From Ref. 416.)

enatiomers have been resolved with 2,3,4,6-tetra-*O*-acetyl-β-D-glucopyranosyl isothiocyanate [240].

C. Amiodarone

Chronic use of amiodarone can result in a number of side effects, particularly photosensitization, disordered thyroid status, peripheral neuropathy, nausea, headache, and nightmares. Assays using simple reversed-phase, ion-pairing reversed-phase [421−423], or normal-phase [424,425] LC have been described. Measurement of the desethyl metabolite may be relevant.

D. Organic Nitrates

Routine monitoring of nitrates by LC is not practicable. However, a thermal energy analyzer, more readily used in GC, has been linked to LC to achieve the sensitivity required to detect the organic nitrates glycerol trinitrate, isosorbide dinitrate, and pentaerythritol tetranitrate [426,427] and nitroglycerine metabolites [428].

E. Other Cardiovascular Drugs

The β-blockers, considered in Sec. VII.B, are the subject of many reports; for reviews see Mehta [429] and Anhoff [386]. Reversed-phase methods for disopyramide [430] and for quinidine with UV detection [431] have been reported, but neither appears to offer any advantage over previously reported procedures. For the sensitive detection of quinidine, fluorescence detection would be preferred (see quinine, an isomer of quinidine in Sec. III.B). Because of the polymorphic metabolism and narrow therapeutic index of procainamide, measurement of N-acetylprocainamide should also be performed. In addition to the assays noted by Anhoff et al. [386], a reversed-phase ion-pair procedure with enhanced specificity would appear appropriate [432].

Ideally, analytical conditions that allow the analysis of several members of a class of compounds are required. Reversed-phase procedures suitable for analysis of lidocaine, mexiletine, disopyramide, and quinidine [433] or procainamide, N-acetylprocainamide, lidocaine, quinidine, disopyramide, and propranolol [434] might be particularly useful.

VIII. SEDATIVES

A. Barbiturates

Phenobarbitone is the barbiturate most frequently monitored and its determination is discussed in Sec. II. Use of barbiturates as sedatives has decreased, and monitoring of levels is now infrequently performed. However, because of the widespread abuse of barbiturates, often resulting in overdosage, procedures that allow the simultaneous identification and quantitation of several barbiturates are necessary. Chromatographic methods for barbiturate analysis have been reviewed [435,436].

Liquid-liquid extraction, particularly with halohydrocarbon solvents, has been favored, although acetonitrile precipitation has also been used [437,438]. Neutral or acidic eluants have been employed; however, the addition of alkali results in a 20-fold increase in sensitivity, and postcolumn addition of alkali has been used to achieve this [439]. Not only does an alkaline medium improve absorbance, but the resultant hypsochromic shift improves specificity by moving the absorption maximum from the low UV otherwise required. At a pH of 13 there is a further hypsochromic shift.

Because of the pH limitations of silica, strongly alkaline eluants cannot be used, but mildly alkaline eluants allow detection at 240 nm or higher [440–444]. PRP-1, which is stable up to pH 13, was used by Gupta et al. [442] with highly alkaline eluants. This procedure identified a number of barbiturates and would be a suitable screening method for clinical toxicological purposes.

High-dose thiopentane, an ultra-short-acting anesthetic, is now used in treatment of traumatic coma. Knowledge of the serum concentrations is useful and a number of LC assays have been specifically designed for determining thiopentane [438,443–448].

B. Benzodiazepines

The term benzodiazepines encompasses a number of related compounds. Monitoring is not usually necessary, although clonazepam (see Sec. II.A) and those used in induction of anaesthesia, such as diazepam, may be measured for monitoring purposes. Even in overdose benzodiazepines are relatively safe, but qualitative and quantitative information may be useful in differential diagnosis and subsequent management. Benzodiazepine analysis has been reviewed [449] and updated [450].

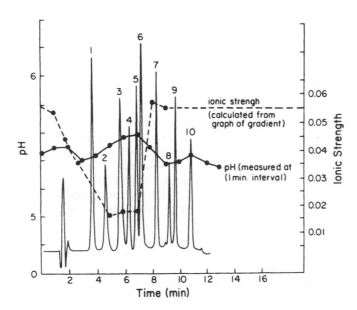

Fig. 10 Representative chromatogram of benzodiazepines showing changes in the ionic strength and pH during the solvent gradient program. Peaks: demoxepam (1), oxazepam (2), *N*-desmethylchlordiazepoxide (3), *N*-desalkylflurazepam (4), *N*-desmethyldiazepam (5), chlordiazepoxide (6), diazepam (7), lorazepam benzophenone (internal standard), prazepam (9), and endogeneous peak (10). (From Ref. 455.)

Benzodiazepines are extensively metabolized, often to active metabolites. Flunitrazepam, which cannot be screened with commonly used enzyme immunoassay procedures, has been satisfactorily estimated by a reversed-phase ion-pair assay with fluorimetric detection [451]. Extraction with diethyl ether is claimed to be most effective for forensic samples [452]. Isocratic reversed-phase procedures with UV detection at 254 nm are suitable for screening [453,454]. A ternary gradient with varying pH and ionic strength allows the simultaneous analysis of eight benzodiazepines; column temperature was maintained at 50°C [455] (Fig. 10). Retention for more than 20 benzodiazepines on four different isocratic LC systems have been published [456].

C. Other Sedatives

These comprise a structurally disparate group of drugs. Cabromal has been determined by reversed-phase [457] and normal-phase [458] chromatography with UV detection at 210 or 205 nm. Chlormethiazole, which can cause respiratory arrest, is toxicologically important and two assays have been described [459,460]. Methaqualone, meprobamate, and ethchlorvynol have each been the subject of reports [461–463].

IX. IMMUNOSUPPRESSIVES

A. Cyclosporin

Azathioprine used to be the mainstay of immunosuppressive therapy, acting through its metabolite 6-mercaptopurine (see Sec. IV.C), but has been supplanted by cyclosporin A, a cyclic undecapeptide used for postoperative transplantation immunosuppression. Toxicity, particularly nephrotoxicity, is a major side effect. Cyclosporin A use has also been associated with hepatotoxicity and an increased incidence of lymphoma. The pharmacology of cyclosporin A has been the subject of recent reviews [464–467].

There are marked differences of opinion as to how and by what procedure cyclosporin A should be measured. Radioimmunoassay is available but uses a tritiated label and cross-reacts with cyclosporin A metabolites; nonetheless, it is a technically simple, although inaccurate, procedure. LC provides an accurate assay [468], although it has been argued that RIA may be more appropriate as the role of metabolites, particularly in nephrotoxocity, is not known. Metabolites can be allowed for by using RIA [469], which is more rapid than LC, but the latter is preferred in hepatic grafts [470]. The status of cyclosporin A assays has been summarized [471–473].

Sampling

Sampling is a major source of variability. Equilibration of cyclosporin A between plasma and red blood cells is temperature- and haematocrit-dependent [474]. Lower plasma concentrations are obtained from samples stored at ambient temperature as opposed to 37°C, and therefore the latter temperature is recommended for plasma [473]. The use of whole blood may obviate this problem [475], although nonspecific quenching in RIA may increase [472].

Assay

Cyclosporin A has minimal UV absorption, mainly at 210 nm or less, and because of the difficulties of obtaining good recovery, accuracy, and chromatography the assays tend to be complex or require a high capital cost for automation. A number of LC procedures have been described [476–490].

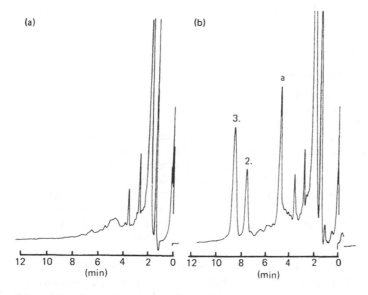

Fig. 11 Determination of cyclosporin A. Extract of (a) drug-free whole blood and (b) sample from patient receiving cyclosporin. Peaks: cyclosporin A metabolite(s) *a*, cyclosporin A (2), and cyclosporin D (internal standard) (3). (From Ref. 485).

Solvent extraction techniques tend to be lengthier than those using solid-phase cartridges, although some semiautomated procedures have been claimed to be very simple [483]. Most liquid-solid extraction methods have a precision better than 7% at 100 µg/liter [484—487], and this can be further improved by using an automated system [488].

Isocratic LC is usually used, although gradient systems have been described [483]; eluants are typically 70% acetonitrile in water at a temperature of 70°C with UV detection at 214 nm. Some authors have added methanol [490] or trifluroacetic acid [489]. A temperature of 50°C has been used [486] to prolong column life, although silica presaturation of the mobile phase via an exhausted analytical column is claimed to be just as effective [473] and allows analysis at 70°C, which ensures that all cyclosporin A has homogenous structural confirmation. Cyclosporin D is the internal standard of choice. When whole blood is the preferred sample there are a number of acceptable methods [483—486,488—490].

Since metabolites, possible active, are measured by RIA, it is surprising that LC has not been more widely used to quantitate them. A red blood cell-bound metabolite has been described [485] (Fig. 11).

X. PSYCHOTROPICS

A. Tricyclic Antidepressants

The necessity for measuring serum concentrations of tricyclic antidepressants has been debated for many years, with no clear recommendations emerging [491].

Sample preparation has been a problem because of the tendency of the compounds to adsorb to glassware; this has been circumvented by silanizing glassware [492], including an alcohol in the extraction solvent, or displacing the analyte by using a large mass of an analog [493]. Extraction with halohydrocarbons may alkylate terminal amines [494]. Hexane alone may be satisfactory [494], but liquid-solid extraction is preferable [495]. The analysis of tricyclic antidepressants has been reviewed [496—499].

An early trend in antidepressant analysis was the use of normal-phase chromatography initially for amitriptyline and nortriptyline [492,500] and subsequently for metabolites and other antidepressants [501—503], and this trend continues [493,504,505]. In a recent review [495], 19 of the 43 quoted procedures for tricyclic antidepressant analysis were normal phase or normal phase ion-pair, an unusually high proportion in comparison with most other applications. However, reversed-phase or reversed-phase ion-pair chromatography may be preferred if maximum flexibility for other procedures is required. Reversed-phase assays tend to include a cationic surfactant in the eluant control tailing [506—512], although this is not invariably required [513,514]. Similarly, in normal-phase chromatography a base, often ammonium hydroxide [492,500], or a cationic surfactant [504] is used to reduce tailing.

Determination of metabolites is particularly relevant as desmethyl metabolites are often pharmacologically active and measurement of the parent drug and the metabolite, for instance, amitriptyline and nortriptyline or imipramine and desmethylimipramine, must be performed. The separation of isomeric 10-hydroxyamitriptyline can be performed by normal-phase [493] or reversed phase [515] methods. Knowledge of the k' values of principal metabolites is essential, particularly as different drugs can be successfully analyzed under the same chromatographic conditions [501,502,513,514]; lack of such information can lead to misinterpretation if the assay is used to screen for antidepressants.

Photometric detection around 210 or 254 nm is usually adequate. Electrochemical detection has been used for determination of trimpramine and its metabolites [508] and mianserin [516]; imipramine and its metabolites may be detected by using fluorescence [517,518].

B. Phenothiazines

Phenothiazines are structurally related to tricyclic antidepressants. Phenothiazines are extensively metabolized, chlorpromazine having 168 theoretical metabolites [519]. There is no definitive evidence of a relationship between plasma concentrations for chlorpromazine, and pharmacological response, partly due to difficulties in scoring schizophrenia symptoms and analyte instability during storage, sample preparation, and analysis [520–523].

Chlorpromazine

Chlorpromazine sulfoxide and 7-hydroxychlorpromazine are principal metabolites and measurement of the latter may be helpful [524].

Early assays used normal-phase systems [502,525] with a sensitivity of 10 μg/liter and UV detection at 250 nm. Use of an electrochemical detector gives a fivefold improvement in sensitivity [526]. As with antidepressants, normal-phase chromatography is widely used and assays said to be normal phase, performed on cyanopropyl packing, have been reported [527,528]. A reversed-phase method has been reported but a complex sample preparation procedure was used [529]. The chromatographic conditions and detector voltage (+0.9v) were identical in two of the procedures [526,528], but no details of metabolite chromatography were given.

Other Phenothiazines

Several assays for thioridazine have reported [530–534]. Normalphase procedures are again favored. Fluorimetric assay following postcolumn oxidation [530] gives good sensitivity once postcolumn extraction of the fluorophore has been performed [533]. Electrochemical detection gives a 20-fold improvement, with a sensitivity of 0.1 ng on column [534]. Solvent extraction yielded 70% recovery; liquid-solid extraction may improve this.

Because of the low concentrations involved, trimeprazine assay has been difficult, but electrochemical detection gives adequate sensitivity [526,535].

C. Haloperidol

Haloperidol is used in the treatment of psychotic syndromes and the treatment of abnormal movements. It is extensively metabolized, with the hydroxy metabolite being active [536]. There is a relationship between plasma concentration and effect [537].

Only recently, with improvements in LC hardware and column technology, could sufficient sensitivity for haloperidol analysis in serum be achieved. Normal-phase and a variety of reversed-phase materials

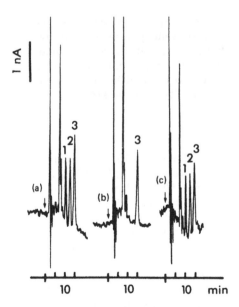

Fig. 12 Chromatograms of haloperidol in standards and patient samples. (a) serum extract spiked with reduced haloperidol (5 ng) (1), haloperidol (5 ng) (2), and chlorohaliperidol (internal standard) (20 ng) (3). (b) Drug-free serum extract. (c) Extract of serum from a patient on haloperidol. (From Ref. 542.)

have been used [538–541]. A sensitivity of 1 µg/liter is achievable with UV detection, which is similar to that obtained with amperometric detection [542]. Such levels are at the lower end of the therapeutic range for senile dementia and can usually be achieved with a precision of better than 10% (Fig. 12).

XI. DRUG SCREENING (CLINICAL TOXICOLOGY)

Drug overdoses account for a significant number of medical admissions to hospital; these range from the trivial to the life-threatening. It is useful to the examining physician and, in some circumstances for medicolegal purposes, to know which, if any, drugs have been taken; sometimes this allows appropriate therapy to be instituted, a differential diagnosis to be clarified, or criminal proceedings to be taken.

An emerging theme in drug analysis is the development of assays capable of being used with little or no modification to determine a number of related drugs. Some examples have been given earlier— for instance, anticonvulsants, benzodiazepines, and β-lactam antibiotics.

A. General Screening

Early attempts were made to provide data bases for the application of LC to general screening [543,544], but although LC has been widely used for quantitative drug analysis, its use in the forensic sciences has been restricted due to perceived technical difficulties [545]. Chromatographic irreproducibility, that is, variable k', between batches of stationary phase and between manufacturers of similar packings is a major limitation [546,547], although interlaboratory comparability can be achieved with standardization on bulk-purched material [546].

Nonaqueous eluants with ionic modifiers have been shown to be flexible, stable systems for chromatography of basic drugs on silica [548,549]. Data are available on 462 compounds, giving details of UV and electrochemical detection response. This system is designed for use with extracts of biological fluids; batch-to-batch variation was compensated for by using an internal standard [549].

A standardized ion-pair extraction procedure and normal- or reversed-phase chromatography have been investigated by Massart and co-workers [550,551] and applied to the determination of such drugs in serum [11]. Micellar chromatography might be potentially useful [30–32].

The problems in using LC for screening have been summarized by de Zeeuw [547] as (a) lack of standardized stationary phases, (b) the necessity for dependence on the UV detector, which is selective, as a "universal" detector, and (c) lack of reliable standardized method of expressing retention. Photodiode array detectors may partly overcome the second objection [552,553]. Detectors with multiple-wavelength monitoring may be appropriate. The use of alkylaryl ketones as retention indices has been shown to be valid, at least for local anesthetics [553], and could satisfy the requirements of (c).

A general problem in chromatographic screening is the difficulty that arises as the number of indexed compounds grows, leading to a greater number of compounds being possible candidates for the assignment of identification. The mean list length approach requires consideration [547].

Systems suitable for identification of drugs within a class have been summarized by Gill [554]; the data base contains a significant amount of previously unpublished data.

B. Drugs of Abuse

The chromatographic analysis of drugs of abuse has been reviewed [435,555], but most assays were not for biological fluids.

Morphine and Other Opiates

Ultraviolet detection of morphine in urine can be readily performed [556,557]. Therapeutic serum levels of codeine can also be determined with UV detection [558,559]. An early procedure for detecting morphine and its metabolites in biological fluids used LC combined with RIA of collected fractions [560]. However, to determine serum morphine levels to the required sensitivity and specificity, an electrochemical detector is required. Many authors have reported assays [561-571]. A typical chromatogram is shown in Fig. 13.

Amperometric detection is usually used, although coulometric detection is more sensitive (0.3 versus 1 µg/liter) [569]; selectivity is better the lower the potential applied. Glassy carbon electrodes are used in amperometry with applied potentials ranging from +0.65 to +1.0 V.

Sample preparation is difficult as morphine is amphoteric. A pH of 8.9 will result in the maximum 80% recovery usually achievable with solvent extraction, with chloroform or 10% butanol in chloroform [561,562,564,570]. Liquid-solid extraction is generally faster; the same pH limitations apply and similar recoveries might be expected [568] although 100% recovery has been claimed [566]. Reversed-phase chromatography with methanol and phosphate buffer [562,567] or ammonium hydroxide [564] has been used; the latter is very similar to reported normal-phase eluants [561,568]. Reversed-phase ion-pairing with anionic surfactants has been claimed to improve the

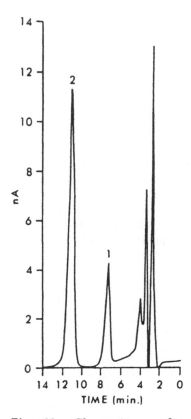

Fig. 13   Chromatogram from a serum sample containing 100 mg/liter morphine.   Peaks: morphine (1) and 5-hydroxyquinoline (internal standard) (2).   (From Ref. 570.)

chromatography [566].   Coefficients of variation of less than 10% at 10 μg/liter can usually be attained.

A number of internal standards have been used, including hydromorphone, nalorphine, dextrophan, and normophine.   Since the last-named compound is a minor metabolite it is an inappropriate choice.

Meperidine has been abused, and a rapid procedure using a cyanopropyl column with detection at 205 nm has been reported [572].   Dextropropoxyphene is used in combination with acetaminophen in Britain and elsewhere, and death from respiratory arrest is not uncommon; one of the few reported procedures uses conditions similar to those for meperidine [573].

Opiate Antagonists

Reversed-phase C_{18} or phenyl columns have been used to determine naloxone or naltrexone with low-UV detection (<220 nm) [574,575] or electrochemical detection [571] with equivalent reported sensitivity. Two of the procedures use benzene for extraction [571,575], and one uses codeine, inappropriately, as the internal standard [571].

Cannabis

Since levels of cannabis and its metabolites in serum are low, LC estimation has been confined to urinary estimation of metabolites. Procedures are aimed at detecting the principal metabolite, 9-tetra-hydrocannabinol carboxylic acid. Concentration on a solid-phase cartridge with reversed-phase chromatography and detection at 214 nm had a sensitivity of 25 µg/liter [576,577]. LC has been combined with RIA [578] with a sensitivity of 1 µg/liter [579]; coulometric detection has a sensitivity of 35 pg on column [580].

Cocaine

There a few reports of screening for cocaine abuse by LC. As with cannabinoids, detection of a urinary metabolite, benzylecogonine, is performed. An early paper used reversed-phase LC to isolate benzylecogonine followed by GC-chemical ionization mass spectrometry [581] with a sensitivity of 1 µg/liter. Reversed-phase chromatography with dual-wavelength UV detection following solvent extraction resulted in a sensitivity of 0.1 µg/liter [582]. Cocaine, nor-cocaine, and benzylnorecogonine can also be determined, as they can by another reported procedure [583]. Methods for analyzing cocaine and its metabolites have been reviewed [584].

C. Other Toxicological Applications

Toxicological analysis for most analytes can be performed by procedures used for drug monitoring, always provided there are not matrix problems such as occur in forensic specimens [452].

Cyanide and thiocyanate in blood have been measured following microdiffusion for cyanide and ultrafiltration or protein precipitation for thiocyanate. Detection is by measuring the fluorescent products of the Konig reaction [585] following separation on an anion exchange column [586].

The chlorinated herbicides 2,4,5-trichlorophenoxyacetic acid and 2,4-dichlorophenoxyacetic acid have been measured with reversed-phase chromatography and UV detection [587,588].

The analysis of urinary metabolites of organic solvents has been the subject of a review [589] which summarizes information on the liquid chromatography of the metabolites of toluene, xylenes, benzene,

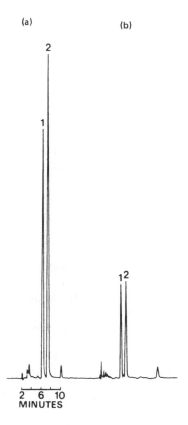

Fig. 14   Chromatograms of benzoyl derivatives of ethylene glycol
(1) and benzyl alcohol (2). (a) aqueous 100 mg/liter ethylene
glycol; (b) plasma extract of 100 mg/liter ethylene glycol. (From
Ref. 590.)

phenol, chlorobenzene, trichlorethylene, all of which are important
in industrial exposure.

   The benzoyl ester of ethylene glycol, with benzyl alcohol as internal standard, can be determined with detection at 237 nm after
reversed-phase chromatography [590] (Fig. 14).

   The mycotoxins from the mushroom *Amanita phalloides*, α-amanitin,
β-amanitin, and phalloidin, can be measured in serum and urine by
gradient chromatography on a reversed-phase column; recovery was
better than 80% and a sensitivity of 10 ng on column was achieved
at 302 nm [591]. An assay for detection of mycotoxins in urine has
been reported [592].

XII. CONCLUSION

This review has concentrated on drugs for which knowledge of the concentration in serum is useful for controlling therapy or enabling effective intervention in overdose. The emphasis in drug assay by LC should be on class analysis to allow determination of analogs and metabolites. Reversed-phase chromatography is most widely used, and for many analytes there are notable methodological similarities. Simultaneous determination of metabolites is of increasing interest, a task for which LC is ideally suited. With the flexibility and cost-effectiveness inherent in the technique, it is inevitable that LC applications in therapeutic drug monitoring and poisoning will continue to grow.

REFERENCES

1. V. Marks, in *Therapeutic Drug Monitoring* (A. Richens and V. Marks, eds.), Churchill Livingstone, Edinburgh, 1981, p. 155.
2. J. E. Strong and R. E. Altman, in *Applied Therapeutic Drug Monitoring*, Vol. 1 (T. P. Moyer and R. L. Boeckx, eds.), American Association for Clinical Chemistry, Washington, 1984, p. 185.
3. J. F. Burdin, in *Applied Therapeutic Drug Monitoring*, Vol. 1, (T. P. Moyer and R. L. Boeckx, eds.), American Association for Clinical Chemistry, 1984, p. 191.
4. J. W. A. Meijer, B. Rambeck, and M. Riedmann, Ther. Drug. Monit. *5*, 39 (1983).
5. A. M. Krstulovic and P. R. Brown, *Reversed-Phase High-Performance Liquid Chromatography, Theory, Practice and Biomedical Applications*, Wiley-Interscience, New York, 1982.
6. P. M. Kabra and L. J. Marton, eds., *Clinical Liquid Chromatography*, Vols. 1 and 2, CRC Press, Boca Raton, Fla. 1984.
7. P. M. Kabra and L. J. Marton, eds., *Liquid Chromatography in Clinical Analysis*, Humana Press, Clifton, N.J., 1981.
8. L. S. Yago and T. J. Good, in *Clinical Liquid Chromatography* P. M. Kabra and L. J. Marton, eds.), CRC Press, Boca Raton, Fla., 1984, p. 197.
9. J. A. F. de Silva, J. Chromatogr. *340*, 3 (1985).
10. G. Hoogewijs and D. L. Massart, J. Pharm. Biomed. Anal. *1*, 321 (1983).
11. G. Hoogewijs and D. L. Massart, J. Pharm. Biomed. Anal. *3*, 165 (1985).
12. B. Widdop, in *Clarke's Isolation and Identification of Drugs*, 2nd ed. (A. C. Moffat, J. V. Jackson, M. S. Moss, and B. Widdop, eds.), Pharmaceutical Press, London, 1986, p. 3.

13. R. D. McDowall, J. C. Pearce, and G. S. Murkitt, J. Pharm. Biomed. Anal. *4*, 3 (1986).

14. R. V. Smith and J. T. Stewart, *Textbook of Biopharmaceutic Analysis*, Lea & Febiger, Philadelphia, 1981, p. 27.

15. P. A. Toseland, in *Clarke's Isolation and Identification of Drugs,* 2nd ed. (A. C. Moffat, J. V. Jackson, M. S. Moss, and B. Widdop, eds.), Pharmaceutical Press, London, 1986, p. 111.

16. K. C. van Horne, ed., *Sorbent Extraction Technology,* Analytichem International, Harbor City, California, 1985.

17. R. W. Giese, Clin. Chem. *29*, 1331 (1983).

18. G. K. Szabo and T. R. Browne, Clin. Chem. *28*, 100 (1982).

19. S. J. van der Wal and L. R. Synder, Clin. Chem. *27*, 1233 (1981).

20. D. B. Campbell, in *Assay of Drugs and Other Trace Organic Compounds* (E. Reid, ed.), Elsevier, Amsterdam, 1976, p. 106.

21. M. D. Osselton, I. C. Shaw, and H. M. Stevens, Analyst *103*, 1160 (1978).

22. M. D. Osselton, M. D. Hammond, and P. J. Twitchett, J. Pharm. Pharmacol. *29*, 460 (1977).

23. L. C. Franconi, G. L. Hawk, B. J. Sandmann, and W. G. Haney, Anal. Chem. *48*, 372 (1976).

24. I. D. Watson, J. Chromatogr. *337*, 301 (1985).

25. G. Ray, W. G. Trawick, and R. E. Mullins, Clin. Chem. *31*, 131 (1985).

26. J. L. Bock and J. Ben-Ezra, Clin. Chem. *31*, 1884 (1985).

27. L. Jack, C. Cunningham, I. D. Watson, and M. J. Stewart, Ann. Clin. Biochem., *23*, 603 (1986).

28. W. Roth and K. Beschke, J. Pharm. Biomed. Anal. *2*, 189 (1984).

29. B. L. Karger, R. W. Giese, and L. R. Snyder, Trends Anal. Chem. *2*, 106 (1983).

30. M. Arunyanart and L. J. C. Love, J. Chromatogr. *342*, 293 (1985).

31. L. J. C. Love, S. Zibas, J. Noroski, and M. Arunyanart, J. Pharm. Biomed. Anal. *3*, 511 (1985).

32. F. J. de Luccia, M. Arunyanart, and L. J. C. Love, Anal. Chem. *57*, 1564 (1985).

33. L. R. Snyder and S. J. van der Wal, Anal. Chem. *53*, 877 (1981).

34. S. H. Curry and R. Whelpton, in *Blood Drugs and Other Analytical Challenges* (E. Reid, ed.), Ellis Horwood, Chichester, U.K., 1978, p. 29.

35. P. Haefelfinger, J. Chromatogr. *218*, 73 (1981).

36. L. Lund, Arc. Neurol. *31*, 289 (1974).

37. H. Kutt and J. K. Penry, Arch. Neurol. *31*, 382 (1974).

38. M. J. Eadie, in *Handbook of Clinical Pharmacokinetics* (M. Gibaldi and L. Prescott, eds.), ADIS Health Science Press, Sydney Australia, 1983, Sec. 4, p. 35.

39. S. D. Sharron, D. Chadwick, A. W. Galbraith, and E. H. Reynolds. Br. Med. J. *1*, 474 (1978).

40. R. F. Adams, Adv. Chromatogr. *15*, 131 (1977).

41. J. C. Kraak and J. P. Crombeen, J. Liq. Chromatogr. *5* (Suppl. 2), 273 (1982).

42. J. T. Burke and J. P. Thenot, J. Chromatogr. *340*, 199 (1985).

43. R. J. Perchalsky and B. J. Wilder, Anal. Chem. *50*, 554 (1978).

44. V. Rovei and M. Sanjuan, Ther. Drug Monit. *2*, 283 (1980).

45. S. Bouquet, P. Aucouturier, A. M. Brisson, Ph. Courtois, and J. B. Fourtillan, J. Liq. Chromatogr. *6*, 301 (1983).

46. E. H. Taylor, D. Sloniewsky, and R. H. Gadsden, Sr., Ther. Drug Monit. *6*, 474 (1984).

47. I. Petters, D.-R. Peng, and A. Rane, J. Chromatogr. *306*, 241 (1984).

48. R. L. Heazelwood and R. W. J. Lemass, J. Chromatogr. *336*, 229 (1984).

49. P. M. Kabra and E. U. Nzekwe, J. Chromatogr. *341*, 383 (1985).

50. J. P. Moody and S. M. Allan, Clin. Chim. Acta *127*, 363 (1983).

51. M. Nakamura, K. Kondo, R. Nishioka, and S. Kawai, J. Chromatogr. *310*, 450 (1984).

52. L. Richard, M. J. Bugugnani, H. Fouye, and J. Batten, Ann. Biol. Clin. (Paris) *43*, 279 (1985).

53. K. Kushida and T. Ishizaki, J. Chromatogr. *338*, 131 (1985).

54. M. J. Eadie, Rev. Drug Metab. Drug Interact. *3/4*, 317 (1981).

55. J. T. Lum, N. A. Vassanji, and P. G. Wells, J. Chromatogr. *338*, 242 (1985).

56. Y. Bergqvist, S. Eckerbom, and L. Funding, Clin. Chem. *30*, 465 (1984).

57. J. A. Christofides and D. E. Fry, Clin. Chem. *26*, 499 (1980).

58. N. Wad, J. Chromatogr. *305*, 127 (1984).

59. A. Kumps, J. Genin-Ramakers, and Y. Mardens, J. Chromatogr. *342*, 469 (1985).

60. J. Cornelis and F. Chaillet, Ann. Biol. Clin. (Paris) *42*, 295 (1984).

61. F. Sarhan, J. M. Ziegler, A. Nicolas, and G. Siest, J. Chromatogr. *183*, 505 (1983).

62. K. Kushida, K. Chiba, and T. Ishizaki, Ther. Drug Monit. *5*, 127 (1983).

63. D. C. Turnell, S. C. Trevor, and J. D. H. Cooper, Ann. Clin. Biochem. *20*, 37 (1983).

64. R. Soto-Otero and G. Sierra-Marcuno, Clin. Chem. *30*, 817 (1984).

65. B. Gerson, F. Bell, and S. Chan, Clin. Chem. *30*, 105 (1984).

66. K. Chan, S. Lok, and R. Teoh, Methods Find. Exp. Clin. Pharmacol. *6*, 701 (1984).

67. M. Riedmann, B. Rambeck, and J. W. A. Meijer, Ther. Drug Monit. *3*, 397 (1981).

68. N. Wad, J. Chromatogr. *338*, 460 (1985).

69. R. Alric, L. Arce-Corrales, J. P. Blayac, and R. Peuch, Methods Find. Exp. Clin. Pharmacol. *6*, 353 (1984).

70. S. J. Soldin and J. G. Hill, Clin. Chem. *22*, 856 (1976).

71. P. B. Kabra, B. E. Stafford, and L. J. Marton, Clin. Chem. *23*, 1284 (1977).

72. J. E. Slonek, G. W. Peng, and W. L. Chiou, J. Pharm. Sci. *67*, 1462 (1978).

73. M. Pesh-Imam, D. W. Fretthold, I. Sunshine, S. Kurmar, S. Terentine, and C. E. Willis, Ther. Drug Monit. *1*, 289 (1979).

74. C. N. Ou and C. L. Rognerud, Clin. Chem. *30*, 1667 (1984).

75. Y. Haroon and D. A. Keith, J. Chromatogr. *276*, 445 (1983).

76. J. C. Kraak, F. Smedes, and J. W. A. Meijer, Chromatographia *13*, 673 (1980).

77. P. M. Kabra, M. A. Nelson, and L. J. Marton, Clin. Chem. *29*, 473 (1983).

78. J. W. Dolan, S. J. van Der Wal, S. J. Bannister, and L. R. Snyder, Clin. Chem. *26*, 871 (1980).

79. L. M. St. Onge, E. Dolar, M. A. Anglim, and C. J. Least, Jr., Clin. Chem. *25*, 1373 (1979).

80. R. C. Williams and J. L. Viola, J. Chromatogr. *185*, 505 (1979).

81. U. Jurgens, J. Chromatogr. *310*, 97 (1984).

82. W. Kuhnz and H. Nau, Ther. Drug Monit. *6*, 478 (1984).

83. A. Nazareth, L. Jaramillo, B. L. Karger, and R. W. Giese, J. Chromatogr. *309*, 357 (1984).

84. W. Roth, K. Beschke, R. Jauch, A. Zimmer, and F. W. Koss, J. Chromatogr. *222*, 13 (1981).

85. W. Roth, J. Chromatogr. *278*, 347 (1983).

86. U. Juergens, Pre-column switching techniques for the determination of antiepileptic drugs and metabolites in body fluids, 2nd International Symposium on Sample Preparation, Freiburg, 1985, (abstract).

87. H. G. M. Westenberg, and R. A. de Zeeuw, J. Chromatogr. *118*, 217 (1976).

88. A. Bugge, J. Chromatogr. *128*, 111 (1976).

89. P. M. Kabra, G. Gotelli, R. Stanfill, and L. J. Marton, Clin. Chem. *22*, 824 (1976).

90. P. J. Helmsing, J. van der Woude, and O. M. van Eupen, Clin. Chim. Acta *89*, 301 (1978).

91. T. Tsuneyoshi, A. Kawamoto, and J. Koezuka, Clin. Chem. *30*, 1889 (1984).
92. B. Rambeck, M. Riedmann, and J. W. A. Meijier, Ther. Drug Monit. *3*, 377 (1981).
93. D. S. Reeves, I. Phillips, J. D. Williams, and R. Wise, eds., *Laboratory Methods in Antimicrobial Chemotherapy*, Churchill Livingstone, Edinburgh, (1978).
94. F. Follath, M. Wenk, and S. Vozeh, J. Antimicrob. Chemother. *8* (Suppl. A), 37 (1981).
95. M. Wenk, S. Vozeh, and F. Follath, Clin. Pharmacokinet. *9*, 475 (1984).
96. J. de Louvois, J. Antimicrob. Chemother. *9*, 253 (1982).
97. M. C. Rouan, J. Chromatogr. *340*, 361 (1985).
98. I. Nilsson-Ehle, Acta Pathol. Microbiol. Scand. Suppl. *259*, 61 (1977).
99. T. T. Yoshikawa, S. K. Maitra, M. C. Schotz, and L. B. Guze, Rev. Infect. Dis. *2*, 169 (1980).
100. I. Nilsson-Ehle, J. Liq. Chromatogr. *6* (Suppl. 2), 251 (1983).
101. P. M. Kabra, P. K. Bhatnagar, M. A. Nelson, J. H. Wall, and L. J. Marton, Clin. Chem. *29*, 672 (1983).
102. P. M. Kabra, P. K. Bhatnagar, and M. A. Nelson, J. Chromatogr. *307*, 224 (1984).
103. N. Kurosawa, S. Kuribayashi, E. Owada, and K. Ito, J. Chromatogr. *343*, 379 (1985).
104. J. d'Souza and R. I. Ogilvie, J. Chromatogr. *232*, 212 (1982).
105. D. B. Haughey, D. M. Janicke, M. Adelman, and J. J. Schentag, Antimicrob. Agents Chemother. *17*, 649 (1980).
106. S. K. Maitra, T. T. Yoshikawa, C. M. Steyn, L. B. Guze, and M. C. Schotz, Antimicrob. Agents. Chemother. *14*, 880 (1978).
107. S. K. Maitra, T. T. Yoshikawa, J. L. Hansen, M. C. Schotz, and L. B. Guze, Am. J. Clin. Pathol. *71*, 428 (1979).
108. L. Essers, J. Chromatogr. *305*, 345 (1984).
109. H. Kubo, T. Kinoshita, Y. Kobayashi, and K. Tokunaga, J. Chromatogr. *227*, 244 (1982).
110. T. G. Rosano, H. H. Brown, J. M. Meola, and C. McDermott, Clin. Chem. *25*, 1064 (1979).
111. J. P. Anhalt and S. D. Brown, Clin. Chem. *24*, 1940 (1978).
112. K. Koshide, R. Tawa, S. Hirose, and T. Fujimoto, Clin. Chem. *31*, 1921 (1985).
113. W. L. Chiou, R. L. Nation, G. W. Peng, and S. M. Huang, Clin. Chem. *24*, 1846 (1978).
114. S. E. Walker and P. E. Coates, J. Chromatogr. *223*, 131 (1981).
115. D. M. Barends, J. S. F. van der Sandt, and A. Hulshoff, J. Chromatogr. *182*, 201 (1980).

116. D. M. Barends, C. L. Zwaan, and A. Hulshoff, J. Chromatogr. *222*, 316 (1981).
117. D. M. Barends, C. L. Zwaan, and A. Hulshoff, J. Chromatogr. *225*, 417 (1981).
118. D. M. Barends, J. S. Blauw, M. H. Smits, and A. Hulshoff, J. Chromatogr. *276*, 285 (1983).
119. L. T. Wong, A. R. Beaubien, and A. P. Pakuts, J. Chromatogr. *231*, 145 (1982).
120. S. J. Kohlepp, M. O. Loveless, P. W. Kohnen, D. C. Houghton, W. M. Bennett, and D. N. Gilbert, J. Infect. Dis. *149*, 605 (1984).
121. L. O. White, A. Lovering, and D. S. Reeves, Ther. Drug Monit. *2*, 123 (1983).
122. W. C. Griffiths, J. F. Belliveau, C. L. Bensted, S. G. Calabresi, and I. Diamond, J. Pharm. Sci. *73*, 1836 (1984).
123. P. J. Claes, R. Busson, and H. Vanderhaeghe, J. Chromatogr. *298*, 445 (1984).
124. A. Mosegaard, P. G. Welling, and P. O. Madsen, Antimicrob. Agents Chemother. *7*, 328 (1975).
125. V. Schwarz, Z. Deyl, and K. Macek, J. Chromatogr. *340*, 401 (1985).
126. M. Edstein, J. Stace, and F. Shann, J. Chromatogr. *278*, 445 (1983).
127. L. C. Patchen, D. L. Mount, I. K. Schwartz, and F. C. Churchill, J. Chromatogr. *278*, 81 (1983).
128. Y. Bergqvist and M. Frisk-Holmberg, J. Chromatogr. *221*, 119 (1980).
129. Y. Bergqvist and B. Domeij-Nyberg, J. Chromatogr. *272*, 137 (1983).
130. E. Pussard, F. Verdier, and M. C. Blayo, J. Chromatogr. *374*, 111 (1986).
131. S. A. Ward, G. Edwards, M. L'E. Orme, and A. M. Breckenbridge, J. Chromatogr. *305*, 239 (1984).
132. M. V. Nora, G. W. Parkhurst, R. W. Thomas, and P. E. Carson, J. Chromatogr. *307*, 451 (1984).
133. J. H. Peters, G. R. Gordon, L. Levy, M. A. Storkan, R. R. Jacobson, C. D. Enna, and W. F. Kirchheimer, Am. J. Trop. Med. Hyg. *23*, 222 (1974).
134. K. Lammintausta, L. Kargas, and R. Lammintausta, Int. J. Clin. Pharmacol. Biopharm. *17*, 159 (1979).
135. M. Edstein, J. Chromatogr. *305*, 502 (1984).
136. M. Edstein, J. Chromatogr. *307*, 426 (1984).
137. H. S. Lee, T. Y. Ti, P. L. Lee, and C. L. Yap, Ther. Drug Monit. *7*, 415 (1985).
138. D. W. Warnock, M. D. Richardson, and A. Turner, J. Antimicrob. Chemother. *10*, 467 (1982).

139. I. Nilsson-Ehle, T. T. Yoshikawa, J. E. Edwards, M. C. Schotz, and L. B. Guze, J. Infect. Dis. *135*, 414 (1977).
140. C. L. Golas, C. G. Prober, S. M. MacLeod, and S. H. Soldin, J. Chromatogr. *278*, 387 (1983).
141. J. W. Mayhew, C. Fiore, T. Murray, and M. Borza, J. Chromatogr. *274*, 271 (1983).
142. P. R. Bach, Antimicrob. Agents Chemother. *26*, 314 (1984).
143. G. G. Granich, G. S. Kobayashi, and D. J. Krogstad, Antimicrob. Agents Chemother. *29*, 584 (1986).
144. R. R. Brodie, L. F. Chasseaud, and L. M. Walmsley, J. Chromatogr. *155*, 209 (1978).
145. L. A. Sternson, T. F. Patton, and T. B. King, J. Chromatogr. *227*, 223 (1982).
146. A. Turner and D. Warnock, J. Chromatogr. *227*, 229 (1982).
147. K. B. Alton, J. Chromatogr. *221*, 337 (1980).
148. F. A. Andrews, L. R. Peterson, W. H. Beggs, D. Crankshaw, and G. A. Sarosi, Antimicrob. Agents Chemother. *19*, 110 (1981).
149. P. T. Mannisto, R. Mantyla, S. Nykanen, V. Lamminsivu, and P. Ottoila, Antimicrob. Agents Chemother. *21*, 730 (1982).
150. S. F. Swezey, K. M. Giacomini, A. Abang, C. Brass, D. A. Stevens, and T. F. Blaschke, J. Chromatogr. *227*, 510 (1982).
151. V. L. Pasucci, J. Bennett, P. K. Narang, and D. C. Chatterji, J. Pharm. Sci. *72*, 1467 (1983).
152. N. R. Badcock, J. Chromatogr. *306*, 436 (1984).
153. L. P. Hackett and L. J. Dusci, J. Chromatogr. *155*, 206 (1978).
154. R. L. Nation, G. W. Peng, V. Smith, and W. L. Chiou, J. Pharm. Sci. *67*, 805 (1978).
155. M. C. Meyer and G. Raghow, J. Pharm. Sci. *68*, 1127 (1979).
156. H. Zia, W. J. Proveaux, J. P. O'Donnell, and J. K. H. Ma, J. Chromatogr. *181*, 77 (1980).
157. A. D. Blair, A. W. Forrey, B. T. Meijsen, and R. E. Cutler, J. Pharm. Sci. *64*, 1334 (1975).
158. R. W. Bury, M. L. Mashford, and H. M. Miles, Antimicrob. Agents Chemother. *16*, 529 (1979).
159. R. B. Diasio, M. E. Wilburn, S. Shadomy, and A. Espinel-Ingroff, Antimicrob. Agents Chemother. *13*, 500 (1978).
160. J. O. Miners, T. Foenander, and D. J. Birkett, Clin. Chem. *26*, 117 (1980).
161. D. W. Warnock and A. Turner, J. Antimicrob. Chemother. *7*, 363 (1981).
162. U. Schwertschlag, L. M. Nakata, and J. Gal, Antimicrob. Agents, Chemother. *26*, 303 (1984).
163. M. Foulstone and C. Reading, Antimicrob. Agents Chemother. *22*, 753 (1982).

164. T. B. Vree, Y. A. Hekster, A. M. Baars, and E. van der Kleijn, J. Chromatogr. *145*, 496 (1978).

165. J. Carlqvist and D. Westerlund, J. Chromatogr. *164*, 373 (1979).

166. H. H. W. Thijssen, J. Chromatogr. *183*, 339 (1980).

167. D. Westerlund, J. Carlqvist, and A. Theodorsen, Acta Pharm. Suec. *16*, 187 (1979).

168. K. Miyazaki, K. Ohtani, K. Sunada, and T. Arita, J. Chromatogr. *276*, 478 (1983).

169. F. W. Teare, R. H. Kwan, M. Spino, and M. MacLeod, J. Pharm. Sci. *71*, 938 (1982).

170. J. T. Rudrick and R. E. Bawdon, J. Liq. Chromatogr. *4*, 1525 (1981).

171. A. Weber, K. E. Opheim, K. Wong, and A. L. Smith, Antimicrob. Agents Chemother. *24*, 750 (1983).

172. D. Jung and N. K. Mahajan, Clin. Chem. *30*, 122 (1984).

173. R. H. Rumble and M. S. Roberts, J. Chromatogr. *342*, 436 (1985).

174. S. J. Soldin, A. M. Tesoro, and S. M. MacLeod, J. Pharm. Sci. *71*, 938 (1980).

175. A. M. Brisson and J. B. Fourtillan, Antimicrob. Agents Chemother. *21*, 664 (1982).

176. T. Annesley, K. Wilkerson, K. Matz, and D. Giacherio, Clin. Chem. *30*, 908 (1984).

177. U. Gundert-Remy and J. X. de Vries, Br. J. Clin. Pharmacol. *8*, 589 (1979).

178. T. L. Lee and M. A. Brooks, J. Chromatogr. *306*, 429 (1984).

179. R. Hildebrandt and U. Gundert-Remy, J. Chromatogr. *228*, 409 (1982).

180. J. H. G. Jonkman, R. Schoenmaker, and J. Hempnius, J. Pharm. Biomed. Anal. *3*, 359 (1985).

181. K. Yamaoka, S. Narita, T. Nakagawa, and T. Uno, J. Chromatogr. *168*, 187 (1979).

182. R. H. Kwan, S. M. MacLeod, M. Spino, and F. W. Teare, J. Pharm. Sci. *71*, 1118 (1982).

183. R. P. Lindberg, R. K. Huupponen, and P. Huovinen, Antimicrob. Agents Chemother. *26*, 300 (1984).

184. F. Salto, J. Chromatogr. *161*, 379 (1978).

185. J. O. Miners, J. Liq. Chromatogr. *8*, 2827 (1985).

186. E. M. McCormick, R. M. Echols, and T. G. Rosano, Antimicrob. Agents Chemother. *25*, 336 (1984).

187. C. Purser, A. Baltar, I. K. Ho, and A. S. Hume, J. Chromatogr. *311*, 135 (1984).

188. R. L. Yost and H. Denendorf, J. Chromatogr. *341*, 131 (1985).

189. J. B. Lecaillon, M. C. Rouan, C. Souppart, N. Febvre, and F. Juge, J. Chromatogr. *228*, 257 (1982).

190. F. Kees and H. Grobecker, J. Chromatogr. *305*, 363 (1984).
191. S. A. Signs, T. M. File, and J. S. Tan, Antimicrob. Agents. Chemother, *26*, 652 (1984).
192. J. Haginaka, H. Yasuda, T. Uno, and T. Nakagawa, Chem. Pharm. Bull. (Tokyo) *31*, 4436 (1983).
193. J. Haginaka, H. Yasuda, T. Uno, and T. Nakagawa, J. Chromatogr. *377*, 269 (1986).
194. J. Haginaka, J. Wakai, H. Yasuda, and T. Uno, J. Chromatogr. *341*, 115 (1985).
195. M. Poe, Science, *194*, 533 (1976).
196. T. B. Vree, Y. A. Hekster, A. M. Baars, J. E. Damsa, and E. van der Kleijn, J. Chromatogr. *146*, 103 (1978).
197. R. W. Bury and M. L. Mashford, J. Chromatogr. *163*, 114 (1979).
198. R. Gochin, I. Kanfer, and J. M. Haigh, J. Chromatogr. *223*, 139 (1981).
199. A. Weber, K. E. Opheim, G. R. Siber, J. F. Ericson, and A. L. Smith, J. Chromatogr. *278*, 337 (1983).
200. L. Nordholm and L. Dalgaard, J. Chromatogr. *305*, 391 (1984).
201. I. D. Watson, A. Shenkin, S. J. McIntosh, and H. N. Cohen, Clin. Chem. *26*, 1791 (1980).
202. S. J. McIntosh, D. J. Platt, I. D. Watson, A. J. Guthrie, and M. J. Stewart, J. Antimicrob. Chemother. *11*, 195 (1983).
203. V. Ascalone, J. Chromatogr. *224*, 59 (1981).
204. R. E. Weinfeld and T. C. Macasieb, J. Chromatogr. *164*, 73 (1979).
205. M. R. Holdiness, J. Chromatogr. *340*, 321 (1985).
206. R. L. Smith and D. D. Walker, J. Chromatogr. *343*, 203 (1985).
207. G. Land and A. Bye, J. Chromatogr. *224*, 51 (1981).
208. F. J. Ryan, M. A. Austin, and J. C. Mathies, Ther. Drug Monit. *6*, 465 (1984).
209. A. Weber, D. Chaffin, A. Smith, and K. E. Opheim, Antimicrob. Agents Chemother. *27*, 531 (1985).
210. D. E. Nix, J. M. De Vito, and J. J. Schentag, Clin. Chem. *31*, 684 (1985).
211. L. I. Harrison, D. Schuppan, S. R. Rohlfing, A. R. Hansen, C. S. Hansen, M. L. Funk, S. H. Collins, and R. E. Ober, Antimicrob. Agents Chemother. *25*, 301 (1984).
212. J.-O. Svensson, A. Muchtar, and O. Ericsson, J. Chromatogr. *341*, 193 (1985).
213. J. L. Uhl and J. P. Anhalt, Ther. Drug Monit. *1*, 75 (1979).
214. P. A. Ristuccia, A. M. Ristuccia, J. H. Bidanset, and B. A. Cunha, Ther. Drug Monit. *6*, 238 (1984).
215. M. A. Pfaller, D. J. Krogstad, G. G. Granich, and P. R. Murray, J. Clin. Microbiol. *20*, 311 (1984).
216. I. N. Bever, P. R. Finley, C. Fletcher, and J. Williams, Clin. Chem. *30*, 1586 (1984).

217. F. Jehl, C. Gallion, R. C. Thierry, and H. Monteil, Antimicrob. Agents Chemother. *27*, 503 (1985).
218. R. J. Hoagland, J. E. Sherwin, and J. M. Phillips, Jr., J. Anal. Toxicol. *8*, 75 (1984).
219. A. Aszalos, J. Liq. Chromatogr. *7*, 69 (1984).
220. S. Eksborg, and H. Ehrsson, J. Chromatogr. *340*, 31 (1985).
221. P. Leff and W. G. Bardsley, Biochem. Pharmacol. *28*, 1289 (1979).
222. D. R. Newell, L. I. Hart, and K. R. Harrap, J. Chromatogr. *164*, 114 (1979).
223. A. G. Ahmed, M. Koenig, and H. H. Farrish, Jr., J. Chromatogr. *233*, 392 (1982).
224. M. Zakaria and P. R. Brown, J. Chromatogr. *230*, 381 (1982).
225. C. M. Egan, C. R. Jones, and M. McCluskey, J. Chromatogr. *224*, 338 (1981).
226. A. G. Bosenquet and E. D. Gilby, J. Chromatogr. *232*, 345 (1982).
227. C. G. Adair, D. T. Burns, A. D. Crockand, Z. R. Resai, and M. Harriott, J. Chromatogr. *336*, 429 (1984).
228. R. W. Hardy, C. Erlichman, and S. J. Soldin, Ther. Drug Monit. *6*, 313 (1984).
229. D. Edwards, A. B. Selkirk, and R. B. Taylor, Int. J. Pharm. *4*, 21 (1979).
230. A. Kono, Y. Hara, S. Eguchi, M. Tanaka, and Y. Matsushima, J. Chromatogr. *164*, 404 (1979).
231. J. den Hartigh, W. J. van Oort, M. C. Y. M. Bocken, and H. M. Pinedo, Anal. Chem. Acta *127*, 47 (1981).
232. G. A. van Hazel and J. S. Kovach, Cancer Chemother. Pharmacol. *8*, 189 (1982).
233. V. R. Tjaden, J. P. Langenberg, K. Ensing, W. P. Bennekom, E. A. de Bruijn, and A. T. van Oosterom, J. Chromatogr. *232*, 355 (1982).
234. S. Eksborg, H. Ehrsson, and A. Lindfors, J. Chromatogr. *274*, 263 (1983).
235. J. W. Gynes, W. D. Ensminger, D. van Harken, J. Niederhuber, P. Stetson, and S. Walker, Clin. Pharmacol. Ther. *34*, 259 (1983).
236. R. G. Buice, P. Sidhu, B. J. Gurley, and H. B. Niell, Ther. Drug Monit. *6*, 113 (1984).
237. G. K. Shiu and T. J. Goehl, J. Chromatogr. *181*, 127 (1980).
238. S. Eksborg, H. Ehrsson, and I. Andersson, J. Chromatogr. *164*, 479 (1979).
239. P. Gil, R. Favre, A. Durand, A. Iliadis, J. P. Cano, and Y. Carcassonne, Cancer Chemother, Pharmacol. *10*, 120 (1983).
240. R. Rahmani, P. Gil, M. Martin, A. Durand, J. Barbet, and J.-P. Cano, J. Pharm. Biomed. Anal. *1*, 301 (1983).

241. T. Masuike, J. Odake, and Y. Takemoto, Yakugaku Zasshi, *104*, 614 (1984).
242. M. J. M. Oosterbaan, R. J. M. Dirks, T. B. Vree, and E. van der Kleijn, J. Chromatogr. *306*, 323 (1984).
243. J. Cummings, J. F. B. Stuart, and K. C. Calman, J. Chromatogr. *311*, 125 (1984).
244. C. A. Riley, W. R. Crom, and W. E. Evans, Ther. Drug. Monit. 7, 455 (1985).
245. A. El-Yazigi and I. Al-Saleh, J. Pharm. Sci. 74, 1225 (1985).
246. I. D. Watson, M. J. Stewart, and Y. Y. Z. Farid, J. Pharm. Biomed. Anal. 3, 555 (1985).
247. A. N. Kotake, N. J. Vogelzang, R. A. Larson, and N. Choporis, J. Chromatogr. *337*, 194 (1985).
248. J. Cummings, J. Chromatogr. *341*, 401 (1985).
249. M. A. van Lancker, L. A. Bellemons, and A. P. de Leenheer, J. Chromatogr. *374*, 415 (1986).
250. M. A. van Lancker, H. J. C. F. Nelis, and A. P. de Leenheer, J. Chromatogr. *254*, 45 (1983).
251. S. Eksborg, J. Chromatogr. *149*, 225 (1978).
252. J. M. van Baal, M. B. van Leeuwen, T. J. Schouten, and R. A. de Abreau, J. Chromatogr. *336*, 422 (1984).
253. C. E. Whalen, H. Tamary, M. Greenberg, A. Zipursky, and S. J. Soldin, Ther. Drug Monit. 7, 315 (1985).
254. L. E. Lavi and J. S. Holcenberg, Anal. Biochem. *144*, 514 (1985).
255. P. Linssen, A. Drenthe-Schonk, H. Wessels, and C. Haanen, J. Chromatogr. *223*, 371 (1981).
256. H. Breithaupt and J. Schick, J. Chromatogr. *225*, 99 (1981).
257. D. C. Sampson, R. M. Fox, M. H. N. Tattersall, and W. J. Hensley, Ann. Clin. Biochem. *19*, 125 (1982).
258. G. J. Peters, I. Kraal, E. Laurensse, A. Leyva, and H. M. Pinedo, J. Chromatogr. *307*, 464 (1984).
259. M. Iwamoto, S. Yoshida, and S. Hirose, J. Chromatogr. *310*, 151 (1984).
260. L. S. F. Hsu and T. C. Marrs, Ann. Clin. Biochem. *11*, 272 (1980).
261. J. L. Cohen and R. E. Brown, J. Chromatogr. *151*, 237 (1978).
262. R. G. Stoller, K. R. Hande, S. A. Jacobs, S. A. Rosenberg, and B. A. Chabner, N. Engl. J. Med. *297*, 630 (1977).
263. W. H. Isacoff, P. F. Morrison, J. Aroesty, K. L. Willis, J. B. Block, and T. L. Lincoln, Cancer Treat. Rep. *61*, 1665 (1977).
264. W. Sadee, Ther. Drug Monit. *2*, 177 (1980).
265. R. G. Buice and P. Sidhu, J. Pharm. Sci. 71, 74 (1982).
266. C. P. Collier, S. M. MacLeod, and S. J. Soldin, Ther. Drug Monit. 4, 371 (1982).
267. R. C. Donehower, K. R. Hande, J. C. Drake, and B. A. Chabner, Clin. Pharmacol. Ther. 27, 63 (1979).

268. Y. Y. Z. Farid, I. D. Watson, and M. J. Stewart, J. Pharm. Biomed. Anal. *1*, 55 (1983).

269. G. Fabre, J. P. Cano, A. Iliadis, Y. Carcassonne, R. Favre, R. Gilli, and J. Catalin, J. Pharm. Biomed. Anal. *2*, 61 (1984).

270. M. J. Stewart, I. D. Watson, Y. Y. Z. Farid, and G. G. Skellern, Ann. Clin. Biochem. *23*, 210 (1986).

271. J. Lankelma, E. van der Kleijn, and F. Raemaekers, Cancer Lett. *9*, 142 (1980).

272. J. Dutrieu and Y. A. Delmotte, Fresenius Z. Anal. Chem. *315*, 539 (1983).

273. J. A. Nelson, B. A. Harris, W. J. Decker, and D. Farquar, Cancer, Res. *37*, 3970 (1977).

274. L. A. Zwelling and K. W. Kohn, in *Pharmacologic Principles of Cancer Treatment* (B. Chabner, ed.), Saunders, Philadelphia, 1982, p. 309.

275. Y. Chang, L. A. Sternson, and A. J. Repta, Anal. Lett. *B11*, 449 (1978).

276. K. J. Himmelstein, T. F. Patton, R. J. Belt, S. Taylor, A. J. Repta, and L. A. Sternson, Clin. Pharmacol. Ther. *29*, 658 (1981).

277. P. A. Reece, J. T. McCall, G. Powis, and R. L. Richardson, J. Chromatogr. *306*, 417 (1984).

278. P. A. Andrews, W. W. Wung, and S. B. Howell, Anal. Biochem. *143*, 46 (1984).

279. K. C. Marsh, L. A. Sternson, and A. J. Repta, Anal. Chem. *56*, 491 (1984).

280. I. S. Krull, X.-D. Ding, S. Braverman, C. Selavka, F. Hochberg, and L. A. Sternson, J. Chromatogr. Sci. *21*, 166 (1983).

281. J. M. Holthuis, F. M. G. M. Romkens, H. M. Pinedo, and W. J. van Oort, J. Pharm. Biomed. Anal. *1*, 89 (1983).

282. T. J. Littlewood, A. L. Hutchings, D. P. Bentley, and B. P. Spragg, J. Chromatogr. *336*, 434 (1984).

283. J. A. Sinkule and W. E. Evans, J. Pharm. Sci. *73*, 164 (1984).

284. C. E. Werkhoven-Goewie, U. A. Th. Brinkman, R. W. Frei, C. de Ruiter, and J. de Vries, J. Chromatogr. *276*, 349 (1983).

285. J. M. Rideout, D. C. Ayres, C. K. Lim, and T. J. Peters, J. Pharm. Biomed. Anal. *2*, 125 (1984).

286. R. J. Strife, I. Jardine, and M. Colvin, J. Chromatogr. *182* 211 (1980).

287. L. Allen, J. Pharm. Sci. *69*, 1440 (1980).

288. P. Farina, G. Marzillo, and M. d'Incalci, J. Chromatogr. *22*, 141 (1981).

289. R. J. Strife, I. Jardine, and M. Colvin, J. Chromatogr. *224*, 168 (1981).

290. B. A. Robinson and F. N. Cornell, Clin. Chem. *29*, 1104 (1983).

291. G. Menge and J. P. Dubois, J. Chromatogr. *310*, 431 (1984).

292. J.-S. Schanche, P. E. Lonning, P. M. Ueland, and S. Kvinnsland, Ther. Drug Monit. *6*, 221 (1984).

293. K. C. Marsh, A. J. Repta, and L. A. Sternson, J. Chromatogr. *187*, 101 (1980).

294. D. L. Gildersleeve, M. F. Tobes, and R. B. Natale, Clin. Chem. *31*, 1979 (1985).

295. I. S. Krull, J. Strauss, F. Hochberg, and N. T. Zervas, J. Anal. Toxicol. *5*, 42 (1981).

296. B. Caddy, O. R. Idowu, and J. F. Stuart, Ther. Drug Monit. *4*, 389 (1982).

297. R. L. Yeager, E. H. Oldfield, and D. C. Chatterji, J. Chromatogr. *305*, 496 (1984).

298. A. K. Done, Paediatrics *62* (Suppl.), 890 (1978).

299. L. F. Prescott, N. F. Wright, P. Pascoe, and S. S. Brown, Lancet *1*, 519 (1971).

300. M. J. Stewart and I. D. Watson, Ann. Clin. Biochem. in press.

301. P. Miskolczi and L. Voreczkey, J. Pharm. Biomed. Anal. 3, 209 (1985).

302. B. E. Cham. D. Johns, F. Bochner, D. M. Imhoff, and M. Rowland, Clin. Chem. *25*, 1420 (1979).

303. R. J. O'Kruk, M. A. Adams, and R. B. Philip, J. Chromatogr. *310*, 343 (1984).

304. R. H. Rumble and R. S. Roberts, J. Chromatogr. *225*, 252 (1981).

305. B. E. Cham. L. Ross-Lee, F. Bochner, and D. M. Imhoff, Ther. Drug Monit. *2*, 365 (1980).

306. S. K. Baker and S. Niazi, J. Pharm. Sci. *72*, 1020 (1983).

307. D. Blair, B. H. Rumack, and R. Peterson, Clin. Chem. *24*, 1543 (1978).

308. D. C. Mays, D. E. Sharp, C. A. Beach, R. A. Kershaw, J. R. Bianchine, and N. Gerber, J. Chromatogr. *311*, 301 (1984).

309. R. A. Brandon, M. J. Eadie, and M. T. Smith, Ther. Drug. Monit. *7*, 216 (1985).

310. D. Dadgar, J. Climax, R. Lambe, and A. Darragh, J. Chromatogr. *342*, 315 (1985).

311. L. Y. Lo and A. Bye, J. Chromatogr. *181*, 473 (1980).

312. M. Rowland and S. Riegelman, J. Pharm. Sci. *56*, 717 (1967).

313. J. Micelli, M. K. Aravind, S. N. Cohen, and A. K. Done, Clin. Chem. *25*, 1002 (1979).

314. J. N. Buskin, R. A. Upton, and R. L. Williams, Clin. Chem. *23*, 1200 (1982).

315. L. I. Harrison, M. L. Funk, and R. E. Ober, J. Anal. Toxicol. *8*, 239 (1980).

316. B. Levine and Y. H. Caplan, J. Anal. Toxicol. *8*, 239–241 (1984).

317. B. Kinberger and A. Holmen, J. Chromatogr. *229*, 492 (1982).

318. N. R. Badcock, A. C. Penna, D. S. Everett, and L. N. Sanson, Ann. Clin. Biochem. *21*, 527 (1984).
319. M. J. Stewart, P. I. Adriaenssens, D. R. Jarvie, and L. F. Prescott, Ann. Clin. Biochem. *16*, 89 (1979).
320. K. Weiner, Ann. Clin. Biochem. *15*, 187 (1978).
321. D. Blair and B. H. Rumack, Clin. Chem. *23*, 743 (1977).
322. J. H. Knox and J. Jurand, J. Chromatogr. *142*, 651 (1977).
323. J. H. Knox and J. Jurand, J. Chromatogr. *149*, 297 (1978).
324. P. I. Adriaenssens and L. F. Prescott, Br. J. Clin. Pharmacol. *6*, 87 (1978).
325. T. Buchanan, P. I. Adriaenssens, and M. J. Stewart, Clin. Chim. Acta *99*, 161 (1979).
326. L. T. Wong, G. Solomonraj, and B. H. Thomas, J. Pharm. Sci. *65*, 1064 (1976).
327. R. A. Horvitz and P. I. Jatlow, Clin. Chem. *23*, 1596 (1977).
328. G. R. Gotelli, P. M. Kabra, and L. J. Marton, Clin. Chem. *23*, 743 (1979).
329. D. Howie, P. I. Adriaenssens, and L. F. Prescott, J. Pharm. Pharmacol. *29*, 235 (1977).
330. C. G. Fletterick, T. H. Grove, and D. C. Hohnadel, Clin. Chem. *25*, 743 (1979).
331. J. C. West, J. Anal. Toxicol. *5*, 118 (1981).
332. B. Ameer, D. J. Greenblatt, M. Divall, D. R. Abernethy, and L. Shargel, J. Chromatogr. *226*, 224 (1981).
333. J. N. Buskin, R. A. Upton, and R. L. Williams, J. Chromatogr. *230*, 443 (1982).
334. S. van der Wal, S. J. Bannister, and L. R. Snyder, J. Chromatogr. Sci. *20*, 260 (1982).
335. C. A. Korduba and R. F. Petruzzi, J. Pharm. Sci. *73*, 117 (1984).
336. F. Demotes-Mainard, G. Vinconn, C. Jarry, and H. Albin, Ann. Biol. Clin. (Paris) *42*, 9 (1984).
337. D. Jung and N. U. Zafar, J. Chromatogr. *339*, 198 (1985).
338. J. E. Mrochek, S. Katz, W. H. Christie, and S. R. Dinsmore, Clin. Chem. *20*, 1086 (1974).
339. R. M. Riggin, A. L. Schmidt, and P. T. Kissenger, J. Pharm. Sci. *65*, 1064 (1976).
340. J. W. Munson, R. Weierstall, and H. B. Kostenbauder, J. Chromatogr. *145*, 328 (1978).
341. P. Surmann, Arch. Pharm. (Weinheim) *313*, 399 (1980) .
342. S. M. Douidar and A. E. Ahmed, J. Clin. Chem. Clin. Biochem. *20*, 791 (1982).
343. H. Court, P. J. Streete, and G. N. Volans, Br. Med. J. *282*, 1073 (1981).
344. M. K. Aravind, J. N. Miceli, and R. E. Kauffman, J. Chromatogr. *308*, 350 (1984).

345. H. Litowitz, L. Olanoff, and C. L. Hoppel, J. Chromatogr. *311*, 443 (1984).

346. J. H. G. Jonkman, R. Schoenmaker, A. H. Holtkamp, and J. Hempenius, J. Pharm. Biomed. Anal. *3*, 433 (1985).

347. A. Shah and D. Jung, J. Chromatogr. *344*, 408 (1985).

348. M. Lalande, D. L. Wilson, and I. J. M. McGilveray, J. Chromatogr. *337*, 410 (1986).

349. R. Ginman, H. T. Karnes, and J. Perrin, J. Pharm. Biomed. Anal. *3*, 439 (1985).

350. D. Pitre and M. Grandi, J. Chromatogr. *170*, 278 (1979).

351. G. C. Kearns and J. T. Wilson, J. Chromatogr. *226*, 183 (1981).

352. J. L. Shimek, N. G. S. Rao, and S. K. Khalil, J. Pharm. Sci. *70*, 514 (1981).

353. B. G. Snider, L. J. Beaubin, D. J. Sears, and P. D. Rahn, J. Pharm. Sci. *70*, 1347 (1981).

354. G. F. Lockwood and J. G. Wagner, J. Chromatogr. *232*, 335 (1982).

355. D. J. Greenblatt, R. M. Arendt, and A. Locniskar, Arzneim. Forsch. *33*, 1671 (1983).

356. A. Ali, S. Kazmi, and F. M. Plakogiannis, J. Pharm. Sci. *70*, 514 (1981).

357. R. I. Ogilvie, Ther. Drug Monit. *2*, 11 (1980).

358. L. Hendeles, M. Weinberger, and G. Johnson, Clin. Pharmacokinet. *3*, 294 (1978).

359. L.-E. Edholm, Eur. J. Respir. Dis. *61* (Suppl. 109), 45 (1980).

360. D. Rowe, I. D. Watson, J. Williams, and D. Berry, Ann. Clin. Biochem., in press.

361. N. Kucharczyk and F. H. Segelman, J. Chromatogr. *340*, 243 (1985).

362. J. Chamberlain, in *Blood, Drugs, and Other Analytical Challenges* (E. Reid, ed.), Wiley, New York, 1978, p. 55.

363. E. C. Lewis and D. C. Johnson, Clin. Chem. *24*, 1711 (1978).

364. J. R. Miksic and B. Hodes, Clin. Chem. *25*, 1866 (1979).

365. J. H. G. Jonkman, R. A. de Zeeuw, and R. Schoenmaker, Clin. Chem. *28*, 1987 (1982).

366. H. H. Farrish and W. A. Wargin, Clin. Chem. *26*, 524 (1980).

367. P. M. Kabra and L. J. Marton, Clin. Chem. *28*, 687 (1982).

368. M. Wenk, B. Eggs, and F. Follath, J. Chromatogr. *276*, 341 (1983).

369. B. J. Starkey and G. P. Mould, Ther. Drug Monit. *6*, 322 (1984).

370. J. L. Bock, S. Lam, and A. Karmen, J. Chromatogr. *308*, 354 (1984).

371. R. Hartley, J. R. Cookman, and S. I. Smith, J. Chromatogr. *306*, 191 (1984).

372. N. R. Scott, J. Chakraborty, and V. Marks, Ann. Clin. Biochem. *21*, 120 (1984).

373. R. Soto-Otero, E. Mendez-Alvarez, and G. Sierra-Marcuno, J. Clin. Chem. Clin. Biochem. *23*, 303 (1985).

374. R. C. Kelly, D. E. Prentice, and G. M. Hearne, Clin. Chem. *24*, 838 (1978).

375. C. A. Robinson, Jr., B. Mitchell, J. Vasilades, and A. L. Siegel, Clin. Chem. *24*, 1847 (1978).

376. C. A. Robinson, Jr., and J. Dobbs, Clin. Chem. *24*, 2208 (1978).

377. G. Lam, S.-M. Huang, M.-G. Lee, R. L. Nation, and W. L. Chiou, Clin. Chem. *25*, 1862 (1979).

378. D. B. Bowman, M. K. Aravind, R. E. Kauffman, and J. N. Miceli, Clin. Chem. *26*, 1622 (1980).

379. N. Weidner, D. N. Dietzler, J. H. Ladenson, G. Kessler, L. Larson, C. H. Smith, T. James, and J. M. McDonald, Am. J. Clin. Pathol. *73*, 79 (1980).

380. R. H. Gannon and R. M. Levy, Am. J. Hosp. Pharm. *41*, 1185 (1984).

381. K. T. Muir, M. Kunitani, and S. Riegelman, J. Chromatogr. *231*, 73 (1982).

382. N. Daoud, T. Arvidsson, and K.-G. Wahlund, J. Pharm. Biomed. Anal. *4*, 253 (1986).

383. P. van Aerde, E. Moerman, R. van Severen, and P. Braeckman, J. Chromatogr. *222*, 467 (1981).

384. T. Zysset, A. Wahllander, and R. Preisig, Ther. Drug Monit. *6*, 348 (1984).

385. P. M. Naish, M. Cooke, and R. E. Chambers, J. Chromatogr. *163*, 363 (1979).

386. M. Anhoff, M. Ervik, P.-O. Lagerstrom, B.-A. Persson, and J. Vessman, J. Chromatogr. *340*, 73 (1985).

387. J. K. Aronson, in *Therapeutic Drug Monitoring* (A. Richens and V. Marks, eds.), Churchill Livingstone, Edinburgh, 1981, p. 404.

388. J. K. Aronson, in *Therapeutic Drug Monitoring* (B. Widdop, ed.), Churchill Livingstone, Edinburgh, 1985, p. 182.

389. R. G. Stoll, M. S. Christensen, E. Sakmar, and J. G. Wagner, Res. Commun. Chem. Pathol. Pharmacol. *4*, 503 (1972).

390. J. G. Wagner, McD. Dick II, D. M. Behrendt, G. F. Lockwood, E. Sakmar, and P. Hees, Clin. Pharmacol. Ther. *33*, 577 (1983).

391. M. H. Gault, L. Longerich, M. Dawe, and S. C. Vasdev, Clin. Chem. *31*, 1272 (1985).

392. J. Plum and T. Daldrup, J. Chromatogr. *377*, 221 (1986).

393. R. Valdes, S. W. Graves, B. A. Brown, and M. Landt, J. Pediatr. *102*, 947 (1983).

394. M. R. Pudeck, D. W. Seccombe, and M. F. Whitfield, N. Engl. J. Med. *308*, 904 (1983).
395. M. R. Pudek, D. W. Seccombe, B. E. Jacobson, and M. F. Whitfield, Clin. Chem. *29*, 1972 (1983).
396. J. M. Hicks and E. M. Brett, Clin. Chem. *29*, 1249 (1983).
397. V. J. Heazlewood, R. L. Heazlewood, L. B. Jellett, Ann. Intern. Med. *100*, 618 (1984).
398. R. W. Yatskoff, P. R. E. Desjardins, and J. G. Dalton, Clin. Chem. *30*, 588 (1984).
399. B. Rosenkranz and J. C. Frolich, Ther. Drug Monit. 7, 202 (1985).
400. A. A. Nanji and D. C. Greenway, Br. Med. J. *290*, 432 (1985).
401. S. W. Graves, B. Brown, and R. Valdes, Ann. Intern. Med. *99*, 604 (1983).
402. M. H. Gault, S. C. Vasdev, and L. L. Longerich, Ann. Intern. Med. *101*, 567 (1984).
403. P. M. Reisert and M. Kistner, Fortschr. Med. *95*, 167 (1977).
404. S. W. Graves, R. Valdes, B. A. Brown, A. B. Knight, and H. R. Craig, J. Clin. Endocrinol. Metab. *58*, 748 (1984).
405. H. Mabuchi and H. Nakahashi, J. Chromatogr. *375*, 75 (1986).
406. E. Wong and K. M. McErlane, J. Chromatogr. *377*, 233 (1986).
407. A. M. Barrett and V. A. Cullum, Br. J. Pharmacol. *34*, 43 (1968).
408. J. Hermansson and C. von Bahr, J. Chromatogr. *227*, 113 (1982).
409. K. Kawashima, A. Levy, and S. Spector, J. Pharmacol. Exp. Ther. *196*, 517 (1976).
410. T. Walle, D. D. Christ, U. K. Walle, and M. J. Wilson, J. Chromatogr. *341*, 213 (1985).
411. M. J. Wilson and T. Walle, J. Chromatogr. *310*, 424 (1984).
412. J. Hermansson and C. von Bahr, J. Chromatogr. *221*, 109 (1980).
413. B. Silber and S. Riegelman, J. Pharmacol. Exp. Ther. *215*, 643 (1980).
414. A. J. Sedman and J. Gal, J. Chromatogr. *278*, 199 (1983).
415. C. Pettersson and G. Schill, J. Chromatogr. *204*, 179 (1981).
416. I. W. Wainer, T. D. Doyle, K. H. Donn, and J. R. Powell, J. Chromatogr. *306*, 405 (1984).
417. G. Schill, I. W. Wainer, and S. A. Barkan, J. Liq. Chromatogr. *9*, 641 (1986).
418. J. Hermansson, M. Eriksson, and O. Nyquist, J. Chromatogr. *336*, 321 (1984).
419. J. Hermansson and M. Eriksson, J. Liq. Chromatogr. *9*, 621 (1986).
420. O. Grech-Belanger, J. Turgeon, and M. Gilbert, J. Chromatogr. *337*, 172 (1985).

421. K. T. Muir, K. A. Kook, C. Stern, and K. M. Gardner, J. Chromatogr. *374*, 394 (1986).
422. J. R. Shipe, Clin. Chem. *30*, 1259 (1984).
423. R. N. Gupta and S. Connolly, Clin. Chem. *30*, 1423 (1984).
424. G. C. A. Storey, D. W. Holt, P. Holt, and P. V. L. Curry, Ther. Drug Monit. *4*, 385 (1982).
425. G. C. A. Storey and D. W. Holt, J. Chromatogr. *245*, 377 (1982).
426. J. Maddock, P. A. Lewis, A. Woodward, P. R. Massey, and S. Kennedy, J. Chromatogr. *272*, 129 (1983).
427. W. C. Yu and E. U. Goff, Anal. Chem. *55*, 29 (1983).
428. A. J. Woodward, P. A. Lewis, M. Aylward, R. Rudman, and J. Maddock, J. Pharm. Sci. *73*, 1838 (1984).
429. A. C. Mehta, Pharm. J. *230*, 191 (1983).
430. C. B. Swezey and J. L. Ponzo, Ther. Drug Monit. *6*, 211 (1984).
431. C. P. Patel, Ther. Drug Monit. *4*, 213 (1982).
432. C. P. Patel, Ther. Drug Monit. *5*, 235 (1983).
433. L. J. Dusci and L. P. Hackett, J. Anal. Toxicol. *9*, 67 (1985).
434. P. M. Kabra, S.-H. Chen, and L. J. Marton, Ther. Drug Monit. *3*, 91 (1981).
435. T. A. Gough and P. B. Baker, J. Chromatogr. Sci. *21*, 145 (1983).
436. R. N. Gupta, J. Chromatogr. *340*, 139 (1985).
437. P. M. Kabra, B. E. Stafford, and L. J. Marton, J. Anal. Toxicol. *5*, 177 (1981).
438. A. Premel-Cabic, A. Turcant, A. Cailleux, and P. Allain, J. Chromatogr. *276*, 451 (1983).
439. C. R. Clark and J. L. Chan, Anal. Chem. *50*, 635 (1978).
440. R. Gill, A. A. T. Lopes, and A. C. Moffat, J. Chromatogr. *226*, 117 (1981).
441. P. C. White, J. Chromatogr. *200*, 271 (1980).
442. R. N. Gupta, P. T. Smith, and F. Eng. Clin. Chem. *28*, 1772 (1982).
443. G. K. Shiu and E. M. Nemoto, J. Chromatogr. *227*, 207 (1982).
444. J. R. Sharman and K. M. Ahern, J. Anal. Toxicol. *7*, 37 (1983).
445. W. Toner, P. J. Howard, J. W. Dundee, and P. D. A. McIlroy, Anaesthesia *34*, 657 (1979).
446. M. Kelner and D. N. Bailey, Clin. Chem. *29*, 1097 (1983).
447. B. Levine, R. Blanke, and J. Valentour, J. Anal. Toxicol. *7*, 207 (1983).
448. C. Salvadori, R. Farinotti, Ph. Duvaldestein, and A. Dauphin, Ther. Drug Monit. *3*, 171 (1981).
449. J. A. F. de Zilva, J. Chromatogr. *273*, 19 (1983).
450. H. Heusler, J. Chromatogr. *340*, 273 (1985).

451. J. P. Weijers-Everhard, J. Wiker, R. Verrijk, H. H. van Rooij, and S. Soudijn, J. Chromatogr. *374*, 339 (1986).
452. H. M. Stevens, J. Forensic Sci. Soc. *25*, 67 (1985).
453. A. S. Wong, J. Anal. Toxicol. 7, 33 (1983).
454. K. Tada, T. Moroji, R. Sekiguchi, H. Motomura, and T. Noguchi, Clin. Chem. *31*, 1712 (1985).
455. G. L. Lensmeyer, C. Rajani, and M. A. Evenson, Clin. Chem. *28*, 2274 (1982).
456. R. Gill. B. Law, and J. P. Gibbs, J. Chromatogr. *356*, 37 (1986).
457. M. Eichelbaum, B. Sonntag, and G. von Unroh, Arch. Toxicol. *41*, 187 (1978).
458. D. R. A. Uges and P. Bouma, Arch. Toxicol. *42*, 85 (1979).
459. R. Hartley, M. Becker, and S. F. Leach, J. Chromatogr. *276*, 471 (1983).
460. C. Kim and J. M. Khanna, J. Liq. Chromatogr. *6*, 907 (1983).
461. R. A. Hux, H. Y. Mohammed, and F. F. Cantwell, Anal. Chem. *54*, 113 (1982).
462. R. N. Gupta and F. Eng, J. High Resolut. Chromatogr. Chromatogr. Commun. *3*, 419 (1980).
463. L. L. Needham and M. M. Kochhar, J. Chromatogr. *111*, 422 (1975).
464. L. D. Bowers and D. M. Canafax, Ther. Drug Monit. *6*, 142 (1984).
465. R. J. Ptachinski, G. J. Burchart, and R. Venkataramanan, Drug Intell. Clin. Pharm. *19*, 90 (1985).
466. W. S. Burkle, Drug Intell. Clin. Pharm. *19*, 101 (1985).
467. R. J. Ptachinski, R. Venkataramanan, and G. J. Burckart, Clin. Pharmacokinet. *11*, 107 (1986).
468. Z. Varghese, M. K. Chan, L. V. Stele, P. Sweny, O. N. Ferando, and J. F. Moorhead, Lancet *1*, 1407 (1984).
469. D. W. Holt and D. J. G. White, Lancet *2*, 228 (1984).
470. W. Woloszczuk, J. Schindler, G. Hamilton, and E. Roth, Lancet *2*, 638 (1984).
471. C. A. Robinson and C. H. Ketchum, Ther. Drug Monit. *5*, 371 (1983).
472. H. F. Schran and E. Abisch, Ther. Drug Monit. *6*, 371 (1984).
473. P. Paull, K. Britton, K. Atkinson, and J. Briggs, Ther. Drug Monit. *6*, 373 (1984).
474. W. Niederberger, M. Lemaire, K. Nussbaumer, G. Maurer, and O. Wagner, Transplant. Proc. *15*, 2419 (1983).
475. M. J. Bennett, K. H. Carpenter, E. Worthy, and J. S. Lilleymann, Clin. Chem. *30*, 817 (1984).
476. W. Niederberger, P. Schaub, and T. Beveridge, J. Chromatogr. *182*, 454 (1980).
477. R. J. Sawchuck and L. L. Cartier, Clin. Chem. *27*, 1368 (1981).

478. K. Nussbaumer, W. Niederberger, and H. P. Keller, J. High Resolut. Chromatogr. Column Chromatogr. 5, 424 (1982).
479. B. Leyland-Jones, A. Clark, W. Kreis, R. Dismore, R. O'Reilly, and C. W. Young, Commun. Chem. Pathol. Pharmacol. 37, 431 (1982).
480. B. D. Kahan, C. T. van Buren, S. N. Lin, Y. Ono, G. Agostino, S. J. Legrue, M. Boileau, W. D. Payne, and R. H. Kerman, Transplantation 34, 36 (1982).
481. G. C. Yee, G. J. Dennis, and S. M. Kennedy, Clin. Chem. 28, 2269 (1982).
482. S. G. Carruthers, D. J. Freeman, J. C. Koegler, W. Howson, P. A. Keawn, A. Laupacis, and C. R. Stiller, Clin. Chem. 29, 180 (1983).
483. H. T. Smith and W. T. Robinson, J. Chromatogr. 305, 353 (1984).
484. R. E. Kates and R. Latini, J. Chromatogr. 309, 441 (1984).
485. G. L. Lensmeyer and B. L. Fields, Clin. Chem. 31, 196 (1985).
486. P. M. Kabra, J. H. Wall, and N. Blanckaert, Clin. Chem. 31, 1717 (1985).
487. D. J. Gmur, G. C. Yee, and M. S. Kennedy, J. Chromatogr. 344, 422 (1985).
488. G. Hamilton, E. Roth, E. Wallisch, and F. Tichy, J. Chromatogr. 341, 411 (1985).
489. R. Garraffo and P. Lapalus, J. Chromatogr. 337, 416 (1985).
490. M. K. Aravind, J. N. Micelli, and R. E. Kauffman, J. Chromatogr. 344, 428 (1985).
491. R. A. Braithwaite, in Therapeutic Drug Monitoring (B. Widdop, ed.), Churchill Livingstone, Edinburgh, 1985, p. 211.
492. I. D. Watson and M. J. Stewart, J. Chromatogr. 132, 155 (1977).
493. P. M. Edelbroek, E. J. M. de Haas,and F. A. de Wolff, Clin. Chem. 28, 2143 (1982).
494. R. Wester, P. Noonan, C. Markos, R. Bible, W. Aksamit, and J. Hribar, J. Chromatogr. 209, 463 (1981).
495. P. W. D. Hughes, S. J. McIntosh, I. D. Watson, and M. J. Stewart, in Proceedings of the 21st International Meeting of the International Association of Forensic Toxicologists (N. Dunnett and K. J. Kimber, eds.), International Association of Forensic Toxicologists, London, 1985, p. 41.
496. T. R. Norman and K. P. Maguire, J. Chromatogr. 340, 173 (1985).
497. B. A. Scoggins, K. P. Maguire, T. R. Norman, and G. D. Burrows, Clin. Chem. 26, 5 (1980).
498. R. Gupta and G. Molnar, Drug Metab. Rev. 9, 79 (1979).
499. P. Jatlow, Clin. Biochem. 18, 143 (1985).
500. I. D. Watson and M. J. Stewart, J. Chromatogr. 110, 389 (1975).

501. I. D. Watson and M. J. Stewart, J. Chromatogr. *134*, 182 (1977).
502. I. D. Watson, Proc. Anal. Div. Chem. Soc. *16*, 293 (1979).
503. R. B. Moyes and I. C. A. Moyes, Postgrad. Med. J. *53*, 431 (1977).
504. F. A. Beierle and R. W. Hubbard, Ther. Drug Monit. *5*, 279 (1983).
505. T. A. Sutfin, R. D'Ambrosio, and W. J. Jusko, Clin. Chem. *30*, 471 (1984).
506. R. D. Faulkner and C. Lee, J. Pharm. Sci. *72*, 1165 (1983).
507. C. T. Hung, R. B. Taylor, and N. Paterson, J. Pharm. Biomed. Anal. *1*, 73 (1983).
508. R. F. Suckow and T. B. Cooper, J. Pharm. Sci. *73*, 1745 (1984).
509. N. Yufu, M. Itoh, A. Notomi, and H. Nakao, Folia Psychiatr. Neurol. Jpn. *38*, 57 (1984).
510. G. L. Lensmeyer and M. A. Evenson, Clin. Chem. *30*, 1774 (1984).
511. A. Kobayashi, S. Sugita, and K. Nakazawa, J. Chromatogr. *336*, 410 (1984).
512. P. W. Hale and A. Poklis, J. Chromatogr. *336*, 452 (1984).
513. P. P. Rop, A. Viala, A. Durand, and T. Conquy, J. Chromatogr. *338*, 171 (1985).
514. P. P. Rop, T. Conquy, F. Gouezo, A. Viala, and F. Grimaldi, J. Chromatogr. *375*, 339 (1986).
515. J. L. Block, E. Giller, S. Gray, and P. Jatlow, Clin. Pharmacol. Ther. *31*, 609 (1982).
516. R. F. Suckow, T. B. Cooper, F. M. Quitkin, and J. W. Stewart, J. Pharm. Sci. *71*, 889 (1982).
517. P. A. Reece, R. Zacest, and C. G. Barrow, J. Chromatogr. *163*, 310 (1979).
518. T. A. Sutfin and W. J. Jusko, J. Pharm. Sci. *68*, 703 (1979).
519. I. S. Forrest, D. E. Green, and B. P. Udale, Proc. West. Pharm. Soc. *7*, 35 (1964).
520. P. Turano, W. J. Turner, and D. Donato, in *The Phenothiazines and Structurally Related Drugs* (I. S. Forrest, C. J. Carr, and E. Usdin, eds.), Raven, New York, 1974, p. 315.
521. R. O. Friedel, Psychopharmacol. Bull *12*, 63 (1976).
522. J. W. Hubbard, J. K. Cooper, E. M. Hawes, D. J. Jenden, P. R. May, M. Martin, G. McKay, T. van Putten, and K. K. Midha, Ther. Drug Monit. *7*, 222 (1985).
523. R. C. Young and R. W. Nysewander, J. Pharm. Biomed. Anal. *4*, 131 (1986).
524. A. D. MacKay, A. F. Healey, and J. Baker, Br. J. Clin. Pharmacol. *1*, 425 (1974).
525. D. Stevenson and E. Reid, Anal. Lett. *14*(B), 1785 (1981).

526. G. McKay, J. Geddes, M. J. Cowper, and T. S. Gurnsey, Prog. Neuro-Psychopharmacol. Biol. Psychiat. 7, 703 (1983).

527. K. K. Midha, J. K. Cooper, I. J. McGilveray, A. G. Butterfield, and J. W. Hubbard, J. Pharm. Sci. 70, 1043 (1981).

528. J. K. Cooper, G. McKay, and K. K. Midha, J. Pharm. Sci. 72, 1259 (1983).

529. K. Murakami, K. Murakami, T. Ueno, J. Hijikata, K. Shirasawa, and T. Muto, J. Chromatogr. 227, 103 (1982).

530. R. G. Muusze and J. F. K. Huber, J. Chromatogr. Sci. 12, 770 (1974).

531. D. C. Williams III and R. W. Burnett, Clin. Chem. 23, 1139 (1977).

532. L. C. Brookes, F. C. Chao, I. S. Forrest, D. E. Green, K. O. Loeffler, and M. T. Serra, Clin. Chem. 24, 1031 (1978).

533. C. E. Wells, E. C. Juerge, and W. B. Furman, J. Pharm. Sci. 72, 622 (1983).

534. A. L. Stoll, R. J. Baldessarini, B. M. Cohen, and S. P. Finklestein, J. Chromatogr. 307, 457 (1984).

535. G. McKay, J. K. Cooper, K. K. Midha, K. Hall, and E. M. Hawes, J. Chromatogr. 233, 417 (1982).

536. A. Forsman and M. Larsson, Curr. Ther. Res. 24, 567 (1978).

537. P. L. Morselli, G. Bianchetti, G. Tedeschi, and R. A. Braithwaite, in *Therapeutic Drug Monitoring* (A. Richens and V. Marks, eds.), Churchill Livingstone, Edinburgh, 1981, p. 296.

538. M. J. Kogan, D. Pierson, and K. Vereby, Ther. Drug Monit. 5, 485 (1983).

539. A. K. Dhar and H. Kutt, Clin. Chem. 30, 1228 (1984).

540. A. McBurney and S. George, J. Chromatogr. 308, 287 (1984).

541. D. Hequet, C. Jarry, C. Rouquette, and A. Bracket-Liermain, Ann. Biol. Clin. (Paris) 43, 739 (1985).

542. E. R. Korpi, B. H. Phelps, H. Granger, W. H. Chang, M. Linnoila, J. L. Meek, and R. J. Wyatt, Clin. Chem. 29, 624 (1983).

543. P. J. Twitchett and A. C. Moffat, J. Chromatogr. 111, 149 (1975).

544. J. K. Baker, R. E. Skelton, and C.-Y. Ma, J. Chromatogr. 168, 417 (1979).

545. I. S. Lurie, J. Forensic Sci. Soc. 29, 607 (1984).

546. R. Gill, J. Forensic Sci. Soc. 24, 322 (1984).

547. R. A. de Zeeuw, J. Pharm. Biomed. Anal. 1, 435 (1983).

548. R. J. Flanagan and I. Jane, J. Chromatogr. 323, 173 (1985).

549. I. Jane, A. McKinnon, and R. J. Flanagan, J. Chromatogr. 323, 191 (1985).

550. G. Hoogewijs and D. L. Massart, J. Pharm. Biomed. Anal. 1, 435 (1983).

551. M. R. Detaevernier, G. Hoogewijs, and D. L. Massart, J. Pharm. Biomed. Anal. *3*, 331 (1983).

551. A. F. Fell, B. J. Clark, and H. P. Scott, J. Pharm. Biomed. Anal. *1*, 557 (1983).

552. A. F. Fell, H. P. Scott, R. Gill, and A. C. Moffat, J. Chromatogr. *273*, 3 (1983).

553. R. M. Smith, T. G. Hurdley, R. Gill, and A. C. Moffat, J. Chromatogr. *355*, 75 (1986).

554. R. Gill, in *Clarke's Isolation and Identification of Drugs in Pharmaceuticals, Body Fluids and Post-Mortem Material* (A. C. Moffat, J. V. Jackson, M. S. Moss, and B. Widdop, eds.), Pharmaceutical Press, London, 1986, p. 201.

555. T. A. Gough and P. B. Baker, J. Chromatogr. Sci. *20*, 289 (1982).

556. J. O. Svensson, A. Rane, J. Sawe, and F. Sjoqvist, J. Chromatogr. *230*, 427 (1982).

557. B. L. Posey and S. N. Kimble, J. Anal. Toxicol. *8*, 68 (1984).

558. V. Nitsche and H. Mascher, J. Pharm. Sci. *73*, 1556 (1984).

559. R. J. Stubbs, R. Chiou, and W. F. Bayne, J. Chromatogr. *377*, 477 (1986).

560. P. E. Nelson, S. M. Fletcher, and A. C. Moffat, J. Forensic Sci. Soc. *20*, 203 (1980).

561. M. W. White, J. Chromatogr. *178*, 229 (1979).

562. J. E. Wallace, S. C. Harris, and M. W. Peek, Anal. Chem. *52*, 1328 (1980).

563. R. G. Peterson, B. H. Rumack, J. B. Sulliven, Jr., and A. Makowski, J. Chromatogr. *188*, 420 (1980).

564. J. A. Owen and D. S. Sitar, J. Chromatogr. *276*, 202 (1983).

565. A. R. Aitkenhead, M. Vater, K. Achola, C. M. Cooper, and G. Smith, Br. J. Anaesth. *56*, 813 (1984).

566. R. A. Moore, D. Baldwin, H. J. McQuay, and R. E. S. Bullingham, Ann. Clin. Biochem. *21*, 125 (1984).

567. C. Kim and T. Kats, J. Anal. Toxicol. *8*, 135 (1984).

568. G. K. Gourlay, C. F. McLean, G. A. Murphy, and N. R. Badcock, J. Pharmacol, Methods *13*, 317 (1985).

569. J.-O. Svensson, J. Chromatogr. *375*, 174 (1986).

570. H. Vandenberghe, L. M. MacLeod, H. Chinyanga, and S. J. Soldin, Ther Drug Monit. *4*, 307 (1982).

571. H. Derendorf, A. El-Din, A. El-Kousi, and E. R. Garret, J. Pharm. Sci. *73*, 621 (1984).

572. R. C. Meatherall, D. R. P. Guay, and J. L. Chambers, J. Chromatogr. *338*, 141 (1985).

573. H. R. Angelo, T. Kranz, J. Strom, B. Thisted, and M. B. Sorensen, J. Chromatogr. *345*, 413 (1985).

575. L. A. Asali, R. L. Nation, and K. F. Brown, J. Chromatogr. *278*, 329 (1983).

575. M. D. Terry, G. H. Hisayasu, J. W. Kern, and J. L. Cohen, J. Chromatogr. *311*, 213 (1984).
576. M. A. Elsohly, H. N. Elsohly, A. B. Jones, P. A. Dimson, and K. E. Wells, J. Anal. Toxicol. *7*, 262 (1983).
577. D. L. Black, B. A. Goldberger, D. S. Isenschmid, S. M. White, and Y. H. Caplan, J. Anal. Toxicol. *8*, 224 (1984).
578. M. A. Peat, M. E. Dayman, and J. R. Johnson, J. Forensic Sci. *29*, 110 (1984).
579. B. Law, J. Forensic Sci. Soc. *21*, 31 (1981).
580. W. H. Cooke, Clin. Chem. *30*, 1030 (1984).
581. A. P. Graffeo, D. C. K. Lin, and R. L. Foltz, J. Chromatogr. *126*, 717 (1976).
582. P. I. Jatlow, C. van Dyke, P. Barash, and R. Byck, J. Chromatogr. *152*, 115 (1978).
583. M. A. Evans and T. Moriarty, J. Anal. Toxicol. *4*, 19 (1980).
584. J. E. Lindgren, J. Ethnopharmacol. *3*, 337 (1981).
585. T. Toida, S. Tanabe, and T. Imanari, Chem. Pharm. Bull. *29*, 3763 (1981).
586. T. Toida, T. Togawa, S. Tanabe, and T. Imanari, J. Chromatogr. *308*, 133 (1984).
587. L. F. Prescott, J. Park, and I. Darrien, Br. J. Clin. Pharmacol. *7*, 111 (1979).
588. A. D. Fraser, A. F. Ismer, and R. A. Perry, J. Forensic Sci. *29*, 1237 (1984).
589. M. Ogata, Acta Med. Okayama *35*, 385 (1981).
590. R. N. Gupta, F. Eng, and M. L. Gupta, Clin. Chem. *28*, 32 (1982).
591. G. Caccialanza, C. Gandini, and R. Ponci, J. Pharm. Biomed. Anal. *3*, 179 (1985).
592. D. L. Orti, R. H. Hill, Jr., J. A. Liddle, L. L. Needham, and L. Vickers, J. Anal. Toxicol. *10*, 41 (1986).

5

Element-Selective Plasma Emission Detectors for Gas Chromatography

Ahmad H. Mohamad and Joseph A. Caruso   *University of Cincinnati, Cincinnati, Ohio*

I. INTRODUCTION AND BACKGROUND

In 1952 James and Martin [1] published the first paper that experimentally demonstrated the potential of gas-liquid chromatography
(GC). Since then the field has developed extensively, covering a
wide range of applications. New chromatographic column designs
and stationary phases were introduced, primarily to improve separation efficiency. Since the GC detectors widely used today (thermal
conductivity, flame ionization, etc.) respond nonselectively, compounds of interest must be well separated prior to detection. Even
more selective detectors (electron capture, thermionic, photoioniza-
tion, etc.) are not able to distinguish compounds within the same
class.

The requirement for highly efficient separation to achieve satisfactory resolution of various compounds can be somewhat relaxed if
the detector used has the ability to respond selectively to certain
compounds but not to others. This selectivity can be based on the
differences, chemical or otherwise, that exist between those compounds. As an example, two compounds with identical retention times
(which will give a single peak if a universal GC detector is used)
can still be selectively detected if any difference in their elemental
composition is used as a basis for their detection (Fig. 1). Compounds 7 and 8 overlapped when a flame ionization detector was used,
but only 8 was detected with a nitrogen/phosphorus-sensitive detector. To a certain extent gas chromatography-mass spectrometry
(GC-MS) and gas chromatography-infrared spectroscopy (GC-IR)
make use of these differences in chemical composition to achieve selective detection. The question of selective detection has been the
subject of an excellent review by Ettre [2]. It was indicated that
usually as a result of selective detection an additional advantage is
observed, an increase in sensitivity, also shown in Fig. 1.

Atomic emission spectroscopy is one of the most selective, sensitive,
reliable, and versatile techniques available for elemental determinations.
Atomic absorption and fluorescence have similar advantages, since all
three of the techniques exploit differences in the elemental composition of the compounds. However, atomic absorption and fluorescence
are not readily adaptable to multielement tracking and are confined to
the determination of a limited number of compounds or elements [3].

With the development of plasmas as highly energetic and stable
analytical excitation sources capable of exciting intense emissions from
all elements in the periodic table, atomic emission spectroscopy has
additional potential for selective chromatographic detection. A
plasma atomic emission system can be readily coupled to a gas chromatograph. At least in theory, this will enable one to excite all the
elements constituting the eluting compounds and subsequently monitor
the elemental emissions optically.

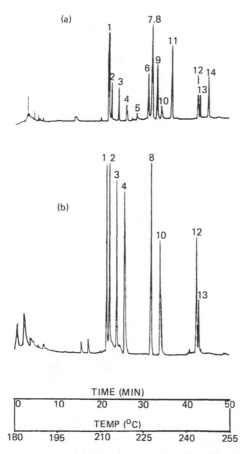

Fig. 1   Chromatograms of a standard steroid mixture detected by
(a) flame ionization detector and (b) nitrogen-sensitive (thermionic)
detector.   Peak 7, pregnanediol; peak 8, pregnenolone.   (From Ref. 81.)

In 1965 the first article describing the use of plasma as an elemental
emission detector for GC was published by McCormack et al. [4].
Two similar works published by Bache and Lisk [5,6] followed shortly
thereafter and further demonstrated the potential usefulness of the
plasma as an element-selective emission detector for GC in the determination of various organic compounds.   Since then, well over 100
articles and reviews have appeared describing applications of plasmas
as GC detectors.   In addition to their excellent selectivity, plasma
emission detectors demonstrate fewer interferences, improved sensitivity, and multielement detection capability (which may lead to

elemental ratioing-empirical formula determinations) and yield more qualitative information than conventional GC detectors. These applications will be discussed later in greater detail. It is perhaps worthy of mention that a commercial plasma emission detector has been marketed by Applied Chromatography Systems.

An important point often overlooked is the fact that the plasma emission detector can be used as a universal detector when the appropriate wavelength is monitored. For example, in the determination of organic compounds, monitoring the carbon emission line allows all organic effluents to be detected. On the other hand, monitoring a sulfur emission line would enable one to detect only sulfur-containing organic compounds. Universal detection and selective detection can be done simultaneously in GC applications by using a polychromator. In this chapter, the development and applications of the plasma atomic emission system as an element-selective detector for GC will be outlined.

II. INSTRUMENTATION

A. Plasma Formation

It is not our intent in this chapter to examine in great detail the theoretical and operational aspects of the various commonly used plasmas. Excellent reviews covering those aspects of the inducively coupled plasma (ICP) [7—11], the microwave-induced plasma (MIP) [12—15], and the direct current plasma (DCP) [16,17] are readily available in the literature. Some discussion, however, will be of help to those less familiar with the field.

The plasma consists of a mass of predominantly ionized gas (usually helium, argon, or nitrogen) at a temperature of 4,000 to 10,000 K. This state can be maintained either directly by an electrical discharge through the gas (DCP) or by suitably coupling a high-frequency radio frequency (ICP) or microwave frequency field (MIP) to the flowing gas stream, typically using 50 to 2000 W power [3]. An analyte molecule entering the highly energetic plasma will be fragmented and excited and will subsequently emit radiation characteristic of its constituent elements. The mechanism of molecular fragmentation and excitation in these plasmas is not yet fully understood. However, the available literature [13,14,18—21] suggests that a multistep sequence of events is involved, probably including some collisional processes leading to ions and possibly recombination with low-energy electrons, resulting in atoms reaching excited states that subsequently emit radiation to return to lower state(s). Direct excitation as a result of collisions between analyte molecules with metastable species or high-energy electrons has been considered, but needs further study.

Various types of DCP devices have been reported, as reviewed by Keirs and Vickers [17]. Two of them are shown in Figs. 2 and 3.

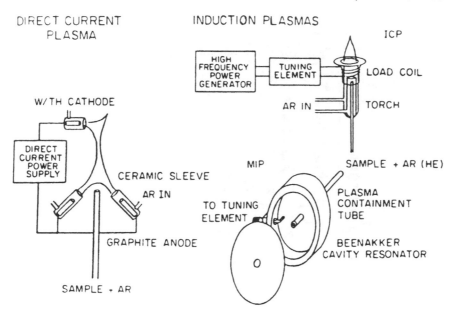

Fig. 2   Plasma sources for elemental emission spectroscopy.   (From Ref. 3).

A number of materials have been used for the fabrication of the anode and cathode. Tungsten and graphite are the popular choices because of their ability to withstand high temperatures and their good conductivity. A plasma gas flow (usually argon) ranges from 5 to as much as 20 liters/min, depending on the design of the DCP device. Power dissipated in the discharge also depends on the design of the DCP device, i.e., how much current at how many volts is used by the device. Typical power used ranges up to a few hundred watts. While the DCP is maintained directly by an electrical discharge through the gas, induction plasmas are maintained indirectly via an induction coil (the ICP) or a cavity resonator (the MIP). Inductive coupling of the high-frequency field to the plasma gas takes place in the induction coil or the cavity resonator as the gas flows through a tube made of quartz, alumina, etc. Design and optimization of these devices are subjects of extensive research (see, e.g., Refs. 7, 8, 10, 12, 14, 22, and 23). Some of the more commonly used cavity resonators for sustaining the MIP are the cylindrical Beenakker TM_{010} cavity (Fig. 2), the 1/4-λ and 3/4-λ Evenson cavities, and the 3/4-λ Broida cavity. The ICP induction coils are usually of the two-turn or three-turn type. Generally these devices are fabricated from copper or brass. High-frequency power generators

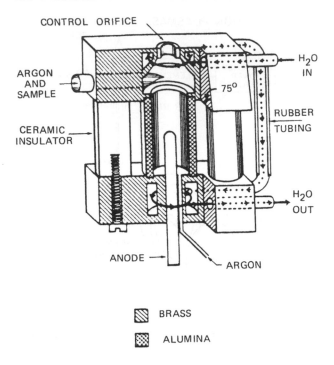

Fig. 3   Cutaway drawing of the anode chamber for direct current plasma. The cathode chamber is the mirror image of the anode chamber. (From Ref. 16.)

(typically 27.12 MHz for the ICP and 2450 MHz for the MIP) are used to supply power to the induction devices via a cable. These generators are designed to provide power efficiently at a certain load impedance (total impedance of the plasma plus coupling device), typically 50 ohms to match that of the generator, otherwise some of the power is reflected back to the generator with possible damage to it.   To compensate for the variation in the load impedance, tuning devices are usually used; variable capacitors and stub tuners are popular choices for ICP and MIP applications, respectively.   The availability of "free-running" generators that automatically compensate for a mismatch between the generator and the load by changing the frequency output minimizes the need for tuning devices.   However, the free-running frequencies generated, if transmitted, may cause interferences with communication bands.   Typical power output of these generators varies from 50 to 500 W for MIP and 0.75 kW to several kilowatts for the ICP.   The power level chosen during an operation is dependent

on many factors and considerations, some of which can only be determined experimentally.

Helium and argon are the two most commonly used plasma support
gases. Nitrogen and air have been used, but less often. DCPs and
ICPs are usually maintained in argon at atmospheric pressure, whereas the MIP uses both helium and argon and is operated from pressures of 1 to 760 torr. The advantage of using helium is the capability of more efficiently exciting analytes such as the difficult-to-excite
nonmetals. Small amounts (less than 5%) of hydrogen, oxygen, or
nitrogen have also been introduced into the plasma as scavenger
gases to prevent excessive carbon deposition on the discharge tube
wall, in the case of MIP.

B. Gas Chromatography-Plasma Detector Interface

As in any gas chromatographic detector, some features are essential
for the plasma to be an acceptable GC detector [14]. These features
are outlined briefly below.

Sensitivity. The demand today is for low detection limits, in the
nanograms (10^{-9} g), to picograms (10^{-12} g) range and even less.
Therefore, the plasma detection system must be able to offer this
sensitivity while not sacrificing its selectivity.

Linearity. This is another important characteristic of a useful GC
detector. The detector's response to the GC effluent must be
proportional to the amount of the effluent reaching it. The
wider the linear dynamic range, the better. In general, the
plasma detection system exhibits linear response of up to five
orders of magnitude.

Low dead volume. Ideally, the volume of the detector (volume between the end of the column packing and the sensing device)
should be at least one order of magnitude smaller than the free
gas volume of the column [14]. A large dead volume will lead
to component dilution, contributing to peak broadening and, therefore, loss of resolution. The design of the interface between the
gas chromatograph and the plasma detector is critical to ensure
that a low dead volume is achieved.

Rapid response. The high-efficiency separation achieved by a GC
column will again be rendered useless if the detector cannot respond quickly to the separated components. Generally, the response must be at least an order of magnitude faster than the
column can separate the components of a mixture [14]. Most detectors, including the plasma, satisfy this requirement.

Some of the interface designs for coupling GC with the plasma detection system are shown in Figs. 4, 5, and 6. Details of those

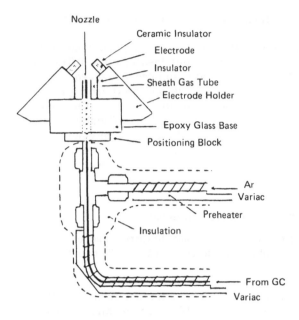

Fig. 4 Diagram of nozzle, sheath gas preheater, and transfer line
from gas chromatograph to DC argon plasma. (From Ref. 32.)

interface designs can be obtained from the original articles. An
auxiliary supply of plasma gas is usually needed since the GC flow
is not sufficient to maintain the plasma by itself, except in the case
of MIP. However, excess solvent is usually vented from the column
instead of being directed to the MIP; an auxiliary gas flow is needed
during this venting step. If directed into the MIP, excess solvent
will cause deposition in the plasma containment tube, degrading it or
even extinguishing the plasma. This is somewhat of a limitation for
the MIP as a GC detector, especially when operated at lower powers
(less than 100 W). Various other interface designs have been presented for the GC-DCP [24], GC-ICP]25], and GC-MIP systems
[26—29]. Generally, these interface designs have taken into consideration the essential features required for a good GC detector as
listed above. In addition, provision is made to prevent analyte condensation along the interface due to a sudden temperature decrease
on leaving the GC oven. This is usually done by using an auxiliary
heater or heating tape along the transfer line.

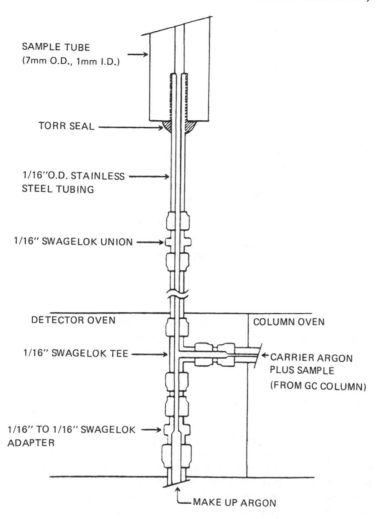

Fig. 5   Interface coupling GC effluent into modified ICP torch.
Note that the sample tube is replaced with a 0.1 mm (inner diameter)
capillary to reduce dead volume.   (From Ref. 71.)

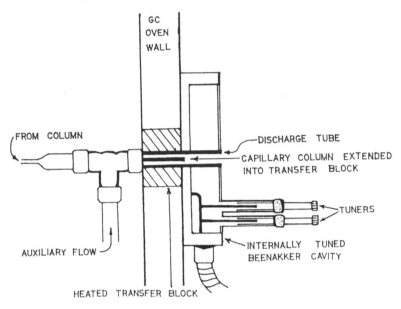

Fig. 6   Low-dead-volume interface for capillary gas chromatography with microwave-induced plasma detection.   The Beenakker cavity was modified to incorporate two internal tuners.   (From Ref. 82.)

C. Optical and Data Handling Systems

As in any atomic emission experiment, some kind of optical system is needed to separate and analyze the emission spectrum obtained from the plasma.   The plasma image is focused onto the entrance slit of a spectrometer with a focusing lens or collimating mirror.   Fiber-optic guides can also be used, but they are not yet popular in commercial plasma instrumentation.   The choice of spectrometer is heavily dependent on the kind of experiment (resolution required, wavelength region, single or multielement monitoring capability).   Detection and data handling systems to be used are again heavily dependent on the kind of experiment and, to a lesser extent, cost, especially with the ready availability of microcomputers that enable both experimental control and data reduction.   A few papers [27, 30-32] can serve as an excellent point of departure to those entering the field.

III. APPLICATIONS

Of the three plasmas, DCP, ICP, and MIP, the MIP, at both reduced and atmospheric pressure, has found the most application as a GC detector. The principal reason is the ability of the MIP to sustain both helium and argon plasmas and to excite emission from metals and nonmetals as well. To date, it has been difficult to form atmospheric helium plasmas by ICP or DCP coupling. Recently, a reduced-pressure helium ICP has been introduced [33,34]. It shows promise as an excitation source for nonmetals; however, its application to actual samples has to be shown. Cost consideration and simplicity of operation favor the MIP as a GC detector.

The application of plasmas as element-selective detectors for GC is not restricted to volatile compounds. Nonvolatile compounds can sometimes be derivatized into volatile ones, thus making them amenable to GC analysis. However, this is done at the expense of losing the elemental ratio information available from the original compound. Delaney and Warren [35] give an excellent overview of the use of element-specific derivatization for GC with MIP detection. In a similar approach, Hagen et al. [36] used a "multielement tagging" method to convert non-chlorine/fluorine-containing compounds into fluorine (and chlorine)-containing compounds, which were subsequently determined by GC-MIP. This approach further demonstrates the flexibility of using plasmas as element-selective detectors for GC. The compounds of interest can be "tagged" with an element of choice so that they can be determined selectively with fewer interferences and possibly with improved sensitivity based on the emission of the tagging element.

In the sections that follow, various applications of plasma GC detectors, especially the MIP, will be discussed. To avoid unnecessary repetition of the discussion of the work accomplished in the subject prior to 1981, the reader is referred to the review by Carnahan et al. [3]. In that article, the use of plasmas as detectors for chromatography is discussed (with heavy emphasis on gas chromatography).

A. Gas Chromatography with Microwave-Induced Plasma Detection (GC-MIP)

Nonmetals

Halogen-containing compounds, including many pesticides, herbicides, and, more recently, the controversial polychlorinated dibenzo-*p*-dioxins (the dioxins), have drawn a great deal of attention because of their toxicity and other ill effects as noted with animals. Mulligan et al. [37] compared an atmospheric-pressure helium MIP sustained in a Beenakker cavity with an electron capture detector (ECD) in the determination of polybrominated biphenyls (PCBs) and related compounds (pesticides and insecticides). Although the MIP detector did not afford as low detection levels as the ECD, it offered the advantage of element selectivity, which can be used to ease the separation

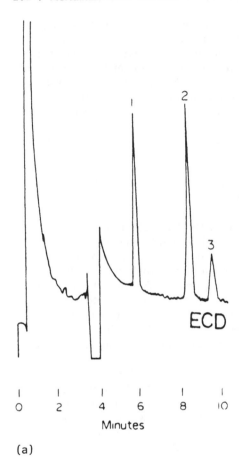

(a)

Fig. 7 Separation of a monochlorobiphenyl isomer mixture and sub-sequent detection by (a) electron capture detector, and (b) flame ionization detector and helium microwave-induced plasma (the chlorine response at 481.0 nm was monitored). Note the absence of matrix re-sponse in the MIP chromatogram. (From Ref. 27.)

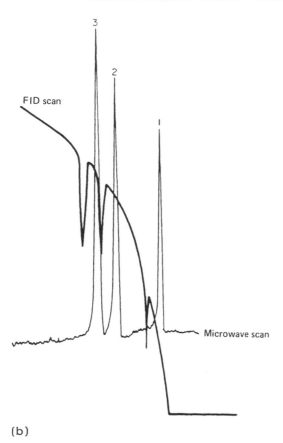

(b)

Fig. 7 (Continued)

requirements of those compounds and thus reduce analysis time. Had ECD been used as the only detector, several of those compounds would be unresolvable because of overlapping peaks.

In two other studies involving detector performance comparisons, the MIP was compared with the flame ionization detector (FID) and the ECD. Chromatograms of monochlorobiphenyl isomers obtained with FID and ECD suffered from matrix responses that caused moderate to severe baseline shifts. The MIP response, on the other hand, was excellent, as Fig. 7 demonstrates [27]. A mixture of alkylmercury compounds and hydrocarbons were detected by both FID and MIP [29]. Even though the FID was more sensitive than the MIP monitored at 193.1 nm (carbon emission), it was shown in the same study that switching to 254 nm (mercury emission) resulted in much improved sensitivity, as shown in Fig. 8. Such versatility makes it possible

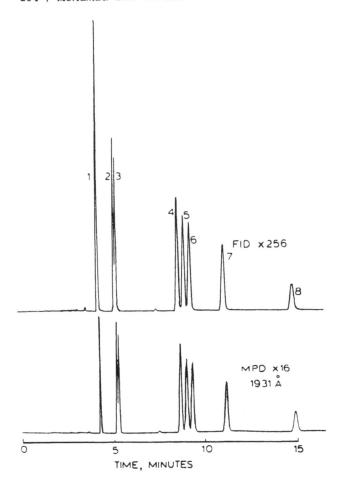

Fig. 8   Separation of (1) dimethylmercury, (2) toluene, (3) per-deuterated toluene, (4) ethylbenzene, (5) *p*-xylene, (6) *m*-xylene, (7) *o*-xylene, and (8) diethylmercury.   Flame ionization and micro-wave-induced plasma detectors were used.   The FID was 50 times more sensitive than the MIP employing the 193.1 nm carbon line; however, the MIP had the same sensitivity as the FID for diethyl-mercury when the 254.3-nm mercury line was used and was twice as sensitive for dimethylmercury.   (From Ref. 29.)

to find an optimum detection condition in terms of specific elemental emission wavelength to be monitored for a particular compound or mixture of compounds.

In other work by Donkin et al. [38], PCBs, DDT, and dieldrin residues in gray seal (*Halichoerus gryphus*) off the northeastern coast of England were determined to assess whether the levels of those pesticides would have harmful effects on the seal's reproduction. The commercial reduced-pressure GC-MIP detection system MPD 850 (Applied Chromatography Systems) was used, monitoring both the chlorine and carbon emissions for the eluting compounds. Carbon/chlorine ratios were determined to aid in identification of the PCBs, DDT, and dieldrin residues, demonstrating the added advantage of multielement monitoring available with the MIP detection system. The MIP detector was used to confirm prior results obtained with ECD. Brenner [39] extended the use of the MPD 850 to the determination of compounds containing C, H, O, N. F. Br, S, and P by monitoring the emission of those elements. About 10 applications involving real samples were presented. Other applications involving the MPD 850 in the determination of nonmetals were reported by Donkin et al. [40] and Hagen et al. [41,42], who determined chlorine-containing (Donkin et al.) and fluorine-containing (Hagen et al.) compounds in various biological samples. Again, in those determinations the excellent selectivity of the MIP detector compared to FID or ECD, reducing the complexity of the chromatograms obtained, was clearly demonstrated. Another important factor is the variation in the ECD response with molecular structure even within different isomers of the same compound. This makes quantification of those isomers difficult at best. Data regarding the MIP response, however, are independent of molecular structure at least in the compounds studied thus far. Bruce and Caruso [43] studied the detection of the dioxins and the pyrethroid class of insecticides. All these compounds contain chlorine or bromine or both (some contain fluorine), which can be used as elemental "handles" to detect the compounds. A laminar flow torch [44] was used in the study instead of a straight quartz tube to obtain a centered, more stable plasma. As a result, the life span of the torch was increased substantially. The torch is an alternative to the tangential flow torch [45]. Detection limits obtained were comparable to those with conventional discharge tubes and significantly better than those with tangential flow torches. Multielement monitoring was also done, using the carbon, chlorine, hydrogen, fluorine, and bromine emissions, and yielded partial empirical formulas of the compounds. A similar study was done by Haas and Caruso [26]; however, a moderate-power MIP (100 to 500 W) was used [46]. Unfortunately, the detection limits were degraded when compared with a low power plasma; this was probably due to several factors, such as the increased helium flow necessary, which reduced analyte residence time in the plasma, and an increase in background emission, which reduced the signal-to-noise ratio.

In perhaps the most comprehensive study on MIP detection for GC to date, Estes et al. [30] determined the detection limits, selectivity ratios, and calibration curves for 29 elements, both metals and nonmetals. The results obtained clearly showed the MIP to be an excellent GC detector, demonstrating good selectivity and detection limits (picogram range) with acceptable linear dynamic response ranges.

Various halogenated compounds in water were determined by Chiba et al. [47,48] and Miller et al. [49]. Fluorine determinations by GC-MIP and fluoride ion-selective electrode (F-ISE) were compared and good agreement was found [47]. However, the GC-MIP allowed monitoring both free and bound fluorine, whereas the F-ISE measures only free fluorine. In the study by Chiba and Haraguchi [48], chlorinated, brominated, and iodinated compounds in water were determined. The plasma discharge tube was heated during those determinations to see any effect on possible analyte condensation on the inside wall of the discharge tube. It was found that for compounds with higher boiling points the response increased with discharge tube heatings, indicating that condensation may have occurred on the inside wall of the discharge tube. Miller et al. [49] derivatized trichloroacetic acid into the corresponding methyl ester to facilitate precolumn trap enrichment in order to analyze parts-per-billion levels of the acid in water. In all the determinations mentioned above, the detection limits obtained were in the picogram (10^{-12} g) range. Derivatization of aliphatic amines and anilines to chlorine-containing compounds was carried out by Delaney and Warren [35]. Determination of the amines then followed, using GC-MIP to monitor the chlorine emission instead of the carbon, hydrogen, or nitrogen elemental emission. This avoided the difficult and tedious precautions needed to prevent interferences from the ubiquitous nitrogen and hydrogen (from moisture in the air), which are difficult to eliminate. Nitrogen has also been known to reduce the effective excitation temperature of the MIP, possibly reducing its ability to excite the maximum emission from the element of interest [55]. In the studies of Cerbus and Gluck [27], Tanabe et al. [31, Goode et al. [50,51], Slatkavitz et al. [52,53], Qing-Yu et al. [54], Koirtyohann [55], and Genna et al. [56], GC-MIP detection of the following nonmetals in various samples was done: the halogens, hydrogen, carbon sulfur, phosphorous, and silicon. Excellent detection limits were obtained (picogram range) and the MIP compared favorably with other GC detectors such as the FID. Koirtyohann [55] extended his study to the effects of nitrogen on emission intensity of various elements excited by the MIP. It was found that a trace amount of nitrogen (0.005 to 0.05%) was helpful in increasing those emissions. However, larger amounts (as in the air) have been found to reduce the effectiveness of the MIP in exciting the elements.

The determination of boron-containing compounds by GC-MIP was done by Krull et al. [57,58] and Sarto et al. [59]. Boron is widely used in GC derivatization procedures to enhance the volatility of many difunctional compounds such as diols and their cyclic boronates. The boron that is introduced can be used to detect the parent compound since boron can be detected with good selectivity and sensitivity [28]. Some of the borated compounds determined were catechol, steroidal carboranes (Krull et al.), carborane silicone polymer, and various diols (Sarto et al.). Selenium and arsenic compounds were determined by both reduced-pressure and atmospheric-pressure MIP [60]. Although the study was only preliminary, at least a few emission lines from each element were considered useful for further analytical study.

In a study by Kei-Wei et al. [61] it was demonstrated that, despite the problem faced by various workers [31,39], it is feasible to design a GC-MIP system capable of maintaining a relatively oxygen-free environment so that oxygenated compounds can be determined reproducibly. Using such a system, four applications involving samples ranging from simple to complex mixtures were presented; two of the samples were cleansing solution and essential oil. In addition to oxygen emission, carbon was monitored simultaneously for universal detection. Elemental ratios were determined and partial empirical formulas calculated.

Finally, it appears that there are only two nonmetals that have yet to be determined by helium/argon MIP (with the exception of the noble gases), namely tellurium and astatine. This is a clear indication of the potential of the helium MIP for successful use in determining nonmetals in a wide variety of compounds.

Metals

Determination of metal-containing compounds by their MIP emission is less common than that of nonmetals. The principal reason is the commercial availability of the more developed argon ICP systems capable of such determinations. With the helium ICP still in its infancy, nonmetals are a "problem" area for argon ICP since the intense emission necessary for low-level work is difficult to obtain. The helium MIP more specifically fills the needs for determining the nonmetals. Both argon and helium MIPs are used in the determination of metal-containing compounds.

Organomercury and organolead compounds are of special interest to many scientists because of their toxicity. Wasik and Schwartz [29] used atmospheric-pressure helium MIP to detect GC effluent containing, among other things, dimethyl- and diethylmercury. Both carbon (193.1 nm) and mercury (254.3 nm) emission lines were monitored for universal and specific detection purposes, respectively. Compared to FID, the carbon emission mode was as much as 50 times less sensitive. However, when the specific mercury emission mode was used, FID and MIP were comparable for diethylmercury detection, and MIP was twice as sensitive for dimethylmercury detection (Fig. 8). Similar mercury compounds were also determined by Olsen et al. [60], using both reduced-pressure and atmospheric-pressure helium MIP. The atmospheric-pressure helium MIP exhibited better sensitivity for organomercury detection. Alkylmercury compounds in seawater were determined with helium MIP by Chiba et al. [62]. Following benzene cysteine extraction, the alkylmercury compounds were separated in OV-101 and Chromosorb columns. Detection limits in the nanogram per liter range with linear responses up to four orders of magnitude were achieved. Argon MIP was utilized by Ballantine and Zoller [63] in the determination of volatile organomercury compounds in the atmosphere. Detection limits obtained were comparable to those of Chiba et al. and in the nanogram range. Methyl ethyl mercury was determined by Qing-Yu et al. [54], using GC-MIP and GC-FID, with comparable results.

Organolead compounds were determined by GC-MIP and GC-FID by Estes et al. [30,64] to compare the two detectors. Thermally unstable trialkyllead was converted to the more stable tetraalkyllead before being determined. It was found that the FID suffered from high carbon backgrounds at least to the extent that the internal reference could not be detected. The MIP, on the other hand, performed satisfactorily when the lead emission was monitored; detection limits were in the low parts-per-billion range (nanograms per milliliter). Iron was determined by Cerbus and Gluck [27], and Bächmann et al. [65] determined nickel, cobalt, and hafnium. The detection limits for iron were impressive: 0.15 pg/sec or 2.5 pg (absolute). Bächmann's work is considered significant since it was one of the few involving inorganic compounds, considering that inorganic gas chromatography is not yet highly developed for trace element analysis.

B. Gas Chromatography-Inductively Coupled Plasma (GC-ICP)

Whereas the popularity of ICP with liquid sample introduction is high, gaseous sample introduction is less common. The use of ICP as a GC detector is even less common. GC is usually employed with organic compounds containing no metals, and these are not readily determinable with the argon ICP as suggested above, although under special conditions requiring a special spectrometer [66] or in the near-infrared region this may be done [67]. Even then, intense emission lines are still lacking from some nonmetals, especially the harder-to-excite halogens [68]. As mentioned above, helium ICP is energetic enough to excite those nonmetals but is not yet widely applied as a GC detector [34].

Determination of metals by GC-ICP should give excellent results, comparable to and possibly better than those obtained with direct solution nebulization into the ICP, since the desolvation and volatilization processes have been accomplished before the analyte enters the ICP by virtue of using GC. Tetramethyl- and tetraethyllead in 0.05 to 0.25 mg/kg of gasoline were easily determined by GC-ICP, as reported by Sommer and Ohls [69]. In a similar study [70] they determined dithiocarbamate complexes of zinc and nickel and compared ICP and FID. ICP was found to be comparable to or better than the FID for GC detection of those complexes. Iron, lead, and tin were determined by both GC-ICP and GC-FPD (flame photometric detector) with detection limits in the low nanogram range, and the ICP demonstrated somewhat better powers of detection (for example, 0.9 ng versus 5 ng for tin) [71].

Despite its shortcomings, the Ar-ICP was used by Windsor and Denton [71–73] for nonmetal determinations. Relatively intense emission lines for oxygen and nitrogen in the near-infrared region were used by Brown and Fry [25,67] for GC-ICP determination and O- and N-containing compounds. Results obtained by both groups varied from reasonable to poor—detection limits ranging from low nanogram to microgram levels.

C. Gas Chromatography—Direct Current Plasma (GC-DCP)

Braman and Dynako [24] in one of the early studies on GC-DCP, used the helium DCP to detect various simple organohalogens. Detection limits in the low picogram range with linear response of at least two orders of magnitude were reported; both were excellent, especially at that time. Selectivity, however, was only fair, ranging from 40:1 to 70:1 for the halogens versus carbon.

More recent work by Lloyd et al. [32] and Uden et al. [74] involving determination of metal complexes and organometallic compounds by argon DCP demonstrated the versatility of this plasma. Compounds of chromium, copper, nickel, palladium, and zinc were determined [32]. The authors described in detail the interface design for GC-DCP and optimization of various parameters to achieve maximum sensitivity. Mixtures of those compounds, separated by GC, were easily detected. Comparison with FID, however, showed that the FID was more sensitive in the determination of hydrocarbons. Carbon emission at 247.8 nm was monitored for the DCP. In contrast, Uden et al. [74] used the same setup to determine methylcyclopentadienyl-manganesetricarbonyl in gasoline and found the DCP to be more sensitive and to give "cleaner" chromatograms than the FID when the manganese 279.8-nm emission was monitored. These two studies illustrate that by using a plasma emission detector, it is possible to choose the best emission line in terms of sensitivity, linearity of response, and absence of interference in the determination of compounds of interest.

D. Elemental Ratios

As pointed out earlier, one of the advantages of the plasma emission detector over other GC detectors is its multielement detection capability. In addition, the energy available in the plasma is capable of causing virtually complete fragmentation and atomization of analyte compounds, consequently yielding an elemental emission signal relatively independent of molecular environment. As a result, elemental ratios and the empirical formula of a compound can be calculated on the basis of its elemental emissions. Together with the retention time data, ambiguous identification of chromatographic peaks can be avoided and peak identification is made much simpler.

In one of the early studies by McLean et al. [75], a mixture of 14 compounds containing a variety of elements was used to test the multielement detection capability of MIP. An FID was used for comparison. Partially unresolved peaks (peaks 6, 7, 8, and 9 in Fig. 9) obtained by using FID were easily distinguishable if different MIP elemental emissions were monitored. Carbon-to-hydrogen and carbon-to-oxygen ratios agreed very well with theoretical values. Empirical formulas of GC effluents determined by both GC-MS and GC-MIP were compared by Brenner [39]; the GC-MS results were more accurate but the MIP results were adequate and better sensitivity was observed. Using experimentally determined empirical formulas and chromatographic retention time data (from which the number of carbons in a hydrocarbon can be approximated), Windsor and Denton [73] accurately determined the molecular formulas of nine hydrocarbons (see Table 1). It is quite conceivable to extend molecular formula determination to other classes of compounds. Studies by Slatkavitz et al. [52,53] of various nonmetal-containing compounds showed that relative errors between experimental and theoretical empirical formulas seldom exceed 3%. Similar results were obtained by Bruce and Caruso [43] in their study of pyrethroids and dioxins, using GC-MIP equipped with various kinds of plasma torches. Chromatograms of six pyrethroids (Fig. 10) monitored at carbon, hydrogen, and fluorine emissions are shown in Fig. 11. Empirical formulas of three of those compounds calculated from such chromatograms are listed in Table 2. Empirical formulas of some representative compounds determined by other workers also are listed in Table 2 to give the reader a feeling for the high potential of this detector to arrive at elemental ratioing/empirical formulas. Haas and Caruso [26] used higher MIP power and compared

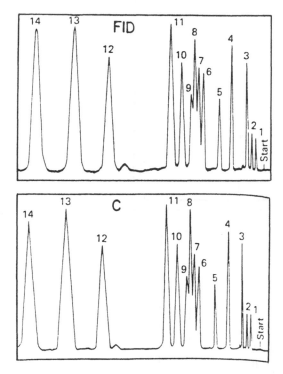

Fig. 9   Element-selective traces on a test sample for carbon, oxygen, chlorine, and sulfur versus a flame ionization detector reference.   Emission wavelengths: C, 247.85 nm; O, 777.19 nm; Cl,
479.45 nm.   Throughout, the peaks are (1) deuteroacetone, (1)
nitroethane, (3) fluorobenzene, (4) toluene, (5) *n*-butyl iodide,
(6) *n*-nonane, (7) chlorocyclohexane, (8) anisole, (9) diethyl disulfide, (10) octan-2-one, (11) bromobenzene, (12) *o*-dichlorobenzene,
(13) *o*-bromotoluene, and (14) *n*-undecane.   (From Ref. 75).

carbon/chlorine ratios determined with and without background correction in their study of chlorinated dioxins and pesticides.   They
reported up to an order of magnitude difference in percent error between background-corrected and non-background-corrected ratios.
By using background correction, accurate C/Cl ratios of the compounds were obtained (less than 1% error); however, the number
of ratios determined was limited.   More ratios were determined by
Qing-Yu et al. [54] using a high-resolution capillary GC; see Fig.
12 and Table 3.   Again, compared to the chromatograms obtained
with FID, those obtained by monitoring appropriate elemental emissions

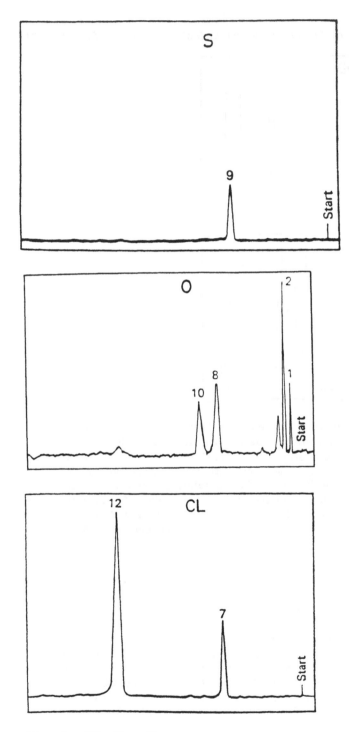

Fig. 9   (Continued)

Table 1   Empirical and Molecular Formulas Derived from GC-ICP

Compound	Retention time (min)	No. of carbons	Empirical formula	Molecular formula
Cumene	6.40	9.4	C_3H_4	C_9H_{12}
Cyclohexene	2.10	7.3	C_3H_5	C_6H_{10}
Ethylbenzene	5.45	8.6	C_4H_5	C_8H_{10}
n-Heptane	1.95	6.6	C_7H_{16}	C_7H_{16}
Isooctane	1.98	6.7	C_4H_9	C_8H_{18}
Methylcyclohexane	2.00	6.8	CH_2	C_7H_{14}
1-Pentane	1.75	5.5	CH_2	C_6H_{12}
o-Xylene	6.85	9.8	C_4H_5	C_8H_{10}
m-Xylene	5.55	8.7	C_4H_5	C_8H_{10}

Source: Ref. 73

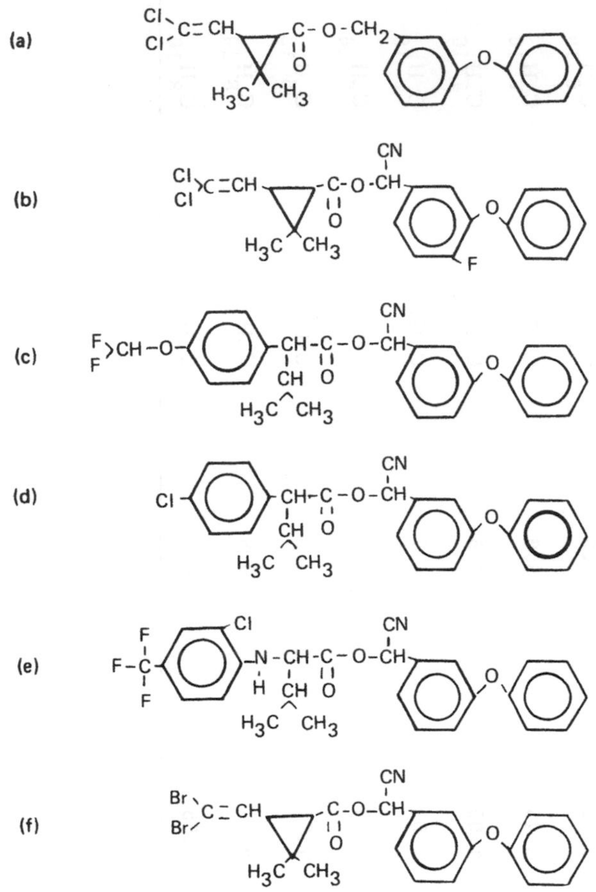

Fig. 10   Structures of permethrin, cyfluthrin, flucythrinate, fenvalerate, fluvalinate, and deltamethrin (the pyrethroids).   (From Ref. 43.)

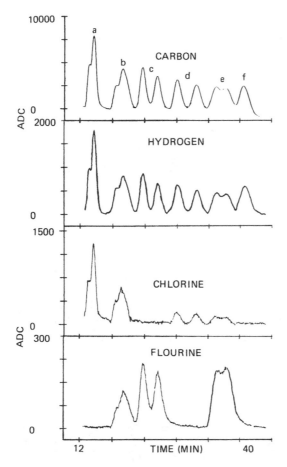

Fig. 11   Background-corrected chromatograms of the pyrethroids taken with the laminar flow torch (LFT) for carbon, hydrogen, chlorine, and fluorine channels.   Peak order: (a) permethrin, (b) cyfluthrin, (c) flucythrinate (two peaks), (d) fenvalerate (two peaks), (e) fluvalinate (double peaks), and (f) deltamethrin. (From Ref. 43.)

Table 2   Empirical Formulae Available From Various Plasma Types

Compound	Empirical formula		DCP/ICP/MIP	Reference
	Theoretical	Experimental		
Undecane	$C_{11}H_{24}$	$C_{11}H_{24.5}$	He-MIP	53
Hexadecane	$C_{16}H_{34}$	$C_{16}H_{34.3}$	He-MIP	53
Ethyl alcohol	C_2H_6O	$C_2H_{5.9}O_{0.96}$	He-MIP	61
tert-Butyl alcohol	$C_4H_{10}O$	$C_4H_{9.8}O_{0.97}$	He-MIP	61
di-n-Butyl ether	$C_8H_{18}O$	$C_8H_{18}O_{1.3}$	He-MIP	61
Methyl benzoate	$C_8H_8O_2$	$C_8H_{7.7}O_{2.2}$	He-MIP	61
Diethyl o-phthalate	$C_{12}H_{14}O_4$	$C_{12}H_{12}O_{3.3}$	He-MIP	61
Methyl isopropyl ketone	$C_5H_{10}O$	$C_5H_{10}O_{1.2}$	He-MIP	61

Compound			Detector	Ref.
7-Ethyl-2-methyl-4-nonanone	$C_{12}H_{24}O$	$C_{12}H_{23}O_{1.0}$	He-MIP	61
Tetrapropylsilane	$SiC_{12}H_{28}$	$SiC_{11.8}H_{28.0}$	He-MIP	53
Dipropyldibutylsilane	$SiC_{14}H_{32}$	$SiC_{13.8}H_{33.9}$	He-MIP	53
1-Chlorobutane	C_4H_9Cl	$C_{3.98}H_{8.97}Cl_{1.00}$	Ar-ICP	73
Iodobenzene	C_6H_5I	$C_{5.92}H_{4.69}I_{1.00}$	Ar-ICP	73
2,7-CDD[a]	C_6H_3Cl	C_6H_3Cl	He-MIP	43
1,2,3,4,6,7,9-CDD[a]	$C_{12}HCl_7$	$C_{11}HCl_7$	He-MIP	43
Permethrin	$C_{21}H_{20}Cl_2$	$C_{21}H_{23}Cl_2$	He-MIP	43
Cyfluthrin	$C_{22}H_{18}Cl_2F$	$C_{24}H_{19}Cl_2F$	He-MIP	43
Deltamethrin	$C_{22}H_{19}Br_2$	$C_{22}H_{19}Br_2$	He-MIP	43

[a]CDD, chlorodibenzodioxin.

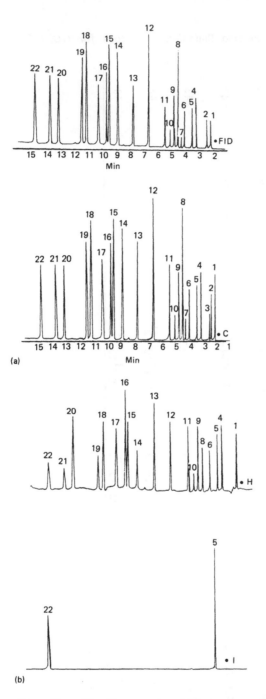

Fig. 12 Simultaneous multidetection chromatogram for C, H, I, Br Cl, F, and S channels and FID chromatogram of a 22-component known mixture (listed in Table 3). Elemental ratios were calculated with the elemental emission intensities obtained at appropriate channels. (From Ref. 54.)

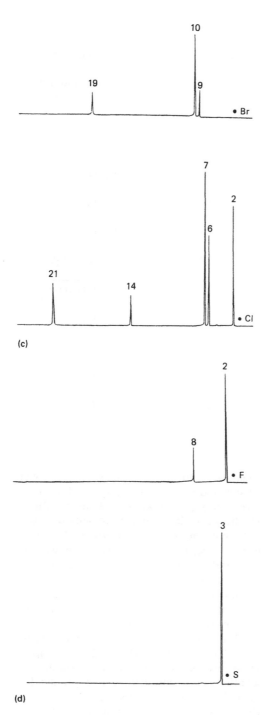

(c)

(d)

Fig. 12   (Continued)

Table 3 Peak Elution Order for 22-Compound Mixture (see Fig. 12)

Peak no.	Compound	Peak no.	Compound
1	$(CH_3)_2CO$	12	C_7H_8
2	$1,1,2\text{-}C_2Cl_3F_3$	13	$CH_3COOC_4H_9$
3	CS_2	14	C_6H_5Cl
4	C_6H_{14}	15	$C_6H_4(CH_3)_2$
5	C_2H_5I	16	$C_4H_9OC_4H_9$
6	$1,2\text{-}C_2H_4Cl_2$	17	$\overline{CO(CH_2)_4CH_2}$
7	CCl_4	18	$C_6H_5CH(CH_3)_2$
8	C_6H_5F	19	C_6H_5Br
9	C_4H_9Br	20	$C_{10}H_{22}$
10	CH_2Br_2	21	$C_6H_4Cl_2$
11	$C_3H_7COOCH_3$	22	C_6H_5I

Source: Ref. 54.

are very helpful in peak identification. In another study by Ke-Wei et al. [61] involving oxygenated compounds, a carefully designed GC-MIP system was used to prevent oxygen leakage. This resulted in very good agreement between the experimentally determined and theoretical empirical formulas of 24 compounds, as shown in Table 2.

The examples cited above are a few representative cases in which the ICP, MIP, and DCP were reported to be capable of exciting elemental emission relatively independent of the molecular environment. As a result, the elemental ratios and empirical formulas of various compounds derived from such elemental emission data were accurate and can be used to simplify gas chromatographic peak identification. This is an added advantage available with plasma emission GC detectors but not with other conventional detectors. Although mass spectrometric detection provides similar information about GC effluents, cost considerations, ease of operation, and sensitivity are pluses for the plasma emission detector, especially the MIP detector.

Finally, a list of elements determined by GC with plasma emission detection is presented in Table 4. It can be seen that various

Table 4 GC-MIP/ICP/DCP Figures of Merit

Element and emission line (nm)	Detection limit ($pg\ sec^{-1}$)[a]	Selectivity versus C	Linearity	DCP/ICP/MIP	Reference
1. Al(I) 396.2	5.0	3.9×10^3	5×10^2	He-MIP	30
2. As(I) 228.8	6.5	4.7×10^4	5×10^2	He-MIP	30
3. B(I) 249.8	3.6	9.25×10^3	5×10^2	He-MIP	30
4. Br(II) 478.6	34	599	5×10^2	He-MIP	30
Br(II) 470.5	6.3	530	10^3–10^5	He-MIP	31
5. C(I) 247.9	2.7	1.00	10^3	He-MIP	30
C(I) 193.1	8.8	1.00	10^3	He-MIP	31
C(I) 247.9	12 ng	1.00	10^3	Ar-ICP	71
6. Cl(II) 479.5	5.4	1000	10^3–10^5	He-MIP	31
Cl(II) 481.0	16	2400	10^4	He-MIP	28
7. Cr(II) 267.7	7.5	1.08×10^5	10^3	He-MIP	30
Cr(II) 267.7	34	—	—	Ar-DCP	32
8. Co(I) 240.7	6.2	1.82×10^5	5×10^2	He-MIP	30
9. Cu(I) 324.7	5.6	—	—	Ar-DCP	32
10. F(I) 685.6	7.5	—	10^4	He-MIP	47
F(I) 685.6	1.8	820	10^3–10^5	He-MIP	31
11. Ge(I) 265.1	1.3	7.57×10^4	10^3	He-MIP	30
12. H(I) 486.1	16	74	5×10^2	He-MIP	30
H(I) 656.3	6.2	—	10^3	He-MIP	31

Table 4   (Continued)

Element and emission line (nm)	Detection limit $(pg\ sec^{-1})^a$	Selectivity versus C	Linearity	DCP/ICP/MIP	Reference
12. H(I) 656.3	5.5 ng	3×10^3	10^3	Ar-ICP	71
13. I(I) 206.2	6.6	530	10^3-10^5	He-MIP	31
I(I) 206.2	24 ng	10^3	10^3	Ar-ICP	71
I(I) 608.2	160	Poor	—	He-DCP	24
14. Fe(II) 259.9	0.28	2.8×10^5	10^3	He-MIP	30
Fe(I) 371.99	5.9 ng	10^3	10^3	Ar-ICP	71
15. Pb(I) 283.3	0.17	2.46×10^5	10^3	He-MIP	30
Pb(I) 217.0	33 ng	3×10^3	10^3	Ar-ICP	71
16. Mn(II) 257.6	1.6	1.11×10^5	10^3	He-MIP	30
Mn(I) 279.83	3 ng	—	—	Ar-DCP	74
17. Hg(I) 253.7	0.6	7.69×10^4	10^3	He-MIP	30
18. Ni(II) 231.6	2.6	6.47×10^3	10^3	He-MIP	30
Ni(I) 341.4	320	—	—	Ar-DCP	32
19. Nb(II) 288.3	69	3.21×10^4	10^2	He-MIP	30
20. N(I) 821.63	1 μg	—	—	Ar-ICP	67
21. Os(II) 225.6	6.3	5.00×10^4	10^3	He-MIP	30
22. O(I) 777.19	300	—	—	He-MIP	61
O(I) 777.19	650 ng	—	—	Ar-ICP	25

23.	Pd(I) 340.4	120	—	—	Ar–DCP	32
24.	P(I) 253.6	3.3	1.06×10^4	5×10^2	He–MIP	30
	P(I) 214.9	0.3	—	—	He–MIP	27
25.	Ru(II) 240.3	7.8	1.34×10^5	10^3	He–MIP	30
26.	Se(I) 204.0	5.3	1.09×10^4	10^3	He–MIP	30
27.	Si(I) 251.6	1.8	2.3×10^3	—	He–MIP	53
28.	S(II) 545.4	25	200	—	He–MIP	76
29.	Sn(I) 284.0	1.6	3.58×10^5	10^3	He–MIP	30
	Sn(I) 284.0	0.9 ng	3×10^4	10^3	Ar–ICP	71
30.	V(II) 268.8	10	5.69×10^4	10^2	He–MIP	30
31.	D(I) 656.1	7.4	194	5×10^2	He–MIP	30

[a]Unless noted otherwise.

plasma emission detectors are indeed capable of detecting ultratrace levels of GC effluents with good selectivity.

IV. FUTURE

It has been shown that the plasma emission detector can be used successfully to detect GC effluents, giving detection power in most cases comparable to if not better than that of conventional GC detectors. Useful applications involving real samples have been demonstrated. Further developments in the form of alternative cavity resonator designs and generators for the MIP [77—80] and ICP sustained in helium [33,34] have been reported. Perhaps a single most significant development would be the commercial introduction of a GC-MIP system operated at atmospheric pressure (as opposed to the reduced-pressure system, which is more difficult to work with). There are many cases where the highly selective plasma emission detector will be very useful, as pointed out in this chapter. Furthermore, with the ability to do empirical formula calculations at nanogram levels, the plasma emission detector may be a competitive alternative to GC-MS.

ACKNOWLEDGMENT

One of the authors (A. H. M) is grateful to the Science University of Malaysia and Public Service Department, Malaysia, who provided financial support in the form of an Academic Staff Training Fellowship. We are also grateful to the National Institute for Environmental Health Sciences (Grant ES-03221) for partial support of this work.

REFERENCES

1. A. J. James and A. J. P. Martin, Analyst 77, 915 (1952).
2. L. S. Ettre, J. Chromatogr. Sci. 16, 396 (1978).
3. J. W. Carnahan, K. M. Mulligan, and J. A. Caruso, Anal. Chim. Acta 130, 227—241 (1981).
4. A. J. McCormack, S. G. Tong, and W. D. Cooke, Anal. Chem. 37, 1470 (1965).
5. C. A. Bache and D. L. Lisk, Anal. Chem. 37, 1477 (1965).
6. C. A. Bache and D. J. Lisk, Anal. Chem. 38, 783 (1966).
7. V. A. Fassel and R. N. Kniseley, Anal. Chem. 46, 1110A (1974).
8. V. A. Fassel and R. N. Kniseley, Anal. Chem. 46, 1155A (1974).
9. L. M. Faires, Am. Lab., 14, p. 16 (November 1982).
10. S. Greenfield, H. M. McGeachin, and P. B. Smith, Talanta 23 1 (1976).

11. V. A. Fassel, Science *202*, 183 (1978).
12. R. K. Skogerboe and G. N. Coleman, Anal. Chem. *48*, 611A (1976).
13. A. T. Zander and G. M. Hieftje, Appl. Spectrosc. *35*, 357 (1981).
14. T. H. Risby and Y. Talmi, CRC Crit. Rev. Anal. Chem. *14*, 231 (1983).
15. J. W. Carnahan, Am. Lab., *15*, p. 31 (August 1983).
16. S. E. Valente and W. J. Schrenk, Appl. Spectrosc. *24*, 197 (1970).
17. C. D. Keirs and T. J. Vickers, Appl. Spectrosc. *31*, 273 (1977).
18. C. I. M. Beenakker, Spectrochim. Acta *32B*, 173 (1977).
19. P. M. Houpt, Anal. Chim. Acta *86*, 129 (1976).
20. K. W. Busch and T. J. Vickers, Spectrochim. Acta *28B*, 85 (1973).
21. P. Brassem and F. J. M. J. Maessen, Spectrochim. Acta *30B*, 547 (1975).
22. C. I. M. Beenakker, Spectrochim. Acta *31B*, 483 (1976).
23. H. A. Dingjan and H. J. DeJong, Spectrochim Acta *36B*, 325 (1981).
24. R. S. Braman and A. Dynako, Anal. Chem. *40*, 95 (1968).
25. R. M. Brown and R. C. Fry, Anal. Chem. *53*, 532 (1981).
26. D. L. Haas and J. A. Caruso, Anal. Chem. *57*, 846 (1985).
27. C. S. Cerbus and S. J. Gluck, Spectrochim Acta *38B*, 387 (1983).
28. B. D. Quimby, P. C. Uden, and R. M. Barnes, Anal. Chem. *50*, 2112 (1978).
29. S. P. Wasik and F. P. Schwarz, J. Chromatogr. Sci. *18*, 660 (1980).
30. S. A. Estes, P. C. Uden, and R. M. Barnes, Anal. Chem. *53*, 1829 (1981).
31. K. Tanabe, H. Haraguchi, and K. Fuwa, Spectrochim. Acta *36B*, 633 (1981).
32. R. J. Lloyd, R. M. Barnes, P. C. Uden, and W. G. Elliot, Anal. Chem. *50*, 2025 (1978).
33. C. J. Seliskar and D. K. Warner, Appl. Spectrosc. *39*, 181 (1985).
34. K. A. Wolnik, D. C. Miller, C. J. Seliskar, and F. L. Fricke, Appl. Spectrosc. *39*, 930 (1985).
35. M. F. Delaney and F. V. Warren, Jr., Spectrochim. Acta *38B*, 399 (1983).
36. D. F. Hagen, J. S. Marhevka, and L. C. Haddad, Spectrochim. Acta *40B*, 335 (1985).
37. K. J. Mulligan, J. A. Caruso, and F. L. Fricke, Analyst, *105*, 1060 (1980).
38. P. Donkin, S. V. Mann, and E. I. Hamilton, Sci. Total Environ. *19*, 121 (1981).

39. K. S. Brenner, J. Chromatogr. *167*, 365 (1978).

40. P. Donkin, S. V. Mann, and E. I. Hamilton, Anal. Chim. Acta *88*, 289 (1977).

41. D. F. Hagen, J. Belisle, J. D. Johnson, and P. Venkateswarke, Anal. Biochem. *118*, 336 (1981).

42. D. F. Hagen, J. Belisle, and J. S. Marhevka, Spectrochim. Acta *38B*, 377 (1983).

43. M. L. Bruce and J. A. Caruso, Appl. Spectrosc. *39*, 942 (1985).

44. M. L. Bruce, J. M. Workman, J. A. Caruso, and D. J. Lahti, Appl. Spectrosc. *39*, 935 (1985).

45. A. Bollo-Kamara and E. G. Codding, Spectrochim. Acta *36B*, 973 (1981).

46. D. L. Haas, J. W. Carnahan, and J. A. Caruso, Appl. Spectrosc. *37*, 82 (1983).

47. K. Chiba, K. Yoshida, K. Tanabe, M. Ozaki, H. Haraguchi, J. D. Winefordner, and K. Fuwa, Anal. Chem. *54*, 761 (1982).

48. K. Chiba and H. Haraguchi, Anal. Chem. *55*, 1504 (1983).

49. J. W. Miller, P. C. Uden, and R. M. Barnes, Anal. Chem. *54*, 485 (1982).

50. S. R. Goode, B. Chambers, and N. P. Buddin, Appl. Spectrosc. *37*, 439 (1983).

51. S. R. Goode, B. Chambers, and N. P. Buddin, Spectrochim. Acta *40B*, 329 (1985).

52. K. J. Slatkavitz, P. C. Uden, L. D. Hoey, and R. M. Barnes, J. Chromatogr. *302*, 277 (1984).

53. K. J. Slatkavitz, L. D. Hoey, P. C. Uden, and R. M. Barnes, Anal. Chem. *57*, 1846 (1985).

54. O. Qing-Yu, W. Guo-Chuen, Z. Ke-Wei, and Y. Wei-Lu, Spectrochim. Acta *38B*, 419 (1983).

55. S. R. Koirtyohann, Anal. Chem. *55*, 374 (1983).

56. J. L. Genna, W. D. McAninch, and R. A. Reich, J. Chromatogr. *238*, 103 (1982).

57. I. S. Krull, S. W. Jordan, S. Kahl, and S. B. Smith, Jr., J. Chromatogr. Sci. *20*, 489 (1982).

58. I. S. Krull, S. W. Jordan, and S. B. Smith, Jr., Anal. Lett. *15*, 1131 (1982).

59. L. G. Sarto, Jr., S. A. Estes, P. C. Uden, S. Siggia, and R. M. Barnes, Anal. Lett. *14*, 205 (1981).

60. K. B. Olsen, D. S. Sklarew, and J. C. Evans, Spectrochim. Acta *40B*, 357 (1985).

61. Z. Ke-Wei, O. Qing-Yu, W. Guo-Chuen, and Y. Wei-Lu, Spectrochim. Acta *40B*, 349 (1985).

62. K. Chiba, K. Yoshida, K. Tanabe, H. Haraguchi, and K. Fuwa, Anal. Chem. *55*, 450 (1983).

63. D. S. Ballantine, Jr. and W. H. Zoller, Anal. Chem. *56*, 1288 (1984).

64. S. A. Estes, P. C. Uden, and R. M. Barnes, Anal. Chem. *54*, 2402 (1982).

65. K. Bächmann, U. Hamm, A. Werner, P. Tschopel, and G. Tolg, in *Developments in Atomic Plasma Spectrochemical Analysis.*, *Proc. Int. Winter Conference, 1980* (R. M. Barnes.), Heyden, London, 1981, p. 361

66. D. D. Nygaard and D. A. Leighty, Appl. Spectrosc. *39*, 968 (1985).

67. R. M. Brown, Jr., S. J. Northway, and R. C. Fry, Anal. Chem. *53*, 934 (1981).

68. S. K. Hughes and R. C. Fry, Anal. Chem. *53*, 1111 (1981).

69. D. Sommer and K. Ohls, Fresenius Z. Anal. Chem. *295*, 337 (1979).

70. K. Ohls and D. Sommer, in *Developments in Atomic Plasma Spectrochemical Analysis.*, *Proc. Int. Winter Conference, 1980* (R. M. Barnes, ed.), Heyden, London, 1981, p. 321.

71. D. L. Windsor and M. Bonner Denton, J. Chromatogr. Sci. *17*, 492 (1979).

72. D. L. Windsor and M. Bonner Denton, Appl. Spectrosc. *32*, 366 (1978).

73. D. L. Windsor and M. Bonner Denton, Anal. Chem. *51*, 1116 (1979).

74. P. C. Uden, R. M. Barnes, and F. P. DiSanzo, Anal. Chem. *50*, 852 (1978).

75. W. R. McLean, D. L. Stanton, and G. E. Penketh, Analyst *98*, 432 (1973).

76. C. I. M. Beenakker, Spectrochim. Acta *32B*, 173 (1977).

77. G. W. Jansen, F. A. Huf, and H. J. DeJong, Spectrochim. Acta *40B*, 307 (1985).

78. J. Hubert, M. Moisan, and A. Ricard, Spectrochim. Acta *34B*, 1 (1979).

79. T. Hanie, S. Coulombe, M. Moisan, and J. Hubert, in *Developments in Atomic Plasma Spectrochemical Analysis, Proc. Int. Winter Conference, 1980* (R. M. Barnes, ed.), Heyden, London, 1981, p. 337

80. K. Cammann, L. Lendero, H. Feuerbacher, and K. Ballschmiter, Fresenius Z. Anal. Chem. *316*, 194 (1983).

81. M. Novotny, M. P. Maskarinec, A. T. G. Steverink, and R. Farlow, Anal. Chem. *48*, 471 (1976).

82. A. H. Mohamad, M. Zerezghi, and J. A. Caruso, Anal. Chem. *58*, 469 (1986).

6

The Use of Retention Data from Capillary GC for Qualitative Analysis: Current Aspects

Lars G. Blomberg   *University of Stockholm, Stockholm, Sweden*

I. INTRODUCTION

Interest has been shown in the use of retention data for qualitative analysis ever since the advent of gas chromatography. Such data are influenced by several factors, and in order to be useful in qualitative analysis, the data should be measured under standardized conditions and reported in a uniform manner. A relatively high insensitivity toward experimental parameters may be achieved when retention is measured under isothermal conditions as "relative retention"; that is, the retention of a compound is compared with that of one or more standard compounds. Such data may be reported as "retention indices."

It is well known that retention data cannot be used as a single basis for solute identification; such data are often used in a negative sense, i.e., as a proof of the absence of a compound. For a strict identification, the use of two independent methods giving positive identification is generally considered necessary. Nevertheless, gas chromatography may be of great utility when retention indices are measured on some different stationary phases. With use of three different types of phases, the probability of peak coincidence in all three columns would be relatively low, at least when the solutes belong to a defined group of compounds. Data from such runs may be quite useful, e.g., in routine analysis and for laboratories that do not have access to instrumentation for positive identification. Moreover, retention indices can be of considerable help in identification studies in combination with supplementary techniques, such as mass spectroscopy and Fourier transform infrared spectroscopy.

Several developments in capillary GC have taken place during recent years with respect to factors that affect precision in retention determinations, and further improvements may be expected in the near future. This may result in a wider use of techniques for "identification" by GC retention indices. An important contribution to this development is the computer-based retention index library that was recently launched by Sadtler Research Laboratories, Philadelphia, Pa. [1]. The recent advances in capillary GC techniques do not, however, per se lead to improvements in reproducibility and precision of retention index determinations. A necessary condition for such improvements is that particular attention is also paid to the sources of errors in retention index determination [2−4].

The errors in gas chromatographic data may be of systematic or random nature [2−6]. Systematic errors have been considered to arise from mass transfer equilibrium effects, fugacity and virial phenomena, solute adsorption at the gas-liquid interface, noninert support materials, and retention of supposedly nonsorbed solutes [7]. Goedert and Guiochon [6] maintain that we can never be sure that all the systematic errors have been identified, since there is, at present, no independent check of the measurements that can be considered with confidence in its accuracy. Random errors may emanate from fluctuations in the experimental parameters and inaccuracy in the estimation of the initial and final events of the chromatographic process, i.e., the injection of the sample and elution of the peak maximum [6].

Concepts such as precision, reproducibility, and accuracy are frequently used in this chapter and some definitions may be necessary. Precision in retention index determinations describes the reproducibility of repeated determinations obtained by one laboratory. When these determinations are performed under controlled and specified conditions, the results can be reproduced by other workers, thus possibly approaching what may be termed accurate values. Such values are thought to result, in an ideal case, when retention is due only to dissolution of the sample in the stationary phase, thus reflecting only the distribution constants between stationary and mobile phase.

The reproducibility of retention indices between different laboratories is generally considered to be within one index unit for non-polar columns and within a few index units for polar columns. Much higher reproducibilities can be obtained within a laboratory; e.g., 0.1 to 0.05 units was reported by Schomburg and Dielmann [5]. Under defined conditions, the reproducibilities are thus quite high; for this purpose, the data need not be accurate. However, in order to improve the reproducibility between laboratories, retention data should approach accurate values. To achieve this, the data must be measured under strictly specified conditions, and the sources of errors should be identified and minimized. This chapter gives an overview of possible sources of error in the determination and calculation of Kováts retention indices. Methods for avoiding these errors, thereby improving laboratory-to-laboratory reproducibility, are discussed. Further, different aspects of the correlation between retention index and molecular structure, especially in connection with prediction of retention indices are considered.

This survey has been directed primarily toward qualitative analysis by capillary GC. The results are, however, also applicable to packed column GC, since the factors determining accuracy in retention indices are largely the same in packed and capillary GC. Further, much of the work compiled here has been performed by packed column GC.

II. RETENTION INDEX SYSTEMS

The use of retention indices in gas chromatography has been treated extensively in the literature and was reviewed by Haken [8]. Several systems for the presentation of normalized retention data have been described, e.g., by Kováts [9], Vanden Heuvel et al. [10], Miwa et al. [11], and Woodford and van Gent [12]. A useful basis for such systems is provided by the linearity of the plot of the log adjusted retention times and carbon numbers of homologous compounds under constant gas chromatographic conditions.

$$\log t'_x = bn + c \tag{1}$$

Here t'_x is the adjusted retention time for a member of the homologous series having n carbon atoms, and b and c are the slope and intercept, respectively, of the straight line plot of log t'_x versus n. Further,

$$t'_x = t_x - t_M \tag{2}$$

where t_x is the retention time, and t_M is the so-called dead time, the "retention time" of an unretained substance. It should be pointed out, however, that for any homologous series of aliphatic organic compounds, there is a strong deviation from linearity of Eq. (1) for the first members of a series [13].

Using *n*-alkanes as standard substances, Kováts [9] suggested an equation for the calculation of retention index by numerical interpolation:

$$I = 200 \frac{\log V_R^0(\text{substance}) - \log V_R^0(nP_z)}{\log V_R^0(nP_{z+2}) - \log V_R^0(nP_z)} + 100z \tag{3}$$

where $V_R^0(nP_z) < V_R^0(\text{substance}) < V_R^0(nP_{z+2})$, V_R^0 is the retention volume and nP_z is the n-paraffin with carbon number z. Retention indices were calculated on the basis of even-carbon n-alkanes since it was considered that there would be an oscillation in the chromatographic properties of successive members of the homologous series. This was experimentally found to be unnecessary, and Kováts redefined the index and proposed an increment of 100 between n-paraffins of general formula C_nH_{2n+2} [14]. The influence of odd or even carbon numbers on the retention of alkanes was later discussed by Vigdergauz and Seomkin [15] and by Kaiser and Rackstraw [16]. Looking at fine details, it was proposed that plots of Eq. (1) showed a zigzag shape in the low molecular weight range.

It was suggested by Ettre [17] that retention volumes could be replaced by adjusted retention times. The well-known equation for Kováts retention index thus reads

$$I = 100\left(n + \frac{\log t'_x - \log t'_n}{\log t'_{n+1} - \log t'_n}\right) \tag{4}$$

where t'_x is the adjusted retention time of the solute, t'_n and t'_{n+1} are the adjusted retention times of the n-alkanes used as standards; $t'_n < t'_x < t'_{n+1}$, I is the retention index, and n is the carbon number of the first alkane used. The standards should enclose the solutes so that extrapolation will not be required [16,18]. Equation (4) can be rewritten using the capacity ratio, k, of the compounds x, n, and n+1; $k = t'_x/t_M$.

Retention index scales are relative and are dependent on the relative retention of the members of the homologous series that are being used as standard substances. Relative retention decreases, i.e., the standards are closer to each other, with increasing temperature and generally also with increasing stationary phase polarity. As a consequence of the lower relative retention on polar columns, the precision of the retention index, for a given precision in time measurement, on such column will be inherently smaller than on nonpolar columns. Further, low k values per se result in low precision (high standard deviation).

As mentioned above, Eq. (1) is not valid for the first few members of a homologous series. It is evident that retention of the homologs in this case cannot be explained by simple additivity. The energy contribution per methylene group to the partial molar free energy of sorption was calculated by Golovnya and Grigoryeva [19–22] for a number of different homologous series. It was found, for nonpolar as well as polar stationary phases, that this energy varied with the increasing number of the member of any homologous series tested

(Figs. 1 and 2) [20]. For n-alkanes, the largest energy contributions were achieved for the first methylene units, thus reflecting the specific van der Waals interactions of two bonded CH_3 groups. Contrary to n-alkanes, the energy contribution for the first methylene units in the polar homologous series is always less than the contribution for the second CH_2 group. It was assumed that the energy of dispersion interaction of a polar functional group, x, with a nonpolar stationary phase decreases considerably when the functional group is directly linked to a CH_3 group.

An increase in chain length by one methylene unit causes, in accordance with the definition, an increase in retention index of 100 units, so that the Kováts indices of standard n-paraffins become 100n, where n is the carbon number. These values of the n-paraffins are by definition independent of the polarity of the stationary phase. Kováts, in extending this concept, stated that "in any homologous series the retention index of higher members increases by 100 index units per methylene group introduced" [14].

$$I = an + b \qquad (5)$$

where n is the number of carbon atoms and a and b are constants for a given stationary phase and temperature. According to the rule mentioned above, a = 100 for any homologous series. This rule was, however, later shown to be an oversimplification [23,24]. Zulaica and Guiochon showed for aliphatic dibasic esters that although the retention index contribution per methylene group in the acid or acyl chain was close to 100, the contribution in the alkyl or alcohol chain was lower. Haken and co-workers [8,24] showed for aliphatic esters that an incremental increase by 100 index units per methylene group would be a chance occurrence. It appears that a slight decrease in "a" is achieved when the functional groups are located at central positions in the carbon chain [25]; thus $1 - ol > 2 - ol > 3 - ol$; formate > acetate > propionate; aldehyde > methylketone > ethylketone. In Eq. (5) "a" is independent of the stationary phase, whereas "b" must be related to solute-stationary phase interactions. Further, because of the deviation from linearity of Eq. (1) for low values of n, the increments, I_{CH_2}, are not constant for lower values of n, i.e., below n = 7.

A. Different Bases for the Retention Index System

The Kováts retention index system is very widely used, which is clearly demonstrated in an extensive review published on the occasion of the 25th anniversary of the index system [26]. This index can thus be applied to all possible combinations of solutes and stationary phases; the reliability is, however, not equally acceptable in all these connections. It seems that the n-alkanes are not the optimal standards

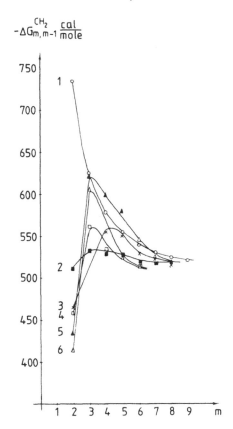

Fig. 1 Change in energy contribution of a methylene unit, $\Delta G^{CH_2}_{(m),(m-1)}$, in R_mX type homologous series, with distance from the functional group, X. Stationary phase, Apiezon L; column temperature, 120°C. The abscissa (m) represents the sequential number of the homologs. Homologous series: 1 (○) R_m—CH_3; 2 (■) R_m—S—CH_3; 3 (x) R_m—NH_2; 4 (□) R_m—O—C—CH_3; 5 (▲) R_m—OH; 6 (△) R_m—O—CH. (From Ref. 20.)

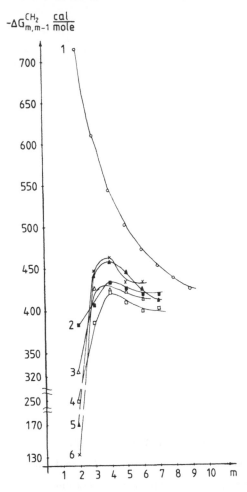

Fig. 2 Change in the energy contribution of a methylene unit, $\Delta G^{CH_2}_{(m),(m-1)}$, in $R_m X$ type homologous series, depending on the distance from the functional group X. Stationary phase: Carbowax 1000 Column temperature: 120°C. The abscissa (m) represents the sequential number of the homologs. Homologous series: 1 (○) R_m—CH_3; 2 (■) R_m—S—CH_3; 3 (△) R_m—O—CH; 4 (□) R_m—O—C—CH_3; 5 (▲) R_m—OH; 6 (x) R_m—NH_2. (From Ref. 20.)

in all instances. Three cases where the use of alternative standards may be motivated will be discussed, these being retention index determinations in connection with (a) polar stationary phases and solutes, (b) temperature programming, and (c) selective detectors.

First, it is desirable that the standards be compatible with the stationary phase. When polar phases are being used, Gibbs adsorption of hydrocarbons is likely to occur at the gas-liquid interface, leading to mixed retention mechanisms and thereby reduced reproducibility of the retention indices. *n*-Alkanes may in such cases show a significant peak asymmetry; further, the relation in Eq. (1) may not be valid. It should be noted that the risk of such surface adsorption may increase with cross-linked polar stationary phases [27], especially in the case of phases that already have a tendency toward such adsorption in the non-cross-linked state, e.g., Carbowax 20M [28].

A further drawback of hydrocarbon standards in connection with polar stationary phases is the reduced retention of *n*-alkanes, which may result in the comparison of low-boiling polar solutes with *n*-alkanes of much higher boiling points. The high-boiling *n*-alkanes may require a significant increase in evaporator temperature, which can result in the formation of artifacts from temperature-sensitive sample components. It may thus be advantageous to use standards that are similar in chemical nature to the sample.

The use of more polar standards has been suggested to improve the precision of retention indices on polar stationary phases [14, 29–31]. *n*-Alkanols were proposed by Gröbler [29]. One drawback of *n*-alkanols may be that they display various solubility parameters such as dispersion, orientation, acidity, and basicity [31] and would thus be sensitive to divergences in stationary phase properties and chromatographic conditions. Good results have been claimed with 2-alkanones [30,32], and *n*-aldehydes [33]; also, *n*-alkyl esters [8, 24,34,35], *n*-alkylbenzenes [36,37], and *n*-alkyl iodides [38] have been investigated. The structure of the *n*-alkylbenzenes, with an aromatic nucleus appended to a relatively short alkane chain, may make these compounds unsuitable as standards since the interaction between the aromatic moiety and the stationary phase may restrict the effect of additional methylene groups [39].

Retention of homologous esters has been extensively investigated by Haken and co-workers in a long series of publications. The suitability of two series of homologous esters—acetates and "symmetrical esters"—for use as so-called secondary reference standards was discussed [39]. Furthermore, saturated methyl esters of fatty acids are being used as standards in the equivalent chain length (ECL) scheme [11]. In this scheme, a scale is constructed on the basis of Eq. (1), where ase member of the fatty acid methyl ester (FAME) series is represented by an ECL value according to its carbon number, e.g., methyl palmitate by 16.000 and methyl stearate by 18.000. The ECL of an unknown compound can be calculated from Eq. (6), which is derived by rearranging Eq. (1):

$$(ECL)_x = \frac{\log t'_x - c}{b} \qquad (6)$$

with t'_x, b, and c as in Eq. (1). The standards used in the ECL system are, however, relatively nonpolar, and they are certainly less compatible with, e.g., cyanosilicones than the unsaturated fatty acid methyl esters, which are often indexed in this system.

The use of n-alkane-1-ols and n-alkane-1-ol acetates as standards for analysis of pheromones was demonstrated by Berezkin and Retunsky [34,35].

General disadvantages of polar reference compounds are that they lack compatibility with nonpolar stationary phases and that it is not possible to determine the index for relatively low-boiling compounds. The latter drawback depends on the fact that polar references show quite high retention on polar phases and, in addition, the first few members of the polar homologous series do not fit Eq. (1), so uncertain indices would be obtained in the first region.

One reason for irregularities in the retention of polar homologs is that when the aliphatic side chain becomes long enough to wrap back on, e.g., a carbonyl group, this will significantly reduce the contribution to retention by the oxygen atom. Such a shielding effect has been demonstrated in several cases; see, e.g., Refs. 24, 32, and 40–42.

In systems where columns of different polarities are coupled in parallel or in series, it would be desirable to select references that are compatible with all the stationary phases involved. It is likely that references of moderate polarity, such as n-propyl ethers, n-alkyl esters, or alkanals, would give the best results in this case [8,39, 43,33].

The second instance when the use of polar reference standards might be justified is the temperature-programmed runs. The distribution constants and thereby the retention indices of different types of compounds are differently affected by changes in temperature, and the compounds will therefore have different dI/dT values. Small irregularities in a temperature program may thus lead to erroneous data. By using reference compounds having structures similar to those of the samples, problems associated with differences in dI/dT values may be minimized.

A third reason for the choice of alternative standards arises when an analysis requires use of a detector that does not respond to n-alkanes. n-Alkyl bromides [44,45] and trichloroacetates [46–49] have thus been used in connection with the electron capture detector. For the flame photometric detector, the use of *n*-alkyl sulfides has been found suitable [50]. Any detector can, however, be used parallel to a flame ionization detector, in which *n*-alkanes are detected sensitively enough.

Two series of multidetector retention index standards, alkyl esters of dimethylphosphine-thioacid and bis(trifluoromethyl)phosphine-thioacid, have been developed by a group at the University of Helsinki (Finland). These investigators demonstrated that, on separation of complex samples, a risk was incurred for coelution of standards and sample components. To circumvent this problem, part of the column effluent was split to a selective detector for standard monitoring, while the major part of the sample was detected by, e.g., a flame ionization detector [51,52]. It is, of course, a fundamental requirement that the standard substances are stable during the analysis.

The retention indices obtained with the alternative standards (secondary standards) can be converted to Kováts indices if the indices in the primary system of the polar standards are known. Kováts pointed out that a homologous series is an ideal secondary reference series when the indices of its consecutive members increase by exactly 100 and are not temperature-dependent [14].

It may be considered that the greatest advantage of the Kováts retention index is that it permits the use of results obtained in other laboratories. This universal role is hindered by different systems for data presentation, including other index systems [26]. In our opinion, it is not necessary to use secondary standards to compensate for deficiencies in the chromatographic system, such as support adsorption and poor temperature and mobile phase flow rate control. In such cases, the remedy should be improved systems. Adsorption at the gas-liquid interface may, however, make the use of polar references necessary if increased reproducibility is desired; the universality of the index would thus have to be sacrificed for the improvement of reproducibility.

Finally, it should be noted that the use of *n*-alkane standards restricts the application of the Kováts index in other chromatographic techniques such as high-performance liquid chromatography (HPLC) and thin-layer chromatography (TLC) where UV or fluorescence detection is widely used.

B. Retention Index in Linear Temperature Programming

For the separation of samples containing components having a wide boiling point range, temperature-programmed gas chromatography is generally necessary. The relationship in Eq. (1) cannot be applied for such runs, but provided the injection temperature is low in comparison with the elution temperature, an approximately linear relationship between the elution temperature of *n*-alkanes and their carbon number may be achieved when the temperature is programmed at a heating rate of $\lambda°$/min. It has been pointed out that linearity can be achieved for a limited range of carbon numbers [53,54]. An expression for the calculation of retention index in temperature-programmed gas chromatography was described by van der Dool and Kratz [55]:

$$I_{TP} = 100 \times n \; \frac{T_{Rx} - T_{Rn}}{T_{Rn+1} - T_{Rn}} \tag{7}$$

Here, the elution temperature, T_R, are used for the calculations, but they can be replaced by the corresponding nonadjusted retention times t_x.

In the temperature-programmed analysis of polycyclic aromatic compounds (PACs), Eq. (7) was applied, with four PACs—naphthalene, phenanthrene, chrysene, and picene—being used as standard substances [56,57]. The first three of these standard substances always occur in PAC samples, and consequently it was not necessary to add standard substances to the samples (t_{picene} could be obtained by calculation using a converging algorithm). This is considered an advantage, since spiking a complex sample with e.g., n-alkanes may lead to coelution problems, as mentioned above.

In splitless and on-column injection techniques, an initial isothermal period is generally used. The effect of the duration of such a period on retention indices was examined by Vassilaros et al. [57]. As might be expected, these effects are related to the difference between injection and elution temperatures. Lee and co-workers [56,57] have shown, however, that high precision in retention data from temperature-programmed GC can be achieved when performed under carefully controlled conditions.

Retention indices of aromatics show a high temperature dependence, dI/dT, on nonpolar columns [58,59]. The use of aromatics as standards for temperature-programmed PAC analysis may thus have a compensating effect on possible small errors in elution temperature. A high precision in linear temperature-programmed retention index determination was also demonstrated by Raymer et al. [32] when using 2-alkanones as standard substances in the analysis of ketones.

In the temperature-programmed determination of retention data, one of the main advantages of the Kováts index is lost, namely the insensitivity to chromatographic conditions. The reason for this is that the temperature-dependent variables, stationary phase polarity (distribution coefficients) and mobile phase flow rate, are involved. Since the polarity varies with temperature, all factors that influence the elution temperature must be standardized. First, the length of the isothermal period and the injection temperature should be established. The second factor is the temperature programming rate, an increase in which results in an increase in elution temperature. Third, for the type of carrier gas chosen, the flow rate must have a standardized value; an increase in flow rate leads to a decrease in elution temperature. Further, provided inlet pressure is kept constant during the chromatographic run, carrier gas flow rates show a decrease during temperature programming, i.e., t_M increases. Such a decrease in flow rate is due to the increase in viscosity of the carrier gas with increasing temperature. The rate of increase in viscosity is different for different gases [60], and thus the choice of carrier gas affects the flow rates. Moreover, optimum flow rates vary considerably for different carrier gases. Fourth, the thickness of the stationary phase film in the column in relation to column diameter must be settled. The importance of this factor may be illustrated by a short discussion of the phase ratio of a capillary column. In a column, the volume of the mobile phase, V_M, is related to the volume of the stationary phase, V_L, by the β value

$$\beta = \frac{V_M}{V_L} \tag{8}$$

The capacity ratio, k, relates the time a solute is retained in the stationary phase to the time spent in the mobile phase,

$$k = \frac{t_x - t_M}{t_M} \tag{9}$$

The distribution coefficient, K, is related to the capacity ratio by the β value,

$$K = k\beta \tag{10}$$

Since K is a constant for a given solute/stationary phase combination at a given temperature, a change in β thus must result in a change in k, according to Eq. (10). In temperature-programmed runs, a change in k results in a change in elution temperature. A fifth factor is the column length, since the use of longer columns results in higher elution temperatures, all other variables being kept constant

For a correlation between isothermal and programmed temperature retention indices, temperature-dependent variables, such as distribution coefficients, carrier gas viscosity, and velocity, must be considered. A method based on such considerations, for the calculation of programmed temperature retention indices from isothermal data, was recently presented [61–63]. This method includes the conversion of programmed retention indices between columns of different dimensions, phase ratios, and temperature conditions.

III. ESTIMATION OF COLUMN DEAD TIME

For the calculation of Kováts retention indices, the column dead time, t_M, of Eq. (2) must be ascertained. Several procedures for estimation of t_M have been presented and have been reviewed by Smith et al. [64]. The methods can be classified in some different categories, the most important being direct measurement, classical methods, statistical, iterative, and miscellaneous. Advantages and disadvantages of these methods will be briefly described. For further details, Refs. 18, 64, and 65 are recommended.

In the direct methods, t_M is achieved by injection of a nonretained substance in the chromatograph. Using helium as carrier gas, the retention on nonpolar phases of a series of permanent gases and light hydrocarbons has been studied [66,67]. Of the gases tested, neon was found to give the lowest retention time. Since the flame ionization detector (FID) does not normally respond to permanent gases, methane is often used for t_M determinations. However, it is evident from Eq. (1) that methane certainly has a definitive retention on the columns, making it unsuitable for this purpose [3,68—70]. There is, however, a wide divergence of opinion on the use of methane as dead time indicator, and methane is thus sometimes recommended for this purpose. Smith et al. [64] stated three factors that may explain such divergences: the experimental conditions, the accuracy of retention time measurements, and the use to which the retention data are put. Retention of methane is particularly noticeable on nonpolar columns and at relatively low temperatures, e.g., 100°C and below. The accuracy of *time* measurement becomes of special importance under the conditions of low methane retention and/or small gross retention times.

Most of the methods for t_M determination are based on Eq. (1), the validity of which for varying numbers of carbons in the n-alkanes has been much disputed. It is clear, however, that the relation is not valid for n-alkanes lower than n-pentane [71,72]. Some authors consider that the equation shows linearity from n-heptane [26]. Rohrschneider [73] maintained that linearity is found from n-nonane, which is in accordance with the values of ΔG^{CH_2} (the energy contribution per CH_2 group) for n-alkanes (Fig. 1).

On the other hand, it was shown that the choice of n-alkanes having longer retention times can lead to errors in calculated t_M values [74]. Such errors may be due to instabilities in mobile phase flow rates and column temperature *during* a relatively long run. In the case of such instabilities, it was thus considered desirable to use relatively low-molecular-weight alkanes, provided retention times can be measured with sufficient accuracy. In the low-k region, however, a difference in retention times of 0.001 min can represent a difference in retention indices of 0.5 unit [75]. Smith et al. [18] recommend that retention times be measured to at least the nearest 0.1 sec unless very long retention times are involved, in which a lower precision in time measurement may be acceptable.

Ashes et al. [76] have shown that more polar homologs could also be used successfully for t_M determinations. The first members of such a series should, however, not be used for this purpose, since their retention values deviate from Eq. (1) [71]; see Fig. 2. The same homologous series could thus be used for determination of t_M and retention index even when alternative standards are used. The advantage of this is, of course, that it is necessary to run only one standard in order to achieve data for the calculation of I.

In the so-called classical methods, a limited number of homologs are generally used. Methods such as those described in Refs. 75–85 belong in this category. Large errors can result from such methods, and they are considered of limited use other than to provide an initial estimate for a more accurate iterative method [64].

Smith et al. [18] have compared some statistical [86,87], iterative [84,85,88,89], and polynomial [84] methods for the calculation of t_M and I. The statistical methods tested were found to provide acceptable estimates of the equation parameters in the majority of cases, although in some instances large errors resulted. It was thus recommended that all homolog combinations be checked to find the best fit. Such a check would make the methods relatively complicated. Of the iterative methods, the procedures of Tóth and Zala [85] and Guardino et al. [88] and the flexible Simplex procedure [89] gave identical results in all cases. Of these procedures, the method of Guardino et al. was the most rapid. A very important point when using this method is that the initial estimate of dead time must be less than the true mathematical dead time or the method will fail [64]. Instructive flowcharts for the calculations are given in Refs. 88 and 64. For the polynomial methods [90], it was concluded that these offer very little improvement. Iterative methods have also been described by Kaiser and Rackstraw [16].

In conclusion, the method of Guardino et al. [88] is strongly recommended. The necessary calculations should be easy to perform now that programmable calculators and microcomputers have become generally available.

IV. RETENTION INDEX AND MOLECULAR STRUCTURE; PREDICTION OF INDICES

Precalculation of retention indices may be of great importance in identification studies. The GC behavior of a compound of interest can be used in choosing its correct structural formula from alternative versions obtained from, e.g., mass spectrometric analysis. Numerous investigators have dealt with the correlation between retention index and molecular structure and the subject has been described in reviews [8,26,91]. Retention prediction may be based on correlation of retention indices with empirical additive schemes, physicochemical parameters, or various topological parameters. Furthermore, studies in this area may give an insight into how different structural features determine the retention index.

A. Empirical Additive Schemes

The relation between structure and retention index was discussed in the early works of Kováts [9,14,92—94]. The difference between I values on nonpolar (squalane) and polar stationary phases, ΔI, was thus considered to reflect the extent of polar interaction between solute and solvent molecules [14]. ΔI was attributed to a zone in the solute molecule that is more strongly retained in polar surroundings than in nonpolar stationary phases, such zones being termed "adhering zones." Kováts stated [14] that if a substance contains more than one adhering zone, then its ΔI value can be calculated by summing the increments for the individual adhering zones. The retention indices are, however, a linear function of the Gibbs energy of solution, which can be represented as the sum of enthalpic and entropic terms. Since the entropy terms depend on the total configuration of the molecule, ΔI cannot be determined in terms of addition [95].

The concept of ΔI values was further treated by Rohrschneider, who presented a method for approximately describing the interaction of a substance with a stationary phase by means of a few selected reference substances, benzene, ethanol, methyl ethyl ketone, nitromethane, and pyridine, and specific Rohrschneider constants of the examined substance [96]. This method allows the calculation of ΔI values and also retention indices [96—104].

B. Correlation with Physicochemical Parameters

For the correlation of retention indices and physicochemical molecular parameters, the problem has been to find variables that are closely correlated with the retention. Further, for a correct prediction of I, it seems to be necessary to consider all relevant retention mechanisms, and thus a corresponding number of variables is required. The interest in correlation of structure with physicochemical properties arises in several areas of chemistry, e.g., in methods for drug design and stereochemistry [105—107]. Methods developed in such connections may also be of interest for the prediction of Kováts indices. The procedure used for studying structure-retention relationships consists of generation of molecular descriptors and the subsequent regression of these descriptors against the retention indices.

It is considered that the retention properties of a solute can be largely expressed by a series of interaction parameters. A method of calculation based on solubility parameters—dispersion, dipole orientation, dipole induction, and hydrogen bonding—was presented by Keller et al. [108]. Lamparczyk et al. [109,110] suggest that the retention indices can be expressed in terms of basic solute-stationary phase electric interaction parameters such as ionization potential (IP), average molecular polarizability (α), and dipole moment (μ). Solubility parameters were applied by Karger et al. [111,112].

The number of carbon atoms, n, can be considered the simplest of the different molecular parameters used for prediction of I; see, Eq. (5). Similarly, molecular weight may also correlate with retention for homologous series [25].

Parameters that account for *dispersive interactions* are molecular refraction, van der Waals volume, molar volume, and polarizability [113—115]; these obviously, largely affect the same factor [25,116]. Differences between them reside in the scaling factors.

Molecular refraction is defined by the Lorenz-Lorentz equation,

$$R_m = \frac{n^2 - 1}{n^2 + 2} \frac{MW}{d} \tag{11}$$

where n is the refractive index, MW the molecular weight, and d the density. The molecular refractivity is considered to be the best "bulk" parameter to use as a measure of polarizability [116].

The van der Waals volume, V_W, is the volume occupied by a molecule, i.e., impenetrable to other molecules, whose thermal energies are at ordinary temperature. The calculation of V_W assumes a knowledge of bond distances, bond angles, contact distances (intermolecular van der Waals radii), and shape characteristics of atoms in various molecular configurations. Values of V_W may be calculated by the method of Bondi [117]. Unfortunately, such calculations do not differentiate between isomers, and an alternative method in which differentiation between alkane isomers is achieved has been published [113]. The correlation between topological indices and van der Waals volume has been demonstrated [118]. The correlation between I and V_W has been studied by Saura-Calixto and Garcia-Raso [113,119,120] and others [121,122].

The molecular volume is the effective volume occupied by 1 mol in the liquid state,

$$V_m = \frac{MW}{d} \tag{12}$$

The molecular electronic polarizability, α, is a measure of the polarizability of the electron cloud of the molecule. Dispersive interactions are mainly due to such electronic polarizability of the solutes [108,115], and therefore α can account for dispersive interactions. The molecular refraction is related to α according to

$$R_m = \frac{4}{3} N_A \pi \alpha \tag{13}$$

where N_A is the Avogadro's number. Values of α can be calculated by an empirical method [123].

Inductive chromatographic effects may be indicated by functions of the refractive index such as n^2 and $(n^2 - 1)/(n^2 + 2)$ [124]. Also, the ratio of R_m and V_m at the boiling point may account for induction forces [102].

Parameters of the dielectric constant, ε, or the permanent dipole moment, μ, can account for dipole-dipole interactions. The Kirkwood function $(\varepsilon - 1)/(\varepsilon + 2)$ has thus been included in the equations for I prediction [124]. Further, μ [124] and μ^2 [102] have been related to I in chromatographic studies. A function called the chromatographic dipole moment has also been suggested [125]. It has been shown, however, that compounds such as dioxane and 1,4-dichlorobutane, which have an overall dipole moment of approximately zero, still behave as polar solutes [112,126]. Further, it was demonstrated by factor analysis that μ and μ^2 may be rather poorly correlated to polar interactions [127]. Energy and charge parameters calculated by quantum chemical methods were found to result in better correlations [127,128].

The total energy of a molecule is a resultant of all attractions and repulsions between nuclei and electrons. Semiempirically, all valence electron molecular orbital calculations generate a total molecular energy presumed to behave in a parallel fashion to the actual total energy. Quantum chemical parameters can be calculated with relative ease by the CNDO/2 method. i.e., the complete neglect of differential overlap [129]. In this approximate method, all two-center overlap integrals vanish, which greatly simplifies the calculations, compared with the full Hartree-Fock-Roothaan equations [105,130]. This type of calculation has been applied to retention predictions by several workers [127,128,131,132]. The total energy was found to correlate with Kováts index for a homologous series of esters [131]. In a subsequent study of alkanes, Garcia-Raso et al. [132] adopted the concept of Takács et al. [133] of separating the retention index into two additive components, molecular and interaction contributions. The molecular contribution was found to correlate with the total energy (E_T) and the binding energy (E_b). Solute-stationary phase interactions were best expressed in this case by the interaction energy, $-\Delta E$, and the energy of the highest occupied molecular orbital, E_{HOMO}. The interaction energy was calculated according to Klopman's equation [134],

$$-\Delta E = \frac{2 \sum C_{HOMO}^2 C_{LUMO}^2 \beta^2}{E_{HOMO} - E_{LUMO}} \tag{14}$$

where C_{HOMO} is the coefficient of HOMO, C_{LUMO} the coefficient of LUMO (the lowest occupied molecular orbital), β the overlap integral, and E_{LUMO} the energy of LUMO. The method allows prediction of the elution of cis/trans isomers, which is of special importance since other methods, e.g., topological, cannot be applied to such tasks (see below). Quantum chemical parameters were also recently applied to predict the retention of various types of amines on columns coated with the dimethylsilicone oil OV-101 [128]. The charge distribution in a molecule may be related to interaction with stationary phases. Quantum chemical charge parameters have been used in combination with E_T for homologous series [128,131,132] and with topological indices [127]. Energy parameters were also applied to retention prediction by Kleinert and Ecknig [135,136].

Other electronic parameters used for correlation with retention indices are the Hansch constant, π [25,137], and the Hammett constant, σ [25,127,137].

Taft-type steric parameters have also been applied to gas chromatography, albeit in the field of stationary phase characterization [138].

In his second rule [14], Kováts proposed a linear correlation between the retention index of two isomers on nonpolar stationary phases and solute boiling points at normal pressure:

$$\partial I \underset{\sim}{\sim} \gamma \partial t_b \tag{15}$$

Values of γ of around 5 for some different homologous series and stationary phases were presented by Saura-Calixto et al. [139,140]. The values of γ may, however, depend on column temperature, and values from 1.8 to 13.9 have been reported for alkenes [13].

When the solutes and the stationary phase form ideal solutions, retention may be a function of solute boiling point:

$$\log V_R = a + bt_b \tag{16}$$

where V_R is the specific retention volume, t_b the boiling point in degrees Celsius, and a and b are constants. Ideal solutions do not exist in chromatography and it thus seems that an equation for prediction of I should also contain terms that account for specific and nonspecific interactions [114]. Bermejo et al. have applied a multiparameter approach of the general form

$$I = m + nt_b + pP + qQ + \cdots \tag{17}$$

where I is the retention index, t_b the boiling point at normal pressure, and P, Q, . . . represent solute parameters that account for different kinds of solute-stationary phase interactions. The equation was applied for precalculation of I for alkylbenzenes [114,141] and linear, branched, and cyclic alkanes [108]. Further, in an attempt to account for the effect of different stationary phases on the retention of aliphatic saturated esters, a factor P_M, which is the stationary phase polarity of the McReynolds scale [97], was introduced in Eq. (17) [124]. For the prediction of I for some olefins on nonpolar stationary phases, an equation was developed that contains the descriptors boiling point, Hansch constant, x moment of inertia, and molecular weight [142].

Roth and Novák [143] used the solution of groups concept for the prediction of I, but this method was found to give unsufficient accuracy for I predictions.

The prediction of retention indices results in approximate values. At present, these may differ from experimental data by at best a few index units. The best results are obtained with homologous series, i.e., when the only difference between substances in a series appears in the skeleton of an alkyl group, especially in nonpolar systems. Prediction of I for solutes that are not homologs but belong to the same chemical family, e.g., aliphatic saturated esters [124], may be somewhat more demanding. Data sets for molecules from different chemical families are more heterogeneous, and molecules with more than one functional group, e.g., an alcohol and a keto group, in the same molecule cannot easily be predicted [127,137].

Equations for the prediction of I are thus established by regression of different descriptor variables against experimental I values. For a given set of solutes, different equations are found for different stationary phases; that is, the coefficients m, n, p, q, . . . in Eq. (17) are different. Further, factors accounting for polar interactions are often included in the equations when polar stationary phases are used. As mentioned above, the polarity, P_M, on the McReynolds scale has been used for this purpose [103,124]. P_M is the sum of the I values of benzene, butanol, 2-pentanone, nitropropane, and pyridine [97]. A drawback of this concept is that it is based on the obsolete stationary phase squalane [98]. Patte et al. [102] calculated solubility factors of stationary phases from the retention indices of five standard solutes: decane, 2-butanone, benzene, 1-butanol, and pyridine. Similarly, Fellous et al. [104] used acetonitrile, methyl iodide, ethyl bromide, and ethanol. Tekler and Takács [101] calculated stationary phase factors from the retention indices of benzene, 2-pentanone, nitropropane, pyridine, and 2-methyl-2-pentanol. Retention polarities for several stationary phases were published by Tarján et al. [144]. Generally, predictions are much less accurate for polar than for nonpolar phases.

The Rohrschneider compounds give index numbers near or below 500, which results in low precision. McReynolds [97] therefore suggested a series of higher-boiling compounds. However, even these compounds often fall in the nonlinear part of Eq. (1); see also Fig. 1.

It was recently pointed out [145] that McReynolds did not fully indicate the conditions used for his index determinations [97], and there was no indication of the reliability of his values, which have been quoted repeatedly in the literature without question. Because of the low thermal stability of squalane, the McReynolds constants are measured at 120°C. In gas chromatography higher temperatures are generally used, since stationary phase polarity may vary with temperature (see Sec. V.B) McReynolds constants may be misleading.

Golovnya and Misharina [146] have shown that the partial molar free energy of solution corresponding to an index unit is different on different stationary phases. Therefore, the *same* values of indices on different stationary phases correspond to different values of free energy and thus different intermolecular interactions with the phase. This problem is avoided when comparison of the GC behavior of a substance of interest on *two* stationary phases is expressed as the difference in partial molar free energy of solution [146]. One of the stationary phases is selected as a reference phase. This phase should have a minimum of selectivity, it must be usable at high temperatures, and it must be chemically homogeneous. Squalane does not meet these requirements [3]. Now, SE-30 is recommended as the reference stationary phase [19,146].

C. Applications of Topology and Graph Theory to the Prediction of Retention Indices

Structural information concerning branching, atomic connections, shape, and size can be classified under the general term *topology*. The structural formula of a compound may thus be viewed as a *molecular graph*, where the vertices represent atoms and the edges represent covalent bonds. Characterization of chemical graphs by graph theory provides numerical data, a topological index, derived from the chemical structure, which can then be used for correlation with chemical properties. To date, more than 20 different indices have been devised in topology and graph theory and, in general, they correlate well with a broad range of physicochemical parameters [147]. The indices may be classified into four groups: topological indices based on the adjacency matrix, topological indices based on the distance matrix, centric topological indices, and indices based on information theory [148].

Several extensive reviews have been published describing the theories and methods of calculation [106,107,147–151]. Information theory indices may be defined in terms of any structural feature of chemical species and thus need not be of topological origin.

By its very nature, there are some important questions that topology should not be expected to address. Most stereoisomers are distinct by virtue of some kind of molecular rigidity. Euclidean geometric properties of the bond graphs are responsible for the distinct character of, for example, enantiomers having a chiral center or E/Z isomers of alkenes. In such cases, the bond graphs of a pair of stereoisomers are topologically equivalent, even when embedded in 3-space.

Topological indices were first applied to retention predictions by Randić [152,153], who used the first-order molecular connectivity index, χ (originally called the branching index [152]). This index accounts for dispersive interactions; if other types of interactions are involved, the use of higher-order indices is necessary.

On the basis of the Randić scheme, a connectivity method has been developed [116] which, in contrast to the original Randić scheme, accounts for heteroatoms and unsaturated bonds. The index is also useful for cyclic compounds [106] and was used by Sabljić for correlation with retention indices of chlorinated alkanes [154] and chlorinated benzenes [155]. The connectivity index has also been applied to 2-ketones [32]; additional references are given in the papers of Sabljić [154,155].

Other types of topological indices are applied to retention prediction in GC are the Wiener number [155—157], the DARC topological system [158—160], and the average distance sum connectivity index J [155]. The latter index is calculated according to Balaban [161]. Further, an information theoretical index, giving mean information content of the distribution of distances $\bar{I}_D^E$, in a graph according to their equality or inequality, has been applied to series of aliphatic hydrocarbons and alkylbenzenes [162,163] and isoalkanes [157]. Being based on the distance matrix, D, of a graph, $\bar{I}_D^E$ is calculated according to Bonchev and Trinajstić [150].

A correlation between the retention index, I, on a squalane coated column and the heats of formation, ΔH_f, of hydrocarbons that was described [162]. Further, it has been demonstrated that ΔH_f of alkylbenzenes may be correlated with the information theory index I_D [164]. It was reported that introducing $\bar{I}_D^E$ into equations for I versus ΔH_f improves the correlations [162].

In several cases, a very high correlation coefficient between first-order connectivity index and retention indices has been obtained, but the elution order has still been incorrectly predicted for a considerable number of compounds [165]. In view of this observation, Sabljić [154] emphasized that both a high correlation coefficient and a correctly predicted elution sequence are necessary before any approach can be usefully and widely applied to predict the retention indices.

An evaluation of the use of topological methods for prediction of I has been presented by Sabljić [154,155]. Equations based on empirical additive schemes and molecular connectivity for the prediction of I for chlorinated alkanes were compared [154,166]. For simple homologous series, e.g., linear monoalkanes, the empirical additive scheme resulted in very good correlations, whereas for branched chlorides the method failed [166]. The topological method gave excellent correlations in all instances. Further, in the empirical approach, twice as many variables are required as in the connectivity method.

Raymer et al. [32] achieved good correlations for a number of ketones when using the zero-, first-, and second-order terms of the connectivity index. Some ketones, however, showed large deviations from predicted I values.

Three different topological/information theory indices, the connectivity index, the Wiener number, and the Balaban index, were compared by Sabljić [155] for the prediction of I for chlorinated benzenes. The connectivity index gave the best results. In conclusion, the general connectivity index, as demonstrated by Sabljić, seems to be a quite promising approach, although since only a limited number of applications have been carried out so far, this method has not yet been thoroughly established. However, an impressive number of retention indices have been published [26]; this large amount of material, e.g., data in the Sadtler index library, has a high potential value and just awaits mathematical processing.

Solute retention may result from the complex interplay of several effects. It is expected that most of these can be adequately related to molecular connectivity terms, but in some cases the use of additional descriptors (physicochemical parameters) may be needed, especially for polar systems and data sets consisting of molecules from different chemical families [127,137].

Buydens et al. [127] combined connectivity parameters with some different quantum chemical charge parameters. Local charges were considered, i.e., the charges in and near the polar functional groups. Good correlations were achieved with both mono- and difunctional derivatives on nonpolar and polar stationary phases. Factor analysis was used to determine the degree of relationship between different parameters.

Papazova et al. [157] combined the Wiener number with some structural elements [167].

Most of the topological indices are related in some way [118,148,149], and in a comparative study it was found that no indices presented so far discriminate isomers uniquely [148,168]. A combined topological index called the superindex, which consists of a number of topological indices, was suggested for complete discrimination of a large number of structures [148,168]. This index has, however, not been tested on chromatographic data.

To summarize this section, topological indices basically reflect molecular shape and size and thereby account for dispersive interactions, which may be sufficient with nonpolar phases. For more polar phases, it seems that the combination of topological indices with "interaction parameters," e.g., molecular energy or charge parameters, may be more rewarding [127,137].

V. INSTRUMENTAL ASPECTS

A. Influence of Injection System on the Retention Index

Both splitless and on-column injection techniques may be used in isothermal analysis when it is desirable to take advantage of the so-called solvent effect for solute focusing. In such systems, the solvent takes an active part in the early stages of the chromatographic process, thus affecting solute retention. Splitless and on-column injections can be very useful for qualitative analysis in temperature-programmed runs, as shown by, e.g., Lee and co-workers [56,57].

Split injection is the most commonly used injection technique for qualitative capillary GC. Certain factors should be taken into considerations with split injection when high precision in retention data is desired. First, it is important that the injector temperature be high enough to provide rapid evaporation of the solutes [169]. Second, it is essential that the time of injection be correctly indicated. Third, syringe introduction and sample injection should be performed rapidly.

B. Influence of Column Temperature on the Retention Index

The influence of temperature on the retention index has been extensively studied. Retention indices of nonpolar solutes on nonpolar phases show an almost linear dependence on column temperature, at least within relatively narrow temperature ranges (about 50°C) [170, 171]. Ashes and Haken [172] studied the retention of the principal McReynolds compounds at temperatures between 90 and 150°C on four polysiloxane stationary phases, OV-1, OV-17, OV-210, and XE-1150, and also on the hydrocarbon squalane. Plots of $\log V_R$ versus T as well as 1/T showed linear relationships for the phases and solutes tested (V_R is the retention volume). An equation was derived by Chovin and Lebbe [173] for the relationship between relative retention and temperature:

$$\log r = a + \frac{b}{T} \tag{18}$$

In gas-liquid chromatography, Antoine-type equations have been used to relate vapor pressure to temperature. Takács et al. [174]

thus applied an equation of the Antoine type to express the relation between I and temperature:

$$I^{\text{stationary phase}}_{\text{substance}} (T) = A + \frac{B}{T + C} \tag{19}$$

where T is the column temperature in Kelvin, and A, B, and C are constants defined by solute and stationary phase. These constants are, however, not easily obtained, since knowledge of the molar heat of evaporation of the substances to be characterized is required. Further, fairly extensive experimentation and calculations are necessary [8]. For alkanes, the parameters A, B, and C correlate with the connectivity index, χ [107]. When a wide temperature range is considered, the retention index shows a hyperbolic temperature dependence [14,173,174]. A plot of Eq. (19), however, generally has a linear section in the temperature interval used in practice in gas chromatography [174], although exceptions may occur, especially with polar stationary phases. When several retention mechanisms, e.g., dispersive, inductive, orientative, and various other types of complexing are involved, it seems unlikely that all of these should be equally affected by an increase in temperature.

The fact that the value of dI/dT varies for different solutes on a particular stationary phase may be used in qualitative analysis [175–178]. For that purpose the dI/dT values must of course, not be lower than the measurement error. Running a sample isothermally at different temperatures may thus give retention data that can contribute to solute identification.

A large effect of temperature on column polarity is often observed with polar stationary phases, as in the retention of unsaturated fatty acid methyl esters on cyanosilicone stationary phases at different temperatures [179]. Another example is the retention of alkanals on Carbowax 1540 [180].

Certain solutes are also sensitive to temperature changes on nonpolar stationary phases. For dI/dT (for a 10° increase), Schomburg et al. [59] found values of 5.76 for decalin and 6.53 for naphthalene on the dimethylsilicone gum OV-1. Similarly, Johansen and Ettre [58] and Ettre [91] found that highly branched paraffins, naphthenes, and aromatics have high dI/dT values on the nonpolar silicone OV-101. The influence of temperature on retention has also been discussed by Grob and Grob [181].

Technical Aspects of Column Temperature

Three factors concerning column oven temperature must be considered. First, the temperature must be regulated to constancy better than 0.1°C [5]. Second, the temperature gradient in the oven should be low [182]. Third, the temperature must be correctly determined.

Temperature control is improved in modern instruments by the use of microprocessors [183]. Further, improved precision is achieved when the platinum resistance thermometer is situated very close to the column [184].

High-performance column thermostats have been available for many years in connection with physicochemical measurements by gas chromatography. Rijks and Cramers [178] found gradients of less than 0.01°C when a Tamson TEV 70 liquid bath was used. In commercial chromatographs, air ovens are most commonly used, and it appears inevitable that air ovens cannot give the same good temperature precision as liquid thermostats. However, some improvement of GC column thermostats has recently been experienced [185,186]. This improvement has become necessary with the introduction of fused silica capillary columns. These have a rather small thermal mass and therefore small local fluctuations in column temperature may lead to peak splitting and broadening [186]. While counteracting peak splitting, the new improved ovens may at the same time afford improved reproducibility in retention determinations.

It is, of course, necessary that the column oven temperature be correctly determined. Some chromatographs now have a simple means of two-point temperature calibration. In our experience, it is advisable from time to time to check the correctness of the temperatures indicated on displays on the chromatographs.

A temperature effect is also caused by the heat generated by the heat of absorption in the stationary phase, and it is thus considered that a compound is accompanied by a heat wave during its travel through the column [187]. Such an effect could be significant in columns of low heat capacity at high sample concentrations.

C. Influence of Carrier Gas on the Retention Index

Carrier gases exhibit nonideal properties and therefore so-called virial effects may occur. Because of the presence of such effects, the use of different carriers may lead to changes in solute retention times and, in special cases, even shifts in elution order [188]. These effects were studied by Desty et al. [189], who considered that a change of carrier gas and/or column pressure may be used as an additional tool in peak identification. The influence of column pressure on I for alkyl benzenes was discussed by Soják et al. [190]. Virial effects are related to column pressures. When an increase in virial effects is desirable, it can be achieved by attaching a choke to the column outlet, and increasing the column inlet pressure [191].

It is considered that the influence of virial effects is small when helium, hydrogen, or nitrogen is used at low column pressures ($p_{inlet} < 1.2$ atm at $p_{outlet} = 1$ atm) [192,193]. The effects, however, may be appreciable at higher pressures or when other carrier gases are used.

The solubility of carrier gas in the stationary phase may influence retention; however, when using helium, nitrogen, hydrogen, or argon as the carrier, errors due to such solubility are negligible [194].

In temperature-programmed runs, as mentioned above, the thermal dependence of carrier gas viscosity is significant. When inlet pressure is kept constant during temperature programming, the carrier gas flow rate will decrease. The rate of such a decrease depends on the nature of the carrier gas [60].

Fluctuations in the carrier gas flow rate, which can give rise to errors in retention index determinations, are likely to occur when conventional flow controllers are used. These are sensitive to temperature and inlet pressure fluctuations. Pressure controllers appear to be relatively less sensitive in this respect [194]. When using commonly available equipment, high precision in flow regulation can be achieved with a two-stage pressure regulator at the gas cylinder head in connection with a precision pressure controller. Another source of error is the outlet pressure if it is equal to ambient pressure, i.e., the influence of weather conditions. Finally, the stability of carrier gas flow rates would be greatly improved if modern gas flow controllers, working according to so-called fluidex principles, were used in gas chromatography [16,195].

Retention indices may be measured by the separate injection of standards and sample; however, such procedures are inappropriate when high precision is desired. Smith et al. [18] recommend that a series of runs be performed using the homolog under appropriate conditions and that three or four homologs (two for a simple analysis) be *added* to the sample to be chromatographed. The sample should be run several times and the means of retention times calculated. With improved constancy of carrier gas flow rate and column temperature in combination with higher precision in time measurement, the practical determination of high-precision retention data can be greatly simplified.

D. Requirements of the Detection System

Detection and readout systems should have a low time constant in order to facilitate a true representation of peaks [4]. Under noise conditions the peak maximum may be difficult to find, and in some integrators this is "solved" by the use of a relatively high time constant or a noise suppression algorithm, which leads to changes in peak shapes and retention time data. The peak maximum can be found by least-squares polynomial interpolation. Second-order regression is often used for this purpose, but it was shown by Kaiser and Rackstraw [16] that fifth-order polynomial regression gives a more true representation of the elution curve and therefore a more correct value of the retention time. According to Kaiser and

Rackstraw [16], the time precision must be at the 10-msec level, i.e., 10 times better than suggested by Smith et al. [18].

   Retention time is generally measured at the peak maximum, which is, in fact, strictly correct only for Guassian peaks; for skewed peaks, the retention time should be obtained from the first moment of the zone [4,194]. In general, capillary GC peaks do not deviate too much from a symmetrical signal.

VI. THE SEPARATION COLUMN IN RETENTION INDEX DETERMINATIONS

For given combinations of solutes and stationary phases, retention data should be readily reproduced with a high degree of precision from column to column and between laboratories. This may be difficult when more than one type of retention mechanism is active. The primary retention mechanism is normally partition between the gas phase and the bulk liquid phase. Contributions may, however, also be made by adsorption at the gas-liquid interface (Gibbs surface adsorption) and adsorption on the solid support surface. Further, the support may have an orientating effect on the molecules of some types of stationary phases. These factors will be considered in some detail.

A. Adsorption Effects

Adsorption on the Solid Support Surface

Adsorption of solutes on the support material often leads to peak tailing and a decrease in retention time with increasing amount of sample. In cases of serious adsorption, retention indices cannot be established in a meaningful way; at moderate adsorption levels, i.e., slightly tailing peaks, useful retention data can be obtained, but the position of the peak maximum in a chromatogram will be slightly shifted by the tailing. Moreover, for high reproducibility, it is very desirable for retention data to be based only on the dissolution of the solutes in the stationary phase. Adsorption should thus not contribute to the retention. Berezkin [196] considered that the main reason for the relatively poor reproducibility of the index between laboratories appears to be adsorption. The contribution of adsorption on the solid support surface to solute retention may increase when the film of stationary phase becomes very thin [197,198].

   The surface of the fused silica capillary tubing that is presently being used as column material is somewhat heterogeneous; different types of adsorptive sites are present. Siloxane bridges, Si—O—Si, in the surface can thus act as proton acceptors, and silanol groups,

Si—OH, may act as proton donors. Further, cracks and crevices in the surface may affect adsorptive activity.

Capillary deactivation methods have been very much improved during recent years. Now, typical polar solutes can be successfully chromatographed when occurring in at least nanogram amounts. However, *n*-amines still show some tailing. Several investigators have attempted to perform separations of solutes at the picogram level. With present-day columns, this can be achieved for nonpolar and moderately polar solutes. The elution of polar solutes in low amounts, however, puts stringent demands on column deactivation, and at present we have no means of attaining the high degree of surface deactivation that would be necessary in these cases. It seems that the remaining active sites in the surface exert a strong attraction, so they are almost covered even at low pressures of solutes.

Adsorption at the Gas-Liquid Interface

Strongly polar stationary phases and nonpolar solutes are relatively incompatible. Such solutes are thus excluded from the bulk of the stationary phase, and an excess surface concentration is thereby formed. The importance of this effect in connection with gas chromatography was first realized by Martin [199]. It was later shown by Pecsok and Gump [200] that surface adsorption effects also occur for polar solutes with nonpolar stationary phases (squalane). After some debate, these results now seem to have been accepted [34,194]. The effects of adsorption at the gas-liquid interface in GC were reviewed by Martire [201].

The occurrence of gas-liquid adsorption in a chromatographic system is a complicating factor in retention index determinations. Retention indices then tend to become concentration-dependent even at relatively low concentrations. However, in order to minimize the influence of Gibbs surface adsorption in the systems mentioned above, the use of very small amounts of solutes is generally recommended [194]. A further factor of importance is, as mentioned above, the cross-link density in immobilized polar stationary phases; too high a density leads to severe hydrocarbon adsorption.

B. Partition Between the Gas Phase and the Bulk Stationary Phase

This is the retention mechanism that in an ideal case should give the "true" retention data. A presupposition for reproducible results is that the solute concentrations lie within the linear part of the solution isotherm; i.e., the sample size must not exceed a maximum permissible level. This upper limit depends on the amount of stationary phase in the column and, for a given phase, the type of solute, i.e., the solubility of the solute in the stationary phase. Further, the maximum permissible sample size is a function of solute concentration in the column.

C. Orientation of the Stationary Phase on the Support Surface

Because of their backbone flexibility, silicones can adopt various orientations at different interfaces [202]. The support surface may thus have an orientating effect on molecules of such stationary phases. The effect may be observed in connection with thin films of stationary phase.

D. Stationary Phase as a Source of Error in Retention Index

A necessary condition for the determination of accurate retention data is of course, that stationary phases with reproducible retention properties be employed. Three factors are critical for the reproducibility of stationary phase properties: (a) synthesis of the stationary phase, (b) preparation of the separation column, and (c) possible modification of the properties due to gas chromatographic use. These factors will be discussed here.

Aspects of Stationary Phase Synthesis

Most stationary phases used today consist of polymers. For reproducibility, it is desirable that these polymers have a narrow and defined chain length distribution. When low-molecular-weight material is present, column bleed-off will cause changes in the retentive properties. Examples of such phases are Carbowax 20M [203,204] and OV-17 [205]; see Figs. 3 and 4. The polymerization of cyclic siloxanes to silicones proceeds via an equilibrium reaction and the product thus also contains monomer residues, which should be removed. It was shown by Rogers and co-workers [206,207] that the methylphenylsilicone stationary phase DC-710 consists of a mixture of cyclics and linear tetramers through heptamers. From the results obtained by Hubball and co-workers [205], it may also be presumed that OV-17 contains low-boiling compounds of this type (Fig. 4). The various oligomers were shown to have different retentive properties [206,207].

SE-33, a dimethylsilicone gum with approximately 1% vinyl substitution, is another example of a phase that exhibits inhomogeneities. It was reported that SE-33 also contains some polydimethylsiloxane of relatively short chain length [208]. It was speculated that the low-molecular-weight material has been added to SE-33 in order to improve flow or other properties. On cross-linking, the vinyl-containing

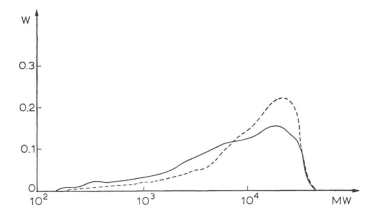

Fig. 3   Molecular weight distribution of Carbowax 20M before (dashed line) and after treatment, 200 temperature program cycles to 220°C when coated on Chromosorb (solid line). (From Ref. 204.)

component of the polymer is immobilized, while the dimethyl-substituted component remains soluble. In column preparation, excellent results have been achieved with the new stationary phase PS-255 (Petrarch Systems Inc., Bristol, Pa.) [209,210], which is reported to be a dimethylsilicone gum with approximately 1% vinyl substitution [211].

Reactions that take place on the silicon substituent groups during synthesis of the phase can be an additional source of stationary phase batch-to-batch variation. Substituents may undergo cleavage from the silicon atom or be otherwise chemically changed. Some years ago, it was noted that SP-2340 (Supelco Inc., Bellefonte, Pa.) and OV-275 (Ohio Valley Specialty Chemicals Inc., Marietta, Ohio) showed infrared adsorption in the carbonyl region [212]. Later, it was shown that a few commercial cyanopropyl-substituted silicone stationary phases contain some carboxamide substitution [213,214]. Some polysiloxanes thus exhibited multiple absorbance in the carbonyl region, inconsistent with their supposed composition. Some of these bands were attributed to residual acetone and disappeared from the spectra when the polymers were vacuum-dried [213]. However, four polymers still showed prominent carbonyl absorbance in the neighborhood of 1675 cm^{-1}. These polymers were SP-2310, SP-2330, and SP-2340 from Supelco and OV-275 from Ohio Valley. Amides may be formed from cyano groups in dilute acid solutions [215], and such a reaction may have occurred during the synthesis of these phases. The formation of carboxamide can be avoided by the use of suitable reaction routes [214].

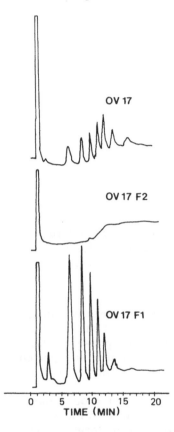

Fig. 4   Packed column gas chromatograms of acetone solutions of OV-17, OV-17 (F1), and OV-17 (F2).   Column: length 45 cm, outer diameter 0.125 in., stainless steel; packed with 5% Dexsil 300 on 80—100 mesh Chromosorb W, DMCS; nitrogen flow rate, 20 ml/min; injector and detector temperatures, 425°C. F_1, low molecular weight material removed by supercritical extraction with CO_2; F_2, residue after supercritical extraction.   (From Ref. 205.)

In order to control the reproducibility of the phases, they must be characterized. Specifications of properties should thus be established.

Another important factor is the procedure for packing the phases. When stored in plastic bottles, phthalates may migrate into the phase. Soda glass may act as a catalyst for the degradation of silicones. Pyrex vials appear to be the safest alternative. Further, some phases, e.g., Carbowaxes, are sensitive to oxygen; these phases should be stored under an inert atmosphere. Moreover, it should be noted that polyethylene glycols are hygroscopic.

Aspects of Column Preparation

In order to coat the capillaries, the stationary phases must first be dissolved in a solvent. It is well known that many types of silicones are somewhat unstable in solution [216–218]. This is especially critical with trifluoropropyl-substituted silicones [219]. It is generally recommended that fresh coating solutions be used.

Immobilization of the stationary phase by in situ cross-linking has turned out to be a fruitful approach in capillary gas chromatography [220, 221]. Several methods have been developed for such immobilization. When high precision in retention determinations is sought, however, it is evident that some of these methods give rise to small shifts in retention properties [222].

The curing of methylvinylsilicone, SE-33, with dicumylperoxide (DCP) has been studied in some detail [208]. Some aromatic material was found to be incorporated in the cross-links. Since the aromatic yield increases with increasing DCP content, it is desirable to keep the DCP content as low as possible. In addition, it was found that the network formed on curing was essentially complete with a DCP content of about 0.50%.

Azo compounds were recommended as free-radical generators for in situ immobilization of silicones. It has been noted that the use of azo-*t*-butane (ATB) as a cross-linking agent may lead to slightly lowered retention indices [223]. This effect may be due to incorporation of hydrocarbon moieties in the stationary phase. Excessive cross-linking of a polar stationary phase may, on the other hand, lead to Gibbs surface adsorption of *n*-alkanes, which may be experienced as an increase in retention indices.

Another method for in situ immobilization of silicones is the heat-curing of silanol-terminated siloxane prepolymers [224–230]. Retentive properties of this type of immobilized stationary phases need to be further studied.

Finally, the columns are often rinsed with solvent after stationary phase immobilization in order to remove residues of radical initiator reaction products and nonimmobilized fragments of the phase. Such a rinsing may result in polarity shifts. It seems that the small changes in column polarity on immobilization deserve further study.

Stationary Phase Stability During Use

Changes in retentive properties of silicone stationary phases may result from changes in the pendant hydrocarbon groups. Such changes may be due to oxidation, especially if the silicone contains groups that are readily oxidized, e.g., bistolyl. Further, some types of substituent groups, e.g., phenyl and tolyl, may be removed from the silicon atom [231,232]. Such cleavages may occur in the presence of acids [233]. Another factor may be the direct reaction of substituent groups with sample constituents, e.g., residues of derivatization reagents.

When silicones are used as stationary phases, the bleed material issuing from the column consists mainly of cyclic siloxanes [234,235]. For stationary phases having a relatively narrow molecular weight distribution, as opposed to, e.g., DC-710, such bleeding should not normally lead to larger shifts in polarity. However, in the case of silicones made by copolymerization of different monomers, such as dimethyl- and cyanopropyl(phenyl)-silanes, the less stable moiety will preferentially bleed off, thereby lowering the polarity [236].

In practical work, changes in retention properties are most often caused by successive deposition of nonvolatiles in the first part of the separation column.

VII. SAMPLE COMPONENTS AFFECTING THE RETENTION OF OTHER COMPONENTS

The partition of a solute between gas and liquid phase in a column is often influenced by the presence of other solutes. This effect may be prominent when a trace component is eluted on the back edge of a major component; the major component will take part in the separation process, thus acting as a "dynamic stationary phase" [237]; compare the influence of the solvent effect as mentioned in Sec. VA. The retention time of the trace component may thereby be drastically changed.

VIII. MULTIDIMENSIONAL COLUMN SYSTEMS

With the advent of fused silica capillary tubing, it has become very easy to mount two or three columns in parallel in one injector. This, of course, greatly simplifies qualitative analysis based on retention indices.

For the separation of complex samples, columns of different polarities coupled in series may be quite useful. By so-called heart cutting, fractions of special interest may be trapped between the first and second column. Such intermediate trapping is necessary when an exact start of the separation on the second column is required. Using such systems, high precision in retention index determinations has been reported by Schomburg et al. [59].

As mentioned above, relatively safe identification can be achieved by running the sample on columns of different polarities. This can be done in practice very elegantly by coupling two columns of different polarities in series, when it is possible to change the solute *residence time* independently in the two columns [238—240]. Such a change in residence time can be obtained by changes in temperature when the columns are mounted in two separate ovens or, more simply, when the mobile phase pressure can be regulated "independently" for the two columns. Systems of this type have attracted some recent interest [179,241]. A tandem system is, of course, useful only when the solutes are retained in a different way on the two columns; thus, such a system cannot offer any advantages for the separation of homologs or completely nonpolar compounds.

IX. COMMENTS ON THE SADTLER STANDARD RETENTION INDEX LIBRARY

Several computerized standard retention index libraries were presented in the earlier days of chromatography. Advances in gas chromatography, however, have motivated the introduction of a new system. A computerized standard retention index library for capillary GC was recently introduced by Sadtler Research Laboratories [1]. At this writing, the system contains 2000 compounds. It was pointed out [1] that "both GC hardware and columns have advanced to the point where it is now practical to measure a retention index library that is highly reproducible and can be used for qualitative analysis of unknown compounds." It is claimed that data can be reproduced within one index unit for nonpolar stationary phases and within a few index units for the polar phases. We consider that much better results can be obtained when means are used to minimize the sources of errors that exist.

For determination and calculation of the index according to Sadtler, improvements could be made at several points. First, column dead time is now based on the retention of methane; as mentioned above, dead time calculations based on homologous series give more accurate values. For the calculation of Kováts index, the method of Guardino et al. [88] is recommended [18,64]. The data for such calculations are available on computer, and it is, in fact, surprising that this possibility has not been utilized. Second, the properties of the columns should be specified in more detail. Four types of stationary phases are employed: immobilized OV-1 and SE-54, Carbowax 20M, and bonded Carbowax 20M. OV-1 is a pure dimethylsilicone gum that can easily be prepared with a high degree of reproducibility. It would, however, also be of interest to know which reagents have been used for the immobilization and in what amounts they have been added. Vinyl substitution in a silicone greatly facilitates cross-linking when peroxides are used for the initiation. Thus, much smaller amounts of peroxide are required for the immobilization of dimethylsilicones containing some vinyl substitution than for those with pure dimethyl substitution. As mentioned above, larger amounts of peroxide would lead to shifts in silicone composition. On these grounds, we would prefer a vinyl-containing dimethylsilicone, e.g., PS-255 (Petrarch), as the nonpolar stationary phase.

In the case of SE-54, a UV spectrum of the phase should be given, and even the type and amount of cross-linking agent should be mentioned. Further, as we reported earlier, the development of an improved version of this phase would be desirable [236].

The choice of Carbowax 20M may lead to some complications, since this phase is not strictly defined; see the molecular weight distribution as shown in Fig. 3. When a column coated with such a phase is used, low-molecular-weight material bleeds off, thereby causing a change in column retentive properties.

Bonded Carbowax 20M is, unfortunately, not standardized. Such bonding can be performed in several different ways, leading to different column properties [242]. One column manufacturer prepares bonded Carbowax according to certain patents [243,244] another manufacturer uses a method developed in our laboratory [245], and so on. We consider the bonded Carbowax 20M to be a good choice because of its chromatographic utility, but the columns used should be better defined.

Low reproducibility of the Kováts index on Carbowax 20M has been reported in several cases [24,246,247]. Furthermore, it was claimed that for n-alkanes, the even carbon number series should have stronger retention than the odd carbon number series [15].

The Sadtler concept also includes a temperature-programmed version. It was pointed out by Curvers [63] that the Sadtler temperature-programmed retention indices can be reproduced only on identical columns with predescribed temperature programming conditions. A considerable improvement would be achieved by adoption of the concept presented by Curvers et al. [61—63] for the calculation of temperature-programmed retention indices.

X. CONCLUDING REMARKS

With the recent developments in capillary gas chromatography technology, the use of GC as a method for accurate qualitative analysis could be feasible in the near future. Computer-based retention index systems facilitate calculations and interpretations. The instrumentation, especially concerning carrier gas regulation and temperature control, has been improved. Future instrumental improvements should include improved flow controllers, improved precision in time measurement (10-msec level), and improved methods for finding the apex of the chromatographic peaks.

Fused silica capillary tubing has made possible simple coupling of columns in parallel or in series. Columns coupled in series offer new possibilities for the separation of complex samples. Further, changing the solute *residence time* on two columns coupled in series is a simple and elegant way to run a sample under a series of different selectivities, when using only two columns of different polarities. Retention data for solute identification are thus generated.

Developments in column technology have increased stability and separation properties. The highest precision in retention indices is obtained with nonpolar columns. Such columns are easy to prepare in a reproducible manner and are quite durable. It is more difficult to prepare polar columns reproducibly and they are also less durable than nonpolar ones. Durability may be a problem when running columns of different polarities in parallel or in series in one oven. Such runs are often restricted by the thermal stability of the polar column. Improvements of polar columns would thus be desirable for the future.

The *n*-alkane standards for the Kováts index are not very compatible with highly polar stationary phases. Most likely, Gibbs surface adsorption of hydrocarbons occurs on the surface of such phases, which causes retention to be rather sensitive to sample size. The use of alternative homologous series, such as so-called secondary standards, for these stationary phases might be a solution to the problem. Uniform reporting of retention data is very desirable, however, and we are therefore disinclined to use systems that cannot be converted into Kováts index values. The degree of surface adsorption seems to be, inter alia, a function of stationary phase hardness.

The recent developments in capillary GC have thus made possible increased precision in retention index determinations. Nevertheless, there are some factors concerning the stationary phases that may act in the opposite direction [248]. First, it seems that the retentive properties of columns marketed under a particular name—immobilized

Carbowax 20M, OV-1701, etc.—are somewhat different when they originate from different manufacturers. As a consequence, the name of the column manufacturer must also be reported together with retention data. Second, fewer chromatographers make their own columns and the market for stationary phases is thus decreasing. At the same time, demands for high stationary phase quality and therefore production costs are increasing. Stationary phase synthesis can, however, be economically sound if it is combined with column manufacturing. There is now a tendency for new stationary phases to be developed by the column manufacturing companies. It seems that the bigger column manufacturing companies will have their own sets of stationary phases; see, e.g., Ref. 249. The trend is thus toward an increased number of stationary phases. In addition, these phases will be available only in the form of coated columns. Of course, competition between companies will lead to the development of new stationary phases, and new improved phases are really needed, but for successful use of the retention index concept, we wish that a few generally useful stationary phases could be at least temporarily established as a basic set. The most obvious choice of stationary phases at present is dimethylsilicone, methyl(phenyl)silicone, Carbowax 20M, cyanopropylsilicone, and trifluoropropylsilicone [250]. The phases should be designed for in situ immobilization.

Finally, it should be noted that the retention scale is very short when one considers the great number of chemical compounds that have been described so far, albeit only a minor part of these are suitable for gas chromatography.

REFERENCES

1. J. F. Sprouce and A. Varano, Int. Lab., 54 (Nov./Dec. 1984).
2. M. Goedert and G. Guiochon, Anal. Chem. *42*, 962 (1970).
3. L. S. Ettre, Chromatographia *6*, 489 (1973).
4. J. A. Rijks, Characterization of Hydrocarbons by Gas Chromatography; Means of Improving Accuracy, Thesis, Eindhoven University of Technology, Eindhoven, Netherlands, 1973.
5. G. Schomburg and G. Dielmann, J. Chromatogr. Sci. *11*, 151 (1973).
6. M. Goedert and G. Guiochon, Anal. Chem. 45, 1188 (1973).
7. R. J. Laub and R. L. Pecsok, *Physicochemical Applications of Gas Chromatography*, Wiley, New York, 1978.
8. J. K. Haken, in *Advances in Chromatography*, Vol. 14, (J. C. Giddings, E. Grushka, J. Cazes, and P. R. Brown, eds.), Marcel Dekker, New York, 1976, p. 367.
9. E. Kováts, Helv. Chim. Acta *41*, 1915 (1958).
10. W. J. A. Vanden Heuvel, W. L. Gardiner, and E. C. Horning, J. Chromatogr. *26*, 387 (1967).

11. T. K. Miwa, K. L. Mikolajczak, F. R. Earle, and I. A. Wolff, Anal. Chem. *32*, 1739 (1960).

12. F. P. Woodford and C. M. van Gent, J. Lipid Res. *1*, 188 (1960).

13. J. Janák and L. Soják, Ber. Bunsenges. Phys. Chem. 77, 205 (1973).

14. E. sz. Kováts, in *Advances in Chromatography*, Vol. 1 (J. C. Giddings and R. A. Keller, eds.), Marcel Dekker, New York, 1965, p. 229.

15. M. S. Vigdergauz and V. I. Seomkin, J. Chromatogr. *158*, 57 (1978).

16. R. E. Kaiser and A. J. Rackstraw, *Computer Chromatography*, Vol. 1, Hüthig Verlag, Heidelberg, 1983.

17. L. S. Ettre, Anal. Chem. *36*(8), 31A (1964).

18. R. J. Smith, J. K. Haken, M. S. Wainwright, and B. G. Madden, J. Chromatogr. *328*, 11 (1985).

19. R. V. Golovnya, Chromatographia *12*, 533 (1979).

20. R. V. Golovnya and D. N. Grigoryeva, Chromatographia *17*, 613 (1983).

21. R. V. Golovnya and D. N. Grigoryeva, Chromatographia *18*, 449 (1984).

22. R. V. Golovnya and D. N. Grigoryeva, J. Chromatogr. *290*, 275 (1984).

23. F. T. Zulaica and G. Guiochon, Bull. Soc. Chim. Fr. *1*, 1242 (1963).

24. J. R. Ashes and J. K. Haken, J. Chromatogr. *101*, 103 (1974).

25. F. Saura-Calixto, A. Garcia-Raso, and J. Cañellas, An. Quim. *79*, 411 (1983).

26. M. V. Budahegyi, E. R. Lombosi, T. S. Lombosi, S. Y. Mészáros, Sz. Nyiredy, G. Tarján, I. Timàr, and J. M. Takács, J. Chromatogr. *271*, 213 (1983).

27. V. G. Berezkin and A. A. Korolev, Chromatographia 20, 482 (1985).

28. D. F. Fritz, A. Sahil, and E. sz, Kováts, J. Chromatogr. *186*, 63 (1979).

29. A. Gröbler, J. Chromatogr. Sci. *10*, 128 (1972).

30. R. G. Ackman, J. Chromatogr. Sci. *10*, 535 (1972).

31. S. J. Hawkes, J. Chromatogr. Sci. *10*, 536 (1972).

32. J. Raymer, D. Wiesler, and M. Novotny, J. Chromatogr. *325*, 13 (1985).

33. U. Heldt and H. J. K. Köser, J. Chromatogr. *192*, 107 (1980).

34. V. G. Berezkin and V. N. Retunsky, J. Chromatogr. *292*, 9 (1984).

35. V. G. Berezkin and V. N. Retunsky, J. Chromatogr. *330*, 71 (1985).

36. L. Mathiasson, J. Å. Jönsson, A. M. Olsson, and L. Haraldsson, J. Chromatogr. *152*, 11 (1978).

37. M. S. Wainwright, C. S. Nieass, J. K. Haken, and R. P. Chaplin, J. Chromatogr. *321*, 287 (1985).

38. G. Castello, G. D'Amato, and E. Biagini, J. Chromatogr. *41*, 313 (1969),

39. J. K. Haken, J. Chromatogr. *99*, 329 (1974).

40. J. R. Ashes and J. K. Haken, J. Chromatogr. *111*, 171 (1975).

41. G. Schomburg, in *Advances in Chromatography*, Vol. 16 (C. J. Giddings and R. A. Keller, eds.), Marcel Dekker, New York, 1968, p. 211.

42. R. U. Luisetti and R. A. Yunes, J. Chromatogr. Sci. *9*, 624 (1971).

43. L. J. Lorenz and L. B. Rogers, Anal. Chem. *43*, 1593 (1971).

44. F. Pacholec and C. F. Poole, Anal. Chem. *54*, 1019 (1982).

45. F. Pacholec and C. F. Poole, J. Chromatogr. *302*, 289 (1984).

46. K. Ballschmitter, Ch. Unglert, and H. J. Neu, Chemosphere *6*, 51 (1977).

47. K. Ballschmitter and M. Zell, Z. Anal. Chem. *293*, 193 (1978).

48. K. Ballschmitter and M. Zell, Z. Anal. Chem. *302*, 20 (1980).

49. T. R. Schwartz, J. D. Pelty, and E. M. Kaiser, Anal. Chem. *55*, 1839 (1983).

50. L. N. Zotov, G. V. Golovkin, and R. V. Golovnya, J. High Resolut. Chromatogr. Chromatogr. Commun. *4*, 6 (1981).

51. J. Enquist, P. Sunila, and U.-M. Lakkisto, J. Chromatogr. *279*, 667 (1983).

52. J. Enquist and A. Hesso, Precision Chromatography as Analytical Method (in Swedish), brochure available from Orion Corp. Ltd., P.O. Box 8, SF-02101 Espo 10, Finland.

53. J. C. Giddings, J. Chromatogr. *4*, 11 (1960).

54. H. W. Habgood and W. E. Harris, Anal. Chem. *32*, 450 (1960).

55. H. van den Dool and P. Dec. Kratz, J. Chromatogr. *11*, 463 (1963).

56. M. L. Lee, D. L. Vassilaros, C. M. White, and M. Novotny, Anal. Chem. *51*, 768 (1979).

57. D. L. Vassilaros, R. C. Kong, D. W. Later, and M. L. Lee, J. Chromatogr. *252*, 1 (1982).

58. N. G. Johansen and L. S. Ettre, Chromatographia *15*, 625 (1982).

59. G. Schomburg, H. Husmann, L. Podmaniczky, and F. Weeke, in *Analysis of Volatiles* (P. Schreier, ed.), de Gruyter, Berlin, 1984, p. 120.

60. L. S. Ettre, Chromatographia *18*, 243 (1984).

61. J. Curvers, J. Rijks, C. Cramers, K. Knauss, and P. Larson, J. High Resolut. Chromatogr. Chromatogr. Commun. *8*, 607 (1985).

62. J. Curvers, J. Rijks, C. Cramers, K. Knauss, and P. Larson, J. High Resolut. Chromatogr. Chromatogr. Commun. *8*, 611 (1985).

63. J. M. P. M. Curvers, Thesis, Qualitative and Quantitative Aspects of Trace Analysis by Capillary Gas Chromatography, University of Eindhoven, Einhoven, Netherlands, 1985.
64. R. J. Smith, J. K. Haken, and M. S. Wainwright, J. Chromatogr. *334*, 95 (1985).
65. L. S. Ettre, Chromatographia *13*, 73 (1980).
66. F. Riedo, D. Fritz, G. Tarján, and E. sz. Kováts, J. Chromatogr. *126*, 63 (1976).
67. J. F. Parcher and D. M. Johnson, J. Chromatogr. Sci. *18*, 267 (1980).
68. V. A. Ezrets and M. S. Vigdergauz, Chromatographia *9*, 205 (1976).
69. M. R. Becerra, E. Fernández Sánches, J. A. Garcia Dominguez, J. Garcia Munoz, and M. J. Molera, J. Chromatogr. Sci. *20*, 363 (1982).
70. M. S. Wainwright, J. K. Haken, and D. Srisukh, J. Chromatogr. *179*, 160 (1979).
71. M. S. Wainwright, J. K. Haken, and D. Srisukh, J. Chromatogr. *188*, 246 (1980).
72. L. Soják, J. Krupčik, and J. Rijks, Chromatographia 7, 26 (1974).
73. L. Rohrschneider, Chromatographia *2*, 437 (1969).
74. J. K. Haken, M. S. Wainwright, and R. J. Smith, J. Chromatogr. *133*, 1 (1977).
75. A. J. Lubeck and D. L. Sutton, J. High Resolut. Chromatogr. Chromatogr. Commun. 7, 542 (1984).
76. J. R. Ashes, S. C. Mills, and J. K. Haken, J. Chromatogr. *166*, 391 (1978).
77. M. L. Peterson and J. Hirsch, J. Lipid Res. *1*, 132 (1959).
78. M. Hafferkamp, in *Chromatographie in der Gasphase* (R. Kaiser, ed.), Part II, Bibliographisches Institut, Mannheim, 1966, p. 93.
79. H. L. Hansen and K. Andresen, J. Chromatogr. *34*, 246 (1968).
80. H. J. Gold, Anal. Chem. *34*, 174 (1962).
81. R. E. Kaiser, Chromatographia 7, 251 (1974).
82. W. K. Al-Thamir, J. H. Purnell, C. A. Wellington, and R. J. Laub, J. Chromatogr. *173*, 388 (1979).
83. A. Tóth and E. Zala, J. Chromatogr. *298*, 381 (1984).
84. J. A. Garcia Dominguez, J. Garcia Muñoz, E. Fernández Sánchez, and M. J. Molera, J. Chromatogr. Sci. *15*, 520 (1977).
85. A. Tóth and E. Zala, J. Chromatogr. *284*, 53 (1984).
86. A. Grobler and G. Balizs, J. Chromatogr. Sci. *12*, 57 (1974).
87. L. Ambrus, J. Chromatogr. *294*, 328 (1984).
88. X. Guardino, J. Albaigés, G. Firpo, R. Rodriguez-Viñals, and M. Gassiot, J. Chromatogr. *118*, 13 (1976).
89. J. A. Nelder and R. Mead, Comput. J. 7, 308 (1965).
90. F. J. Heeg, R. Zinburg, H. J. Neu, and K. Ballschmitter, Chromatographia *12*, 451 (1979).
91. L. S. Ettre, Chromatographia 7, 39 (1974).

92. A. Wehrli and E. sz. Kováts, Helv. Chim. Acta 42, 2709 (1959).
93. E. sz. Kováts, Z. Anal. Chem. 181, 351 (1961).
94. E. Kováts, Helv. Chim. Acta 46, 2705 (1963).
95. G. I. Spiakovskii, A. I. Tishchenko, I. I. Zaslavskii, and N. S. Wulfson, J. Chromatogr. 144, 1 (1977).
96. L. Rohrschneider, J. Chromatogr. 22, 6 (1966).
97. W. O. McReynolds, J. Chromatogr. Sci. 8, 685 (1970).
98. L. S. Ettre, Chromatographia 7, 261 (1974).
99. Zs. Szentirmai, G. Tarján, and J. Takács, J. Chromatogr. 73, 11 (1972).
100. Zs. Szentirmai, G. Tarján, L. Békési, J. Gajári, and J. M. Takács, J. Chromatogr. 119, 333 (1976).
101. V. Tekler and J. M. Takács, J. Chromatogr. 202, 179 (1980).
102. F. Patte, M. Etcheto, and P. Laffort, Anal. Chem. 54, 2239 (1982).
103. R. Fellous, D. Lafaye de Micheaux, L. Lizzani-Cuvelier, and R. Luft, Anal. Chim. Acta 154, 191 (1983).
104. R. Fellous, L. Lizzani-Cuvelier, and R. Luft, Anal. Chim. Acta 174, 53 (1985).
105. L. B. Kier, Molecular Orbital Theory in Drug Research, Academic Press, New York, 1971.
106. L. B. Kier and L. H. Hall, Molecular Connectivity in Chemistry and Drug Research, Academic Press, New York, 1976.
107. A. T. Balaban, A. Chiriac, I. Motoc, and Z. Simon, Steric Fit in Quantitative Structure-Activity Relationships, in Lecture Notes in Chemistry, No. 15, Springer Verlag, Berlin, 1980.
108. R. A. Keller, B. L. Karger, and L. R. Synder, in Gas Chromatography 1970 (R. Stock, ed.), Institute of Petroleum, London, 1971, p. 125.
109. H. Lamparczyk and A. Radecki, Chromatographia 18, 615 (1984).
110. H. Lamparczyk, Chromatographia 20, 283 (1985).
111. B. L. Karger, L. R. Snyder, and C. Eon, Anal. Chem. 50, 2126 (1978).
112. B. L. Karger, L. R. Snyder, and C. Eon, J. Chromatogr. 125, 71 (1976).
113. F. Saura-Calixto and A. Garcia-Raso, Chromatographia 15, 521 (1982).
114. J. Bermejo and M. D. Guillen, Chromatographia 17, 644 (1983).
115. J. Bermejo and M. D. Guillen, J. High Resolut. Chromatogr. Chromatogr. Commun. 7, 191 (1984).
116. M. Charton, in Topics in Current Chemistry (F. L. Boschke, ed.), Vol. 114, Springer Verlag, Berlin, 1983, p. 107.
117. A. Bondi, J. Phys. Chem. 68, 441 (1964).
118. I. Motoc and A. T. Balaban, Rev. Roum. Chim. 26, 593 (1981).

119. F. Saura-Calixto and A. Garcia-Raso, Chromatographia *14*, 596 (1981).

120. F. Saura-Calixto and A. Garcia-Raso, Chromatographia *15*, 771 (1982).

121. J. Kuśmierz, E. Maliński, W. Czerwiec, and J. Szafranek, J. Chromatogr. *331*, 219 (1985).

122. J. Macek and E. Smolková-Keulemansová, J. Chromatogr. *333*, 309 (1985).

123. K. J. Miller and J. A. Savchik, J. Am. Chem. Soc. *101*, 7206 (1979).

124. J. Bermejo and M. D. Guillen, J. Chromatogr. *318*, 188 (1985).

125. M. Gassiot-Matas and G. Firpo-Pamies, J. Chromatogr. *187*, 1 (1980).

126. R. P. W. Scott, J. Chromatogr. *122*, 35 (1976).

127. L. Buydens, D. L. Massart, and P. Geerlings, Anal. Chem. *55*, 738 (1983).

128. K. Osmialowski, J. Halkiewicz, A. Radecki, and R. Kaliszan, J. Chromatogr. *346*, 53 (1985).

129. J. A. Pople, D. P. Santry, and G. A. Segal, J. Chem. Phys. *43*, 5129 (1965).

130. C. J. Roothaan, Rev. Mod. Phys. *23*, 69 (1951).

131. F. Saura-Calixto, A. Garcia-Raso, and M. A. Raso, J. Chromatogr. Sci. *22*, 22 (1984).

132. A. Garcia-Raso, F. Saura-Calixto, and M. A. Raso, J. Chromatogr. Sci. *302*, 107 (1984).

133. J. Takács, Zs. Tálas, I. Bernáth, Gy. Czakó, and A. Fischer, J. Chromatogr. *67*, 203 (1972).

134. G. Klopman, J. Am. Chem. Soc. *90*, 223 (1968).

135. Th. Kleinert and W. Ecknig, J. Chromatogr. *315*, 75 (1984).

136. Th. Kleinert, W. Ecknig, and J. Novák, J. Chromatogr. *315*, 85 (1974).

137. L. Buydens, D. Coomans, M. Vanbelle, D. L. Massart, and R. van den Driessche, J. Pharm. Sci. *72*, 1327 (1983).

138. E. Chong, B. de Briceno, G. Miller, and S. Hawkes, Chromatographia *20*, 293 (1985).

139. F. Saura-Calixto, A. Garciá-Raso, and P. M. Déya, J. Chromatogr. Sci. *20*, 7 (1982).

140. F. Saura-Calixto, A. Garcia-Raso, J. Cañellas, and J. Garcia-Raso, J. Chromatogr. Sci. *21*, 267 (1983).

141. J. Bermejo, J. S. Canja, O. M. Gayol, and M. D. Guillen, J. Chromatogr. Sci. *22*, 252 (1984).

142. R. H. Rohrbaugh and P. C. Jurs, Anal. Chem. *57*, 2770 (1985).

143. M. Roth and J. Novák, J. Chromatogr. *258*, 23 (1983).

144. G. Tarján, Á. Kiss, G. Kocsis, S. Mészáros, and J. M. Takács, J. Chromatogr. *119*, 327 (1976).

145. T. J. Betts, J. Chromatogr. *354*, 1 (1986).

146. R. V. Golovnya and T. A. Misharina, J. High Resolut. Chromatogr. Chromatogr. Commun. *3*, 4 and 51 (1980).

147. D. H. Rouvray, in *Chemical Applications of Topology and Graph Theory* (R. B. King, ed.), Elsevier, Amsterdam, 1983, p. 159.

148. A. T. Balaban, I. Motoc, D. Bonchev, and O. Mekenyan, in *Topics in Current Chemistry* (F. L. Boschke, ed.), Vol. 114, Springer Verlag, Berlin, 1983, p. 21.

149. N. Trinajstić, *Chemical Graph Theory*, Vols. 1 and 2, CRC Press, Boca Raton, Fla. 1983.

150. D. Bonchev and N. Trinajstić, J. Chem. Phys. *67*, 4517 (1977).

151. D. Bonchev, Chemometrics Series, Vol. 5, *Information Theoretic Indices for Characterization of Chemical Structures*, Research Studies Press, Wiley, Chichester, U.K., 1983.

152. M. Randić, J. Am. Chem. Soc. *97*, 6609 (1975).

153. M. Randić, J. Chromatogr. *161*, 1 (1978).

154. A. Sabljić, J. Chromatogr. *314*, 1 (1984).

155. A. Sabljić, J. Chromatogr. *319*, 1 (1985).

156. D. Bonchev, Ov. Mekenjan, G. Protic, and N. Trinajstić, J. Chromatogr. *176*, 149 (1979).

157. D. Papazova, N. Dimov, and D. Bonchev, J. Chromatogr. *188*, 297 (1980).

158. J. É. Dubois, D. Laurent, and A. Aranda, J. Chim. Phys. *70*, 1608 (1973).

159. J. É. Dubois, D. Laurent, and A. Aranda, J. Chim. Phys. *70*, 1616 (1973).

160. J. É. Dubois and J. Chrétien, J. Chromatogr. Sci. *12*, 811 (1974).

161. A. T. Balaban, Chem. Phys. Lett. *89*, 399 (1982).

162. F. Saura-Calixto, A. Garcia-Raso, and J. Garcia-Raso, J. Chromatogr. *322*, 35 (1985).

163. D. Bonchev and N. Trinajstić, Int. J. Quantum Chem., Quantum Chem. Symp. *12*, 293 (1978).

164. O. Mekenyan, D. Bonchev, and N. Trinajstić, Int. J. Quantum Chem. *18*, 369 (1980).

165. R. Kaliszan and H. Lamparczyk, J. Chromatogr. Sci. *16*, 246 (1978).

166. F. Morishita, Y. Terashima, M. Ichise, and T. Kojima, J. Chromatogr. Sci. *21*, 209 (1983).

167. N. Dimov and D. Papazova, Chromatographia, *12*, 720 (1979).

168. D. Bonchev, O. Mekenyan, and N. Trinajstić, J. Comp. Chem. *2*, 127 (1981).

169. F. Vernon, J. Chromatogr. *179*, 165 (1979).

170. L. Podmaniczky, L. Szepesy, K. Lakszner, and G. Schomburg, Chromatographia *20*, 591 (1985).

171. L. Podmaniczky, L. Szepesy, K. Lakszner, and G. Schomburg, Chromatographia *20*, 623 (1985).

172. J. R. Ashes and J. K. Haken, J. Chromatogr. *84*, 231 (1973).

173. P. Chovin and J. Lebbe, in *Journées Int. d'Etude des Methodes de Separation Immédiate et de Chromatographie 1961* (J. Tranchant, ed.), G.A.M.S., Paris, 1962, p. 90.

174. J. Takács, M. Rockenbauer, and I. Olácsi, J. Chromatogr. *42*, 19 (1969).

175. L. S. Ettre and K. Billeb, J. Chromatogr. *30*, 1 (1967).

176. R. A. Hively and R. E. Hinton, J. Gas Chromatogr. *6*, 203 (1968).

177. N. Dimov, T. Petkova, and D. Shopov, J. Chromatogr. *74*, 165 (1972).

178. J. A. Rijks and C. A. Cramers, Chromatographia 7, 99 (1974).

179. P. Sandra, F. David, M. Proot, G. Diricks, M. Verstappe, and M. Verzele, J. High Resolut. Chromatogr. Chromatogr. Commun. *8*, 782 (1985).

180. F. Saura Calixto and A. Garcia Raso, J. Chromatogr. *216*, 326 (1981).

181. K. Grob and G. Grob, Chromatographia *17*, 481 (1983).

182. M. Goedert and G. Guiochon, Anal. Chem. *45*, 1180 (1973).

183. M. L. Lee, F. J. Yang, and K. D. Bartle, *Open Tubular Column Gas Chromatography*, Wiley, New York, 1984.

184. G. M. Ogle, S. W. S. McCreadie, and D. F. K. Swan, J. Autom. Chem. 7(3), 130 (1985).

185. F. Rowland, Int. Lab. *12*, (4), 104 (1982).

186. F. Munari and S. Trestianu, J. Chromatogr. *279*, 457 (1983).

187. R. P. W. Scott, Anal. Chem. *35*, 481 (1963).

188. A. Goldup, G. R. Luckhurst, and W. T. Swanton, Nature (London) *193*, 333 (1962).

189. D. H. Desty, A. Goldup, G. R. Luckhurst, and W. T. Swanton, in *Gas Chromatography 1962* (M. van Swaay, ed.), Butterworths, London, 1962, p. 67.

190. L. Soják, J. Janák, and J. A. Rijks, J. Chromatogr. *138*, 119 (1977).

191. R. P. W. Scott, in *Gas Chromatography 1958* (D. H. Desty, ed.), Butterworths, London, 1958, p. 189.

192. R. J. Laub and R. L. Pecsok, *Physicochemical Applications of Gas Chromatography*, Wiley, New York, 1978, p. 30.

193. R. J. Laub, Anal. Chem. *56*, 2115 (1984).

194. J. R. Conder and C. L. Young, *Physicochemical Measurement by Gas Chromatography*, Wiley, Chichester, U.K., 1979.

195. R. E. Kaiser, Inst. Chromatographie, Bad Dürkheim, GFR, personal communication, 1986.

196. V. G. Berezkin, J. Anal. Chem. USSR *40*(3), Part 2, 459 (1985).

197. J. Krupčik, E. Matisová, J. Garaj, L. Soják, and V. G. Berezkin, Chromatographia, *16*, 166 (1982).

198. E. Matisová, J. Krupčik, and J. Garaj, Chromatographia *16*, 169 (1982).

199. R. L. Martin, Anal. Chem. *33*, 347 (1961).

200. R. L. Pecsok and B. H. Gump, J. Phys. Chem. *71*, 2202 (1967).

201. D. E. Martire, in *Progress in Gas Chromatography* (J. H. Purnell, ed.), Interscience, New York, 1968, p. 93.

202. M. J. Owen, Ind. Eng. Chem. Prod. Res. Dev. *19*, 97 (1980).

203. Gy. Vigh, Á. Bartha, and J. Hlavay, J. High Resolut. Chromatogr. Chromatogr. Commun. *4*, 3 (1981).

204. J. Hlavay, Á. Bartha, Gy. Vigh, M. Gazdag, and G. Szepesi, J. Chromatogr. *204*, 59 (1981).

205. E. F. Barry, P. Ferioli, and J. A. Hubball, J. High Resolut. Chromatogr. Chromatogr. Commun. *6*, 172 (1983).

206. J. A. Nieman and L. B. Rogers, Sep. Sci. *10*, 517 (1975).

207. F. J. van Lenten, J. E. Conway, and L. B. Rogers, Sep. Sci. *12*, 1 (1977).

208. E. M. Barrall II, R. Hawkins, A. A. Fukushima, and J. F. Johnson, J. Polym. Sci. Polym. Symp. *71*, 189 (1984).

209. K. Grob and G. Grob, J. High Resolut. Chromatogr. Chromatogr. Commun. *6*, 133 (1983).

210. K. Grob, J. Chromatogr. *299*, 1 (1984).

211. Silicon Compounds (1984), catalog available from Petrarch Systems, Inc., Bristol, PA 19007.

212. H. Heckers, K. Dittmar, F. W. Melcher, and H. O. Kalinowski, J. Chromatogr. *135*, 93 (1977).

213. L. G. Sarto, Jr., Applications of Fusion Reaction Gas Chromatography and Pyrolysis Gas Chromatography to the Characterization of Polysiloxanes and Poly-m-Carboranyl-Siloxanes, Thesis, University of Massachusetts, Amherst, 1982.

214. B. A. Jones, J. C. Kuei, J. S. Bradshaw, and M. L. Lee, J. Chromatogr. *298*, 389 (1984).

215. F. C. Schaefer, in *The Chemistry of the Cyano Group* (Z. Rappoport, ed.), Interscience, London, 1970, p. 258.

216. K. Grob, J. High Resolut. Chromatogr. Chromatogr. Commun. *1*, 93 (1978).

217. K. Grob and G. Grob, J. High Resolut. Chromatogr. Chromatogr. Commun. *1*, 221 (1978).

218. A. Venema, L. G. J. v.d. Ven, and H. v.d. Steege, J. High Resolut. Chromatogr. Chromatogr. Commun. *2*, 69 (1979).

219. Yu. A. Yuzhelevskii, E. V. Kogan, A. L. Klebanskii, and O. N. Larionova, J. Gen. Chem. USSR *34*, 2838 (1964).

220. L. G. Blomberg, J. High Resolut. Chromatogr. Chromatogr. Commun. *5*, 520 (1982).

221. L. G. Blomberg, J. High Resolut. Chromatogr. Chromatogr. Commun. *7*, 232 (1984).

222. B. W. Wright, P. A. Peaden, M. L. Lee, and T. J. Stark, J. Chromatogr. *248*, 17 (1982).

223. R. C. M. de Nijs, Chrompack, Middelburg, Netherlands, personal communication, 1984.
224. C. Madani, E. M. Chambaz, M. Rigaud, J. Durand, and P. Chebroux, J. Chromatogr. *126*, 161 (1976).
225. L. Blomberg and T. Wännman, J. Chromatogr. *168*, 81 (1979).
226. L. Blomberg and T. Wännman, J. Chromatogr. *189*, 159 (1979).
227. M. Verzele, F. David, M. van Roelenbosch, G. Diricks, and P. Sandra, J. Chromatogr. *279*, 99 (1983).
228. K. Grob and G. Grob, J. Chromatogr. *347*, 351 (1985).
229. W. Blum, J. High Resolut. Chromatogr. Chromatogr. Commun. *8*, 718 (1985).
230. S. R. Lipsky, in *The Science of Chromatography* (F. Bruner, ed. Elsevier, Amsterdam, 1985, p. 257.
231. G. Rutten, A. van de Ven, J. de Haan, L. van de Ven, and J. Rijks, J. High Resolut. Chromatogr. Chromatogr. Commun. 7, 607 (1984).
232. G. Rutten, J. de Haan, L. van de Ven, A. van de Ven, H. van Cruchten, and J. Rijks, J. High Resolut. Chromatogr. Chromatogr. Commun. *8*, 664 (1985).
233. H. A. Schroeder, Polym. Prep. *13*, 764 (1972).
234. K. Grob and G. Grob, J. High Resolut. Chromatogr. Chromatogr. Commun. 5, 349 (1982).
235. S. Hoffmann, L. Blomberg, J. Buijten, K. Markides, and T. Wännman, J. Chromatogr. *302*, 95 (1984).
236. S. Schmidt, S. Hoffmann, and L. G. Blomberg, J. High Resolut. Chromatogr. Chromatogr. Commun. *8*, 734 (1985).
237. D. R. Deans, Anal. Chem. *43*, 2026 (1971).
238. R. E. Kaiser, L. Lemig, L. Blomberg, and R. I. Rieder, J. High Resolut, Chromatogr. Chromatogr. Commun. *8*, 93 (1985).
239. R. E. Kaiser, R. I. Rieder, L. Leming, L. Blomberg, and P. Kusz, J. High Resolut. Chromatogr. Chromatogr. Commun. *8*, 580 (1985).
240. R. E. Kaiser and R. I. Rieder, Labor Praxis, *10*(3), 184, 186, 188 (1986).
241. J. V. Hinshaw, Jr. and L. S. Ettre, Chromatographia *21*, 561 and 669 (1986).
242. J. K. G. Kramer, R. C. Fouchard, and K. J. Jenkins, J. Chromatogr. Sci., *23*, 54 (1985).
243. S. E. Berger, U.S. Patent 4,136,080 (1979).
244. S. E. Berger, U.S. Patent 4,151,154 (1979).
245. J. Buijten, L. Blomberg, K. Markides, and T. Wännman, J. Chromatogr. *268*, 387 (1983).
246. F. Vernon and J. B. Suratman, Chromatographia *17*, 597 (1983).
247. J. K. Haken, J. D. Openshaw, W. E. Sharples, B. G. Madden, and F. Vernon, J. Chromatogr. *267*, 1 (1983).

248. L. G. Blomberg, Trends Anal. Chem., in preparation.
249. M. F. Mehran, W. J. Cooper, R. Lautamo, R. R. Freeman, and W. Jennings, J. High Resolut. Chromatogr. Chromatogr. Commun. *8*, 715 (1985).
250. J. A. Yancey, J. Chromatogr. Sci. *24*, 117 (1986).

7

Retention Indices in Reversed-
Phase HPLC

Roger M. Smith   *Loughborough University of Technology,
Loughborough, England*

I. INTRODUCTION

High-performance liquid chromatography (HPLC) is now a widely adopted technique with a multiplicity of applications. Its suitability for quantitative analyses have been confirmed by both inter- and intralaboratory comparisons. However, the reproducibility of the retention times determined in different laboratories or on different instruments can be very poor, and if literature methods are repeated the retention times often differ from the published values. These problems have so far severely limited the application of HPLC for qualitative analyses as it is necessary for each chromatograph and column to be calibrated with all the standards of interest. One cause of these difficulties is that retentions in HPLC are very sensitive and can be altered by even small changes in the stationary or mobile phases or in the operating conditions, although this sensitivity also provides the versatility that has made HPLC such a popular method for a wide range of samples.

   Similar poor interlaboratory reproducibility was also faced in gas-liquid chromatography (GLC), and it was realized that as many of the problems arose from factors that would alter the retentions of all analytes, i.e., eluent flow rate and proportion of liquid phase, relative methods of recording retentions would give more reproducible results. This led to the development of a number of methods for recording retentions based on the comparison of analyte retentions with individual standard compounds or with series of standards. Of these, only the Kováts retention index scale [1] has been widely adopted. This method expresses the retentions of the analytes as interpolated retention index (RI) values between n-alkanes chromatographed under the same conditions. Basing the method on a series of homologs has the advantage that a wide range of analyte retentions can be determined by using a single set of standard compounds. The value of this scale is reflected in numerous reviews of its application [2–4], the latest of which, celebrating 25 years of use, contains 1380 references [4]. The indices have been employed in many areas, from structure determination and analyte identification to the physicochemical basis of gas chromatography and as a reference scale for test compounds used to determine the selectivity of stationary phases. In recent years improvements in the reproducibility of stationary phases have prompted the publication of extensive compilations and data bases of the retention index values of a wide range of compounds [5], including drugs of therapeutic and forensic interest measured on selected phases [6–8]. These can be used for identification purposes by either manual or computer-based comparisons with experimental values.

Alternative series of retention standards have been proposed for specialized applications, such as the methyl esters [9] or alkanones [10,11] for polar stationary phases, or homologous series containing elements that can be detected by specific detectors such as the bromoalkanes [12] or *n*-alkyl trichloroacetates [13] for the electron capture detector. In a review of the nomenclature of chromatography, Ettre [14] has recommended the term "retention indices" should be reserved for indices based on the *n*-alkanes calculated according to the method of Kováts and that to avoid confusion alternative or modified terms should be used for index values from other systems.

The absence so far of a widely used corresponding retention index method that could be used for the identification of drugs by HPLC was noted by de Zeeuw [15]: "A generally acceptable method to express retention in a standardized form does not yet exist." He also noted that relative retention times were not useful in HPLC because of the irreproducibility of stationary phase materials.

This chapter discusses the developments so far in the provision of reference methods for reporting retentions in reversed-phase separations in HPLC and their applications. Although individual internal standards are widely used for specific assays, the chapter will concentrate on proposals for retention indices analogous to the systems used in GLC, but will also describe suggestions that individual members of sets of standards could be used to determine relative retentions over a range of elution conditions. Most of this work has appeared since 1979, reflecting the fact that until that time column materials and instrumentation were often insufficiently reproducible for accurate measurements.

II. RELATIVE RETENTION METHODS IN CHROMATOGRAPHY

In HPLC direct experimental retention times or retention volumes are easy to measure but these values are very dependent on the experimental conditions, including the nature of the stationary phase, composition of the mobile phase, flow rate of the mobile phase, internal dimensions of the column and connecting tubing, and column temperature as well as the relative proportions of the stationary and mobile phases. It is difficult to reproduce these experimental conditions exactly in different laboratories or even on different columns or instruments in the same laboratory.

The conventional method of recording retentions has been as capacity factors (k') calculated as the adjusted retention time of the analyte ($t_R - t_0$) divided by the retention time of an unretained compound (t_0, often referred to as the column void volume or column dead time):

$$k' = \frac{t_R - t_0}{t_0} \tag{1}$$

In theory, this relative retention method should compensate for changes in the eluent flow rate and column dimensions. However, the value of t_0 is small and even very small absolute errors have a large effect on the results. In practice, the accurate measurement of t_0 is difficult and interlaboratory reproducibility is poor. This is partly because the measured values often also include any extra-column dead volume of the instrument in the detector, connecting tubing, and injector, which will differ between experimental systems. More important, there is no agreement on a standard method for the measurement of the void volume and frequently the method used is not reported. Many methods have been proposed based on either different "unretained" analytes or physical measurement of the volume of the column, but these methods often given different results for the same column [16,17]. Of particular interest in the present context, it has been proposed that the linearization of the retention times of homologous series can be used, as the zeroth homolog should be eluted at t_0 [18,19]; however, Kaiser and Rackstraw [20] have suggested that in many cases the retention data may be insufficiently accurate for reliable calculations. The true value of the column void volume may well be different for different analytes, as their ability to gain access to various parts of the column will differ because of size exclusion effects [17]. A simple void volume marker sample may therefore never be suitable and in practice a compromise method may be used, but the method used should always be reported or defined.

Because of the problems with capacity factors, in recent years there has been considerable interest in alternative methods for standardizing the recording of retention measurements in HPLC so that results from different laboratories can be compared or can be related to other physical properties. The first approach was to use relative retention times or relative capacity factors (equivalent to relative adjusted retention times) in which the analyte is compared to a reference compound with a similar retention. These values should be largely independent of the factors that have a proportional effect on both analyte and standards, such as flow rate and column dimensions, but they cannot compensate for differences in the selectivity of the system caused by differences in temperature or composition of the stationary and mobile phases. In practice, problems arise because individual laboratories may select different standard compounds for the assay of the same analyte, thus preventing comparison.

In order to obtain comparability between separations it has been proposed that the relative reference standards for all assays should be selected from a limited set of compounds such as a homologous series (Sec. III). Different members of such a set of universal reference compounds could cover a wide range of retentions.

Of wider application, and forming the major part of this chapter, is the use of homologous or pseudohomologous series of compounds as the basis of retention index scales in a similar manner to the use of retention index scales in GLC (Sec. IV). Retention indices have the advantage that one set of standards can be used to provide a single continuous scale over a wide range of separation conditions and that the results from separations under similar but different conditions should be comparable.

III. UNIVERSAL REFERENCE STANDARDS IN HPLC

In an attempt to reduce the number of different compounds that are used as internal standards, it has been suggested that the standards should always be taken from a single homologous series and the member of the series with the most appropriate retention should be selected. Kikta and Stange [21] proposed that the alkylaryl ketones (phenones) could be used in this way for both normal and reversed-phase separations as they are reasonably stable to most common eluents and cover a wide range of elution conditions. Previously, Kikta and Grushka [22] had shown that their retention increases systematically with carbon number in a reversed-phase separation. However, these compounds have not been widely used, although an example of the selection of myristophenone as an internal standard for separations on cyano-bonded silica and ODS-silica columns has been noted [23]. Sets of alkylaryl ketones from acetophenone to octadecaphenone for use as internal standards have now also become commercially available [24,25].

More recently, Verzele et al. [26] proposed that anilides of fatty acids could be versatile internal standards, although the interest was primarily in quantitative rather than qualitative analyses. By using anilines with different substituents on the aromatic ring to prepare the amides, a series of standards can be prepared with adsorption chromophores similar to the analyte of interest. Thus both analyte and standard could readily be detected at the same wavelength, improving selectivity. The different substituents also would enable the internal standard with the optimum retention compared to the components of the sample to be selected.

Experimentally, the easiest relative measurements to determine are relative retention times, as these can often be directly calculated by most electronic integrators. However, the thermodynamically more correct values are relative adjusted retention times (or relative capacity factors), as they should be directly related to the relative distribution coefficients of the two compounds in the separation. Ettre [27] has compared the discrepancies between these two methods and, as expected, they are most significant at low retention times.

It is also difficult to obtain reproducible retention results in thin-layer chromatography (TLC) because of differences in the extent of elution and the retention activity of the plates. In this case multiple internal standard compounds have been used to give corrected retentions (R_f values), which have been used for the identification of drugs of forensic interest [28]. A similar concept has also been applied to the reversed-phase separation of the barbiturates [29]. Although this method required additional calculations compared to other methods of recording retention, it gave better discrimination and more positive identification than relative measurements based on a single internal standard or retention indices, probably because of the close similarity between the standards and the analytes.

IV. RETENTION INDEX SYSTEMS IN HPLC

The Kováts retention index value of an analyte in GLC is calculated by comparing the adjusted retention time (or volume) of the analyte to the corresponding retentions of the homologous n-alkanes that elute just before and after the analyte [Eq. (2)] [1]. The result is expressed as an interpolated value on a scale in which the retention indices of the n-alkanes are 100 times their carbon number (C_n) (i.e., hexane = 600).

$$\text{Retention index} = 100 \times n + \frac{100(\log t'_R - \log t'_{Rn})}{\log t'_{Rn+1} - \log t'_{Rn}} \qquad (2)$$

where t'_R, t'_{Rn}, and t'_{Rn+1} are, respectively, the adjusted retention times of the analyte, the n-alkane with the next shorter retention containing n carbons, and the next higher alkane with n + 1 carbons.

Alternatively, the index value can be derived from a semilog graphical plot or, if the relationship between the carbon number and $\log t'_{Rn}$ is linear, by using a linear correlation on a mainframe or microcomputer.

From the earliest studies in reversed-phase HPLC, it has been recognized that in this method there is also a linear relationship between the adjusted retention times of the members of a homologous series and their carbon numbers:

$$\log k' = A \times C_n + B \qquad (3)$$

However, because even in many reversed-phase separations specific interactions can occur between functional groups in the analyte and uncapped silanols on the column packing, concern has been expressed that it might be necessary to remove these silanophilic interactions in order to achieve a truly systematic behavior before it would be possible to establish a universal retention index scale [30]. This reflects the concern expressed earlier that the irreproducibility of the stationary phase was a problem in qualitative analyses [15].

This linear relationship has been examined in detail by a number of workers for a wide range of compounds. Typical studies have included examinations of alkylbenzenes, alkyl bromides, and alkyl disulfides on nine different stationary phases with a wide range of eluents [31]. Other workers have examined homologous series of aliphatic alkanes, alkanols, benzyl ethers, esters, aromatic amides, and ketones. In some cases the first few members of the homologous series gave anomolous retentions [32]. Vigh and Varga-Puchony [33] demonstrated that the linear relationship between carbon numbers and retentions for n-alkanols, alkanals, and alkan-2-one dinitrophenyl-hydrazones remained valid over a wide range of temperature.

Melander et al. [34], in a similar study with alkylbenzenes suggested the possibility of establishing a hydrophobic retention index system for the characterization of biological substances and prediction of properties in quantitative structure-activity relationships (QSAR) studies. In further work they observed that the methylene group selectivity was largely independent of the make of the packing material so that relative changes in retention of analytes even on different columns would reflect interaction differences [35]. They also suggested that their observations could lead to the introduction of an indexing systems to HPLC for identification and Qsar studies of biologically active compounds.

The most comprehensive study of the relationship between carbon number and retention has been carried out by Colin and Guiochon and co-workers. They initially demonstrated the linearity of the relationship for series of n-alkanes, n-alkylbenzenes, and fatty acid methyl esters on pyrocarbon columns [36]. In a series of papers, they extended the work to ODS-bonded silicas, including an examination of the effect of solvent composition [37]. However, precise work showed that the methylene selectivity ($\alpha = k'_{n+1}/k'_n$) for a range of different homologous series—n-alkanes, n-alkyl chlorides, n-alkanols, alkan-2-ones, and methyl esters of fatty acids—changed by small but significant amounts at a particular length of the carbon chain of the analyte [38] (Fig. 1). They suggested that this occurred when the length of the analyte chain (C_{15-16} for ODS-silica) was similar to the length of the bonded alkyl chains on the silica surface. It therefore

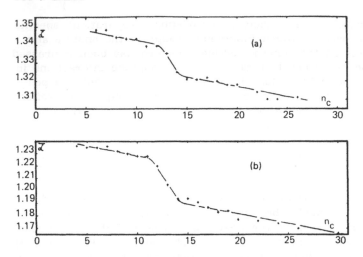

Fig. 1 Mean selectivity $\log k'$ versus number of carbon atoms on C_{14} bonded phase: (a) 90:10 methanol-water; (b) pure methanol. (From Ref. 38.)

corresponded to the point at which the analyte could not completely enter the stationary phase. However, this effect is small and on plots of $\log k'$ against carbon number cannot be easily observed.

They also confirmed earlier studies by Berendsen and co-workers [18], who found that an accurate determination of the void volume was needed to obtain a truly linear relationship. Berendsen had also suggested that linearization of the unadjusted retention times of a homologous series of compounds provided the most accurate determination of the column void volume, but Wainwright et al. [19] found that this method was unreliable.

This work has led to the introduction of a number of different homologous series of compounds as the basis of retention index scales for HPLC. In most cases the indices of analytes were calculated in a similar manner to the Kováts indices but in some papers other calculation methods have been used.

A. *n*-Alkanes

By analogy with GLC, the first choice for a fundamental retention index scale would be the *n*-alkanes. These were often used in the early days of HPLC, almost without comment, as a direct extrapolation of their application in GLC [39,40]. However, although they are free from specific interactions with the stationary phase, these standards are not particularly suitable for most HPLC applications because they lack a chromophore and can only be detected with the relatively insensitive refractive index detector. In addition, their low polarity means that on most reversed-phase chromatographic systems they are strongly retained and would be eluted much later than many analytes of interest, even with relatively nonpolar samples such as fatty acid esters [41].

More recently, they have been applied in the study of column interactions because they are not susceptible to specific interactions with the stationary phase [42,43]. In practice, for the majority of their study, Morishita and co-workers [42] used the more readily detected *n*-alkylbenzenes as marker compounds with predetermined retention indices based on the alkane scale (typical retention index values: benzene, 269; toluene, 342; ethylbenzene, 407; up to 1-phenylnonane, 1063).

B. Alkylbenzenes

The alkylbenzenes have the advantage of being more readily detected than the *n*-alkanes, but they are still very nonpolar. As noted above, they have been used as marker compounds for the *n*-alkane scale [42]. In work closely related to some of the studies using retention indices, Jandera [e.g., 44,45] used a series of homologous *n*-alkylbenzenes as standards for the derivation of interaction indices but did not calculate retention indices.

C. Alkan-2-ones

As part of a series of studies of drug identification, Baker et al. [46] investigated the use of relative retention values and dual-wavelength ultraviolet detection. This led Baker and Ma [47] to propose the use of the alkan-2-ones as the basis of a liquid chromatography retention index scale. The calculation of the retention indices was based on Eq. (2), using the ketones instead of the alkanes. They recognized the limitations of the *n*-alkanes and showed that the alkan-2-ones ranging from acetone (RI = 300) to nonadecan-2-one (RI = 1900) were detectable by ultraviolet spectroscopy. They demonstrated a linear relationship between the carbon numbers and log k' over a wide range of methanol—pH 7.0 phosphate buffers (Fig. 2) and acetonitrile—pH 7.0 phosphate buffer eluents, using a μ-Bondapak C_{18} column. Similar results but over a more limited range were also obtained for a cyano-bonded column. The retention indices of a series of test compounds were virtually constant over the range 20 to 90% methanol on the ODS column (Fig. 3), but different values were obtained on changing to the eluents containing acetonitrile or to the cyano-silica column (i.e., for phenacetin on the ODS-silica with methanol-buffer, RI = 530, and with acetonitrile-buffer,

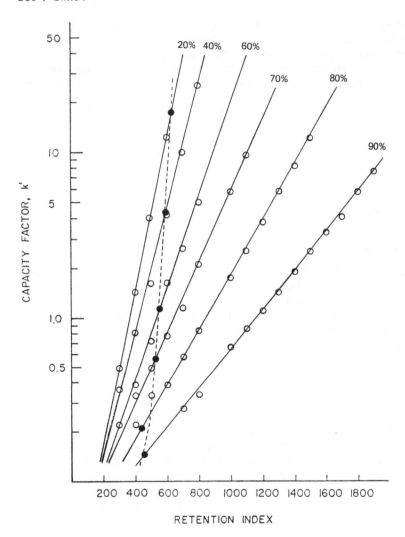

Fig. 2 Effect of solvent composition on capacity factor: (○) 2-keto-
alkanes; (●) phenacetin. μBondapak C_{18} column, pH 7.0, 0.025 M
NaH_2PO_4 aqueous buffer with varying methanol content. (From Ref.
47.)

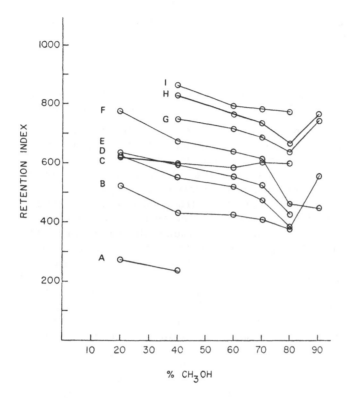

Fig. 3 Variability of the retention indices of selected drugs: (A) aspirin; (B) caffeine; C, acetophenone; D, phenobarbital; E, phenacetin; F, chlordiazepoxide; G, methaqualone; I, androsterone. μBondapak C_{18} column, pH 7.0 0.025 M NaH_2PO_4 aqueous buffer with varying methanol content. (From Ref. 47.)

RI = 526; and on the cyano column with methanol-buffer, RI = 679; and with acetonitrile-buffer, RI = 591).

These results demonstrated that the retention indices were much more robust than capacity factors to changes in the proportion of the organic modifier in the eluent but were sensitive to changes in the selectivity. Subsequently, numerous studies have used these standards and kits of the standard compounds are now commercially available [48]. They have been used primarily to predict retentions and in investigations of drug properties and QSAR studies.

A single report has suggested that instead of the alkan-2-ones, their corresponding 2,4-dinitrophenylhydrazones might be more suitable retention index standards because of their strong chromophores and hence ready detection [49]. However, no reports of their applications have appeared.

D. Alkylaryl Ketones

Because it was felt that the chromophores of the alkan-2-ones (e.g., acetone: λ_{max} = 265 nm, log ε = 1.2 mol^{-1} cm^{-1}) were rather weak and therefore relatively concentrated solutions of standards were needed for ready detection, alternative homologous series were examined as possible retention index standards [50].

A set of criteria were used to guide the selection [50]. To ensure detectability, the compounds in any alternative homologous series of standards must be aromatic or include a strong chromophore. The least retained member of the series should be reasonably polar so that the retentions of the standards will bracket the retention of as many analytes of interest as possible, including common drugs and natural products. The standards also must be commercially available so that they can be readily adopted by different laboratories. These criteria excluded many amides and esters, as often only two or three homologous were available. Because mobile phases in HPLC may be reactive or may be held at a pH between 3 and 8, the standards should be chemically stable in common mobile phase components and should not contain any groups that would be ionized by buffers or suffer specific interactions on the silica surface.

From these studies Smith [50] proposed that the homologous alkylaryl ketones (phenones) (e.g., acetophenone: λ_{max} = 247,279 nm, log ε = 4.1, 3.1 mol^{-1} cm^{-1}) might be suitable standards. These compounds were proposed previously as universal reference standards (see above) and were also briefly examined in other laboratories [51]. A reasonable range of compounds from acetophenone (RI = 800) to octanophenone (RI = 1400) were readily available [50]. More recently, higher homologs up to octadecaphenone have also been marketed specifically for HPLC calibration [24,25], following the proposals of Kita and Stang [21].

There was a linear relationship between the carbon numbers of the ketones and log k' on ODS-silica for a wide range of methanol-water eluent compositions (Fig. 4) [50]. Subsequently, a similar linear relationship was demonstrated for their separation in binary and ternary eluents containing acetonitrile and tetrahydrofuran (THF) [52—54].

As with the alkan-2-ones, the retention indices of a number of test samples based on the alkylaryl ketone scale were virtually constant over a range of eluent compositions (Fig. 5) [50]. Although acetophenone is less polar and more highly retained than the C_3 to C_6 alkan-2-ones, the two ketone scales overlap and in many cases either

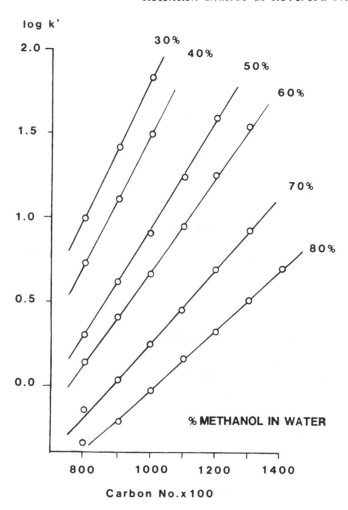

Fig. 4   Capacity factors for alkylaryl ketones compared with retention indices (carbon number × 100), using different percentages of
methanol as eluent.   (From Ref. 50.)

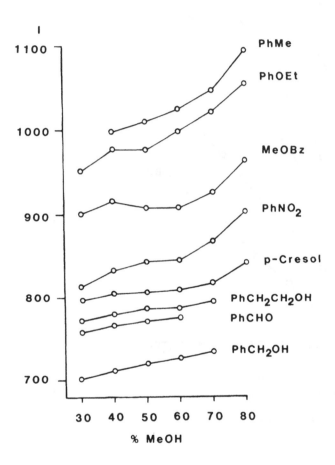

Fig. 5 Variation of retention index values of test compounds with eluent composition. Me, Methyl; Et, ethyl; Ph, phenyl; Bz, PhCO. (From Ref. 50.)

could be used. However, the methylene increments for the two
scales differ and the slopes of their linear relationship are not parallel [50]. For some polar analytes with retention indices less than 800,
the linear alkylaryl ketone scale has to be extrapolated.

During these studies it was demonstrated that the retention indices
determined with this scale were virtually independent of the exact
value measured for the column void volume. Altering the experimental value by ±15% caused large changes in the capacity factors of
analytes but not change in the retention indices unless the retention
times were very short [52]. They therefore offered a method for recording retention values that should be largely independent of the
accurate measurement of the column void volume.

The alkylaryl ketone retention index scale has been used primarily
as the basis of analyte identification in forensic and drug studies
and for the comparison of column/eluent selectivity. Studies were
carried out using C_2, C_{22}, and phenyl-bonded silica columns [55,56]
and polystyrene-divinylbenzene column materials [57,58]. Usually,
on this last material, acetophenone was consistently eluted more
rapidly than expected from a linear relationship between carbon number and log capacity factor.

E. Esters

Vonach and Schomburg [39] used the methyl esters of *n*-fatty acids
as a retention index scale to compare the effect of silver ions on unsaturated fatty acids in a similar manner to ester scales used in GLC.
In a study of the separation of phospholipids and triglycerides,
Compton and Purdy [59] derived retention indices based on the total
acyl portion of triglycerides to correlate structure with retention.
Thus the triglyceride of dodecanoic acid had an index value of 3600
and that of hexanoic acid had an index value of 1800 [59]. However,
this method has not been taken up generally for lipid studies.

Antle et al. [60] used a retention index-type scale based on the
carbon number of the parabens (*p*-hydroxybenzoate esters) to compare the retention properties of different columns. They worked with
a gradient elution of methanol-water or acetonitrile-water and used
methylparaben = 8 and butylparaben = 11 as normalizing compounds
to determine the retention indices.

A different system was used by Wilken [61], who devised a retention scale based on the dialkyl phthalates. Analytes were assigned
"retention indices" derived from the carbon number of the alkyl
group times 10 (i.e., dimethyl phthalate = 10 and diethyl phthalate =
20).

A retention index scale based on (α, ω-phenylcarboxylate esters
(i.e., methyl 3-phenylpropionate) has been studied by Simpson [62]
to examine the effect of modifiers in the mobile phase, but so far
no details are available.

Other acid derivatives such as the N-alkylbenzamides or acyl
anilides might make suitable series of standards, but few of these
compounds are available commercially. Wells and Clark [63] showed
a linear relationship for the N-alkylbenzamides, although the first
members of the series were anomolous, and Dufek [64,65] has examined the separation of a series of homologous N-alkylphthalimides
in a range of solvents. In overall polarity these aromatic amides are
similar to the alkylaryl ketones, but so far none has been used to
derive retention indices.

F. Polycyclic Hydrocarbons

In early studies with aromatic hydrocarbons, Popl et al. [66] proposed that a retention scale could be developed for HPLC on alumina
using polycyclic hydrocarbons as standards. On the basis of experimental adsorption energies, they defined the following series of
standards and corresponding index values: benzene, I = 10; naphthalene, I = 100; phenanthrene, I = 1000; and benz[a]anthracene,
I = 10,000. This was subsequently extended to benzo[b]chrysene,
I = 100,000 [67]. In each case log I corresponds to the number of
aromatic rings. A closely related scale has been used by Karlesky
and co-workers [68,69]: benzene, 10, naphthalene, 100; anthracene,
1000; 1,2-benzathracene, 10,000; and 1,2,3,4-dibenzanthracene,
100,000, and they noted that the differences were not a serious
problem in comparisons.

Although these scales are not based on a homologous series of
standards or a direct carbon number relationship and are thus not
true retention index scales, they have been repeatedly referred to
as such in the literature and are sufficiently closely related in concept to be included in this review.

Using these scales, the retention index (as log I) of an analyte is
calculated in an analogous manner to the Kováts indices by interpolation between standards from Eq. (4):

$$\log I_x = \log I_n + \frac{\log R_x - \log R_n}{\log R_{n+1} - \log R_n} \tag{4}$$

where $\log I_x$ represents the analyte, $\log I_n$ is the index value of the
standard eluting immediately before the analyte, and $\log R_x$, $\log R_n$,
and $\log R_{n+1}$, respectively, are the elution volumes (or retention
times) of the analyte and the standards containing n and n + 1 rings
that are eluted immediately before and after the analyte.

As the standards only represent a pseudoscale, it is not possible
to use a least-squares correlation, although graphical interpolation
methods can be used. For compounds that elute before benzene an

extrapolation method is used based on the first two standards. The first of these methods was subsequently extended to silica columns and to reversed-phase separations [67] and has formed the basis of extensive studies of the aromatic hydrocarbons on amino- and C_{18}-bonded silica columns [70].

A similar scale has been proposed by Lee and co-workers [71,72] for use in GLC, with benzene, naphthalene, phenanthrene, chrysene, and picene as the standards.

G. Misuse of Retention Indices

The current literature is often confused because of the misuse of the term retention indices in cases where the values are not relative measurements compared to a scale of closely related compounds. Often, retention index has been used as a synonym for retention capacity factor [73–76]. It has also been used instead of interaction index, although in this case it was only used in the title [77]. In a recent paper retention indices have been defined as linear interpolated values between the unrelated compounds 2-ethylhexyl phthalate and benzoic acid [78].

V. APPLICATIONS OF RETENTION INDICES IN HPLC

The ability to report and record reproducible retention values that are largely or wholly independent of the source of the data or of small differences in the chromatographic conditions has a number of applications. The primary role is as a method for identification in which the retentions of analytes measured in one laboratory can be compared with results in another laboratory or with a data base of retention values. This leads naturally to attempts to predict retentions based on an incremental approach, in which the different parts of the analyte molecule contribute to the overall retention. This also has a more specialized application as a method for comparing chromatographic behavior with other properties of the analyte, including partition coefficients, and for QSAR studies relating chromatographic retention to biological activity.

The retentions of analytes in a particular column and eluent system can also be used to characterize the properties of the system and enable comparisons of the retentive power and selectivity of different eluents and stationary phases.

The retention indices for a wide range of compounds have now been determined, although a number of different index scales have been used and values in one system may not be directly relatable to results in another (Appendix 1). As might be expected, these compounds are mainly drawn from the major areas of application of HPLC in pharmaceutical, forensic, and food analysis, but a large number of studies have also examined polycyclic aromatic hydrocarbons (PAH) because of their environmental importance.

A. Identification Using Retention Indices

Because retention indices are relative measurements compared to the retention of standard compounds run under the same experimental conditions, they should be more robust than retention times or capacity factors to small differences in the experimental conditions of the separation such as might occur between laboratories or operators. In particular, changes in the flow rate of the eluent, the column dimensions or the design of the instrument and tubing should have a corresponding effect on both analytes and standards. Even differences in the proportion of organic modifier in the eluent, which would have a major effect on the capacity factors, should have little effect on the retention indices as long as the selectivity of the eluent is unchanged. This was confirmed by Baker and Ma [47] and by Smith [50], who found only small differences in the retention indices of test compounds on altering the proportions of the mobile phase components over wide ranges.

This stability led Smith et al. [79—81] to carry out a series of studies of the standardization of retention measurements and to compare the robustness of different methods of recording retention values. These detailed studies examined a number of drug separations of forensic importance. In each case it was of particular importance to identify experimental parameters that would have to be closely defined to obtain reproducible results in different laboratories. The aim was to be able to compile data bases of retention values for drug identification that could be used in a similar manner to the libraries of GLC retention index data.

The retention indices of barbiturates on an ODS-Hypersil column on elution with methanol-aqueous phosphate buffer pH 8.5 (40:60) were virtually independent of small changes in the eluent composition, the column temperature, or the ionic strength of the eluent (Fig. 6) [82]. However, because the barbiturates were partially ionized, the retention indices were susceptible to changes in the pH, and the composition of the buffer had to be closely specified for interlaboratory comparisons. The most important variable in the separation was the make of the stationary phase; different makes had a large effect on the capacity factors and also altered the retention indices [83] but batch-to-batch variations within the ODS-Hypersil had a negligible effect.

In a second study, the HPLC of local anesthetic drugs with an amine-modified eluent of methanol-water-1% aqueous phosphoric acid-n-hexylamine (30:70:100:1.4) was examined [84,85]. The retention indices of the drugs were again much more robust than their capacity factors to small changes in the experimental conditions, and

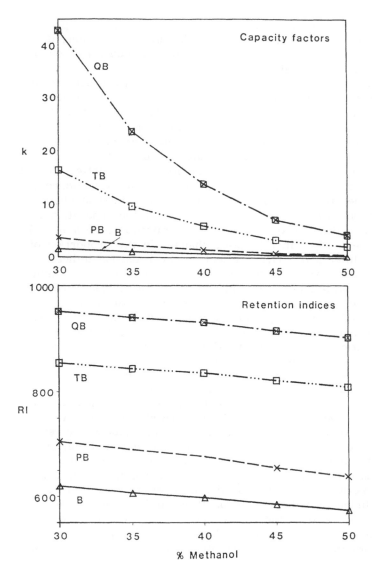

Fig. 6 Variation of capacity factors and retention indices of barbiturates with percentage of methanol. Mobile phase: methanol-
buffer (pH 8.5). B, Barbitone; PB, phenobarbitone; TB, talbutal;
QB, quinalbarbitone. (British Crown copyright. Courtesy of the
Controller of Her Majesty's Stationery Office.)

as long as more than a minimum proportion of the amine was present any changes had little effect on the retention indices. Again, differences in the make of stationary phase had a marked effect on both retention indices and capacity factors.

In both of these studies the group of analytes were closely related and any changes in the conditions would have a similar influence on all the analytes. This was particularly noticeable for the barbiturates when it was found that relative retention times compared to a barbiturate internal standard appeared to be even more robust than the retention indices [29]. The internal standard would also be susceptible to any changes in pH and would therefore provide better compensation than the neutral alkylaryl ketone retention index standards.

To test whether the retention indices were also robust in eluents containing organic modifiers other than methanol, the separation of a group of diuretic drugs with acetonitrile-1% aqueous acetic acid (35: 65) was examined [86]. In this case, changes in the proportion of the organic modifier had a larger effect on the retention indices and the separation was more susceptible to the column temperature and make of packing material. However, the retention indices were still more robust than capacity factors. Because of the diversity of structural types in this assay, no individual analyte would have made a suitable internal standard for direct relative measurements [86].

Further confirmation of the value of retention indices for drug identification has come from two interlaboratory collaborative studies. In the first, Baker et al. [87] distributed seven typical drug compounds to six laboratories, which carried out the separation on their normal make of ODS-silica column but using a common eluent composition. The retention times of the drugs showed very large variations (e.g., androsterone, 758%) although the intralaboratory reproducibility was good. Relative retention times compared to hexan-2-one were better but still very variable [androsterone, relative standard deviation (r.s.d.) 41%]. The retention indices based on the alkan-2-ones were somewhat better with an average r.s.d. of 12.6%, but these included data determined on three different makes of stationary phase. If only the results on μBondapak C_{18} were compared, the variation was reduced to 3.6%, emphasizing the need to specify experimental conditions carefully, in particular the make of packing material.

In the second collaborative study, organized by Gill et al. [29], the separation of the barbiturates, using the method already examined in detail in intralaboratory studies [82,83], was carried out in 10 laboratories, all using columns prepared from a single batch of the same make of packing material. The reproducibility of the retention indices, r.s.d. 0.57 to 1.94%, in this interlaboratory trial was much better than that of the capacity factors, r.s.d. 10.5 to 13.9%. The different methods for recording the retentions were compared by using discrimination numbers (DNs), which were calculated as the number of retention windows, each 2 standard deviations wide, that could be fitted into a defined chromatographic retention range. This represents the maximum number of analytes that could be positively identified within this range. Whereas retention times (DN = 10) and capacity factors (DN = 16) were relatively inefficient, retention indices (DN = 34) and relative adjusted retention times (DN = 44) were much better. The best results were obtained by using a set of internal barbiturate standards to give corrected capacity factors (DN = 64) in a similar manner to the correction of retentions in TLC [28]. As noted earlier, the close structural similarity of the standards and analytes means that in this assay internal standards can also correct for changes in the ionization of the samples, whereas similar factors may not apply in other analyses.

These studies have demonstrated that retention indices can be determined with high reproducibility, if the conditions are carefully controlled. They can largely compensate for differences in the operating conditions in different laboratories, the exact proportions of the components of the eluent, or column temperature. However, retention index values determined on different makes of column packing material will often differ, but, as discussed later, this effect can be used to characterize the stationary phase properties. Clearly, for accurate interlaboratory comparisons separations must be carried out on the same make of stationary phase.

As well as these comparative studies, retention indices have been used for the identification of a wide range of drugs and natural products. As noted earlier, Baker and Ma [47] demonstrated that the retentions based on the alkan-2-one scale of a number of drugs including phenacetin, phenobarbitone, chlordiazepoxide, and androsterone were virtually independent of the composition of the mobile phase. This work was extended to a study of the characterization of drugs used in the management of arthritis [88]. To enhance the identifications and eliminate confusion with possible coeluting compounds, the use of retention indices was combined with UV absorbance ratio measurements of the peaks.

Retention indices based on the alkan-2-ones have also been used by Culberson et al. [89] to standardize the identification of depsides from lichens as part of a chemotaxonomic study.

As well as the detailed studies of drug identification, Smith suggested that retention indices based on the alkylaryl ketones could be used to identify extracts of natural products such as the flavor components in spices, which are often difficult to study by GLC. Retention index values were reported for the gingerols, shogaol, paradol, and sesquiterpenes from ginger and grains of paradise [90,91], eugenol and methyl eugenol from pimento berries [92], and curcuminoids from turmeric powder [93].

Retention indices based on the alkylaryl ketones have also been used to identify mycotoxins in a separation using an acetonitrile-water gradient elution [94]. The retention values were calculated from the retention times of the analyte (T_m) and the standards that elute just before (T_{p1}) and just after the analyte (T_{p2}), using a linear interpolation according to Eq. (5) in which z is the number of carbons in the first standard and Δz the difference in carbon atoms between the standards:

$$RI = \frac{(T_m - T_{p1})\Delta z \times 100}{T_{p2} - T_{p1}} + 100z \qquad (5)$$

This linear rather than logarithmic calibration is similar to the calculation of retention indices in programmed temperature GLC. In a similar application, the long-term reproducibility of the alkylaryl ketones has led to their adoption as a standard scale to characterize peaks in a fingerprint separation of gliadin cereal proteins on a acetonitrile-water gradient (Fig. 7) [95]. This work was carried out as part of a chemotaxonomic procedure to identify varieties of wheat. The retentions were expressed as a "hydrophobicity index" or interpolated carbon numbers based on the length of the ketone side chain (e.g., butyrophenone = 4.0).

One of the main applications of the polycyclic hydrocarbon-based retention indices has been for the identification of polynuclear aromatic hydrocarbons in environmental samples. Following preliminary studies [96], Wise and co-workers reported retention indices for over 80 hydrocarbons on an amino-bonded silica column in normal-phase separations and on two different ODS-silica columns in a reversed-phase separation as part of the development of a sequential characterization procedure [97]. About the same time, a similar study to evaluate the use of different column materials for hydrocarbon separations was carried out by Chmielowiec and George [98]. These methods were used by Murray et al. [99] for the estimation and identification of hydrocarbons in environmental samples and crude oils. This topic has recently been reviewed in detail by Sander and Wise [70], and retention indices for over 150 aromatic hydrocarbons have been reported on mono- and polymeric C_{18}-bonded silica columns.

B. Retention Prediction and Structure-Activity Relationships

The relationship between the structures of analytes and their retention indices has been investigated in a number of laboratories. These studies have taken two main directions, although they often overlap and one study may use both techniques. In the first approach the retention indices of analytes have been predicted by incremental additions to the retention index of a core compound, using structural relationships including connectivity index values, Hansch constants [100], Rekker fragmental constants [101], or experimental octanol-water partition coefficients (log P) values.

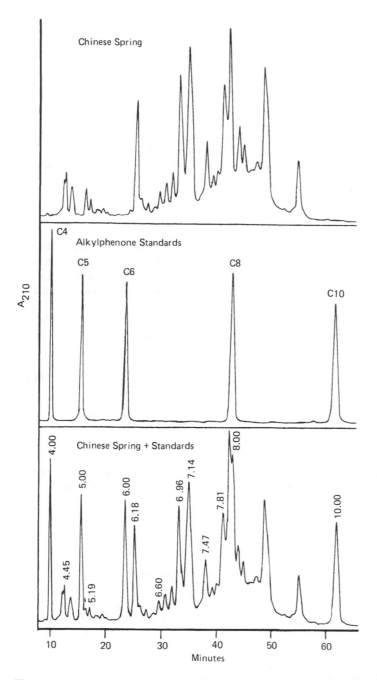

Fig. 7 Reversed-phase HPLC separation of Chinese Spring gliadin, a mixture of alkylphenone standards, and a mixture of the two samples on a SynChronpak RP-P (C_{18}) column at 70°C with a gradient from 25 to 50% CH_3CN(+0.1% TFA) during 55 min with a total run time of 65 min. (From Ref. 95.)

The second approach has correlated experimental retention indices with these structural factors and/or biological activity to examine the effect of structural changes. In both of these studies the retention indices provide a wide and uniform scale for recording retentions, enabling different compounds and methods to be compared. Previously, studies were carried out in many laboratories based on changes in capacity factors, and Kaliszan [102] has comprehensively reviewed the use of chromatography for the examination of quantitative structure-activity relationships.

Retention Prediction and Structure

Once Baker had confirmed that retention indices based on the alkan-2-ones could be used as a reference scale for the retention measurements, he investigated the possibility of predicting retention indices of drugs from their structures or physical properties. Much of this work was recently surveyed [103].

From basic relationships, Baker [104] demonstrated that the retention index of a drug (I_x) could be predicted directly from its calculated or experimental octanol-water partition coefficient:

$$I_x = 200 \log P + 342 \qquad (6)$$

This general relationship between retention index and $\log P$ was examined by comparing values for a group of barbiturates, anthranilic acid analogs, and propranolol analogs [104]. Within each drug group the correlation was good, but the three groups showed different relationships (Fig. 8). It is therefore preferable to calculate the retention index of a core compound ($I_{Reference}$) in conjunction with Hansch π_x values for any structural differences:

$$I_x = 200\pi_x + I_{Reference} \qquad (7)$$

This equation was used to calculate predicted retention indices of 17 barbiturates based on the experimental value for barbital and the results were compared with experimental indices. With the exception of the sulfur-containing derivatives, they gave a close correlation with an average error of only ±24 units [104]. Similar studies based on reference compounds gave comparable accuracy for anthranilic acid derivatives (±22 units) and propranolol analogs, although the predictions in the latter case were less accurate [104]. The same method had also been used to predict the retention of steroids (±18 units) used in the management of arthritis to locate possible coeluting compounds [88].

However, when this approach was extended to a study of the narcotic analgesics, based on morphine as the core compound, the

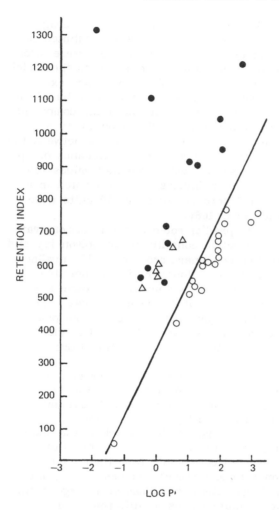

Fig. 8   Correlation between observed retention indices and octanol-water partition coefficients. (○) Barbiturates; (●) propranolol analogs; (△) anthranilic acid analogs. The retention index measurements were all made with a mobile phase of a pH 7.0 buffer and 40% methanol. The log P values for the propranolol series were obtained at pH 7.0; the anthranilic acid series at pH 8.0; and the barbiturate measurements were made in the free acid form. (From Ref. 104.)

correlations were poorer than in the earlier studies, with average errors of ±100 units [105]. These larger errors were attributed to the structural rigidity of the analgesics, and hence there were often differences in the conformation of substituents compared to the model compounds used to determine the π values. This study was extended to an examination of a related series of nortropanes based on fentanyl. The results indicated that the retentions of the compounds with the smaller N-substituents did not follow the lipophilicity model. There was also a clear difference in the retention indices between the α and β isomers of the N-substituted 3-propanilidonortropanes of up to 130 units and often a poor correlation with predicted values [105]. Stereochemical differences in retention indices were also noted in a study of azabicycloalkanes, and differences of about 70 units were found between different conformers [106].

The advantage of being able to predict retentions was demonstrated in a study of the urushiols, the active constituents of poison ivy and poison oak [107]. Many urushiol congeners are not readily available and all are highly toxic and unstable. However, it was possible to predict the retention of the unavailable compounds from their structural differences from a limited number of reference samples. The identifications were confirmed by preparative isolation of the peaks and GC-MS. Calculated retention indices were also used to identify metabolites of imipramine formed in the presence of fungal organisms [108].

Baker [109] also compared a series of drugs and their glucuronides and showed that in each case the formation of the glucuronide caused a similar decrease in the retention index value of −244 (±31 units), confirming the additive effect of chemical groups on retention.

Other groups have carried out similar studies based on the alkan-2-ones. Shalaby et al. [110] estimated the retention indices of a series of about 40 nitrogen-bridged compounds from their log P values and found a close correlation with experimental values. Magg and Ballschmiter [111] determined the retention indices of 11 ergopeptines and found that although the absolute values on different columns differed markedly, changes in the structure of the compounds such as the replacement of an isopropyl by a phenyl group or the addition of an alkyl group caused equivalent changes in the retention indices. In a study of depsides from lichens it was possible to predict the expected retention indices on the alkan-2-one scale of previously unidentified homologs and to aid chemotaxonomic comparisons [89].

Using the n-alkane scale, Morishita et al. [42] predicted the retention indices for substituted benzenes by using cumulative increments for the different substituent groups (alkyl, nitro, phenolic hydroxyl, and amino) and positional corrections derived from model compounds. Wise et al. [112] examined the relationship between the shapes of over 100 polycyclic aromatic hydrocarbons and their retention indices

on the polycyclic hydrocarbon scale. Hasan and Jurs [113] compared
a number of molecular descriptors for the calculation of retention
indices for PAH and concluded that a combination of factors, including the number of rings, the largest and smallest axes, and molecular
connectivity, gave the best correlation with experimental results.

Karlesky and co-workers [69] reported a correlation between the
retention indices, based on their modified polycyclic hydrocarbon
scale, of polynuclear aromatic hydrocarbons on four amino bonded
columns and a trigonal additivity calculation using the electronic interactions and structural features of the analytes.

In a largely theoretical study, Lamparczyk [114] derived electrical
interaction indices from the published capacity factors of polycyclic
aromatic hydrocarbons. He suggested that these could be used as a
universal scale for GLC, TLC, and HPLC to predict the retention
properties of the solute as index values. This work was an extension of an earlier study that related these interaction indices to
Kováts retention indices [115].

Quantitative Structure-Activity Relationships

It has been shown that there is a close relationship between the
lipophilicity of a compound and its biological activity. The latter is
usually expressed as the octanol-water partition coefficient (log P),
which is often closely related to the liquid chromatographic retention
in a reversed-phase partition separation. There should therefore
also be a relationship between retention indices and biological activity.

As part of the studies described earlier, in an extension of their
earlier work on prediction [104], Baker and co-workers tested this
hypothesis for a number of groups of compounds. Although the retention indices on the alkan-2-one scale showed a better correlation
with biological activity for propranolol analogs and barbiturates, log P
values correlated more closely for anthranilic acid analogs [116]. On
this scale 90% of biologically active compounds would be expected to
have retention indices between 350 and 1100 with 800 as the optimum
value [103].

Retention indices were employed as a guide to biological activity in
a study of 4-hydroxyquinoline-3-carboxylic acids as they could cover
a wide range of properties, whereas extremes of the log P values are
difficult to determine by shake flask methods [117]. It was found
that the retention indices correlated closely with respiratory inhibition. Although the relationship become second order at the limits of
the range, the retention indices were better than values derived from
Hansch constants or molar refraction. Retention indices have been
used to derive the log P values of 4-anilidopiperidines, but these
showed poor correlation with their analgesic activity [118]. Schultz
and co-workers [119,120] similarly generated log P_{OW} values from retention indices for a wide range of nitrogen-containing compounds

and found a close relationship with toxicity towards *Tetrahymena pyriformis*.

Other workers have found a good correlation between retention indices and biological activity, including the effect of quinazolinesulfonamide derivatives on cyclic nucleotide phosphodiesterases [121] and the antihypertensive activity of quinazoline derivatives [122]. Shalaby et al. [110] estimated partition coefficients for a series of bridged nitrogen compounds from experimental retention indices and compared them with experimental partition values as part of a study of biological activity. They also showed a close correlation between the retention indices and log P values for a series of synthetic quinazolines [123] and (in unpublished work) have extended the study to the comparison of retention indices on different columns.

Brent et al. [124] compared different chromatographic methods for determining lipophilicity and reexamined the reported retention index data for the barbiturates [116]. They concluded that their direct HPLC method with a deactivated stationary phase gave the best results, suggested that retention indices based on a single scale would suffer from variations in different stationary phases, and noted that changing the methanol content can alter the order of elution [47].

The hydrophobicity of hydrocarbons is an important factor in their transport in the environment and their toxicological activity. Using reversed-phase TLC, Bruggeman et al. [125] determined retention indices compared to *n*-alkylbenzenes as a guide to log P values. They also noted that the indices could be used directly in QSAR studies and suggested that this retention index method could also be used in HPLC.

Govers and co-workers [126,127] compared a range of chromatographic and descriptive methods, including retention indices based on the polycyclic hydrocarbons, Hansch and Rekker calculations, and molecular connectivity, for the prediction of log K_{OW} values for PAH and found that all the methods gave close correlations with partition coefficients. Published retention indices based on the polycyclic hydrocarbons have also been compared with Weiner's number (W) for absorption chromatography and gel permeation chromatography [128].

Polychlorinated biphenyls (PCBs) are another group of compounds of environmental interest. They have been examined by Brodsky and Ballschmiter [129], who correlated retention indices based on the alkan-2-ones with published log P values and aqueous solubilities.

A wide-ranging correlation was carried out by Schulze et al. [130] between retention indices based on the *n*-alkanes and log P values for a number of different groups of analytes, but they found that specific interactions caused deviations from an overall direct relationship.

C. Characterization of Stationary Phase/Mobile Phase Combinations

Because retention indices provide an independent reference scale based on relative retentions, they can be used to compare changes in the selectivity of different chromatographic combinations of stationary and mobile phases. However, unlike GLC, which uses a very limited range of mobile phases with very similar selectivity properties, in HPLC changes in either phase can markedly alter the overall retention and selectivity. In addition, as noted earlier, one of the major problems in achieving reproducibility in HPLC has been differences in the selectivity of nominally equivalent stationary phases and the lack of a method for reliability characterizing these variations. It is also difficult to compare the different modes of HPLC because of the wide variations in selectivity in HPLC that can be generated by using packing materials with different bonded groups and by using different organic modifiers in the mobile phase.

In GLC, probably the most widely used practical application of Kováts indices has been as the basis of the Rohrschneider [131,132] and McReynolds constants [133], which are used to quantify the activity of liquid phases toward different analytes. The retention indices of a set of test compounds on the liquid phase under study are compared to the retention indices determined with a squalane stationary phase, which has been defined as the liquid phase with zero activity [Eq. (8)]. The test analytes were chosen to reflect different types of interactions.

$$\Delta I_{analyte} = RI_{test\ phase} - RI_{Squalane} \qquad (8)$$

It was felt that a similar selectivity measurement method could be developed for HPLC based on a retention index scale. Using the alkylaryl ketone scale, Smith [55] showed that the retention indices of a set of test compounds, containing similar structural groups to the McReynolds test compounds, were significantly different on columns with different alkyl or phenyl bonded phases. As there is no clearly defined and reproducible low-activity stationary phase that could be used as a reference column (equivalent to squalane in GLC), it was suggested that the pure partition system hexane-water could be used as a reference liquid-liquid separation. The theoretical retention values of the test compounds and retention index standards in this system could be calculated from their corresponding hexane-water partition coefficients. However, this part of the concept has not been taken further.

The set of column test compounds were used to compare six different makes of ODS-silica and three mobile phases containing methanol, acetonitrile, or THF as the modifier [52]. The analysis of the retention indices of the test compounds by a principal component multivariate comparison found considerable differences between the different mobile phases. The different brands of ODS-silica column packing materials could also be distinguished and they could be displayed as the contributions to the first two principal components (Fig. 9). Columns prepared from batches of the same packing material, ODS-Hypersil (columns H_1 and H_2), were very similar, but those of a very different material, ODS-Zorbax (Z), were clearly different. The same relationship could also be demonstrated by using a hierarchical cluster analysis diagram [134].

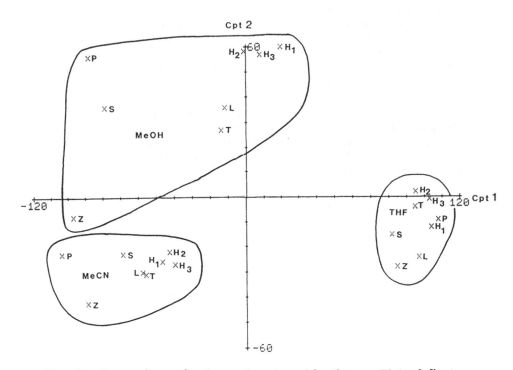

Fig. 9 Comparison of column-eluent combinations. Plot of first and second components from the multivariate analysis of the retention indices of six reference compounds separated on eight columns with 70:30 methanol-water; 50:50 acetonitrile-water; and 40:60 tetrahydrofuran-water as eluents. Columns H_1 and H_2, ODS-Hypersil 5 μm; H_3, ODS-Hypersil 3 μm; L, Lichrosorb RP-18; S, Spherisorb ODS; T, Techsil ODS; Z, Zorbax ODS; P, Partisil 10 ODS. (From Ref. 52.)

From these results the test mixture was refined so that the compounds did not duplicate responses. Three of the members of the final set of column test compounds, toluene, 2-phenylethanol, and p-cresol, and the alkylaryl ketone retention index standards come from different Snyder interaction groups [135] and can thus reflect the different interaction properties of electron donation, electron acceptance, and dipole interaction. Nitrobenzene was also included in the set as it appeared to reflect charge transfer interactions on a phenyl-bonded column.

As well as the detailed comparison of different batches and makes of ODS-silica, this concept has been used to compare a wide range of column-eluent combinations, and some typical values for the column test compounds are given in Appendix 2. As well as the selectivity differences between C_2, C_{18}, and C_{22} alkyl [55] and phenyl [56] bonded stationary phases, the indices have been used to characterize polystyrene-divinylbenzene polymer column packing materials [57,58]. In addition, during the examination of the effect of using different makes of packing materials on the analysis of drugs, the selectivity of the columns was monitored by using the column test compounds [83—86]. Because the retention indices of the column test compounds can also define separation conditions, the values were also determined during the examination of retention indices of the flavor components of spices [90—93]. They reflected the different selectivity between methanol-water (70:30), which could not resolve [6]-shogaol and [8]-gingerol, and acetonitrile-water (60:40), which gave a clean separation [90].

This ability to compare the retention and selectivity of different eluents by using a common retention index scale has been widely used. One of the earliest studies used retention indices based on the n-alkanes or methyl esters to examine the effect of adding silver ions to the mobile phase on the retentions of unsaturated hydrocarbons or unsaturated fatty acid esters [39]. Large changes of −14 to −54 units were observed.

Comparisons of methanol-, acetonitrile-, and tetrahydrofuran-containing eluents on different stationary phases showed that the eluent changes were greater than differences between column materials [52]. This work was extended to the changes in a triangle of iso-elutropic ternary eluents [53] and the effects of different proportions of acetonitrile [54], tetrahydrofuran [54], or methanol [50] as binary eluents on ODS-columns. These studies agreed with earlier work by Schoenmakers et al. [136], who found that changing the proportion of acetonitrile or THF had a larger effect on the selectivity or relative retentions than changing the proportion of methanol.

One concept behind these studies to characterize column-mobile phase combinations by using the retention indices of the column test compounds was the need to be able to independently define the conditions for official or standard HPLC methods. Ideally, the description of the apparatus used in these methods should be as general as possible and any suitable make of packing material should be acceptable as long as it can meet prescribed selectivity and efficiency criteria. Often the specification of a particular brand name is not permitted. In addition, the retention indices of the column test compounds can serve as a sensitive test of the reproducibility of batches of a packing material, and in the study of the barbiturates, batches of ODS-Hypersil were found to give virtually identical values [83].

As well as these studies based on the column test compounds, comparisons of different packing materials using retention indices have been carried out in a number of laboratories. Pesek and co-workers [137,138] used the retentions of the alkylaryl ketones to study the retention properties of new types of allyl-bonded stationary phases. The retention indices of the ergopeptines on a number of different column packing materials were recorded by Magg and Ballschmiter [111] using the alkan-2-one scale.

Retention indices based on the polycyclic hydrocarbons have been used to compare the separation ability and reproducibility of a number of different commercially available bonded silica columns [98] and ODS and amino-bonded silica columns [97], and the results have been summarized by Sander and Wise [70]. These indices have also been used by Thomson and Reynolds [139] to compare the retentions of aromatic hydrocarbons on five novel bonded columns designed to emphasize charge transfer interactions. To compare a number of different alkyl- and aryl-bonded silica columns, Antle et al. [60] compared the retentions of a number of test compounds, using retentions indices based on the alkyl parabens. In a recent study Figge and co-workers [43] compared retention indices based on the n-alkanes for six homologous series of test compounds in order to measure the silanophilic interactions on five polymer-coated silica and alumina columns.

Because retention indices provide a stable independent reference scale largely unaffected by changes in the operating conditions, they can be used to examine compounds whose chromatographic retention is very susceptible to small changes in conditions. The N-alkyl-anilines have often been reported to give poor chromatograms and be very susceptible to the make of packing material and separation conditions. However, Smith et al. [140] found that if the pH was controlled with a pH 8.5 buffer the retention indices were very reproducible even on different makes of packing material, suggesting an absence of specific silanophilic interactions between the columns and the neutral anilines.

The study was extended to PS-DVB polymer column materials; retention indices of the column test compounds were compared with homologous series of phenylalkanols and aromatic and aliphatic amines and the effects of pH and different organic modifiers in the eluents were investigated [56].

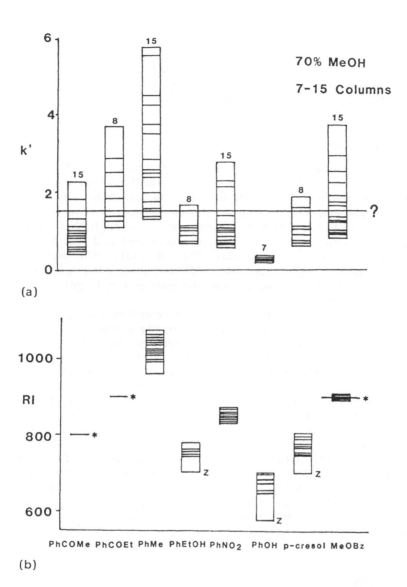

(a)

(b)

PhCOMe PhCOEt PhMe PhEtOH PhNO₂ PhOH p-cresol MeOBz

Fig. 10   Comparison of (a) capacity factors (k') reported in the
literature and (b) calculated retention indices (RI) for compounds
measured on a range of ODS-silica columns with 70:30 methanol-
water mobile phase.   The numbers below the bands indicate the number of columns used.   Asterisks indicate defined points used for
the calculation of RI (800 for acetophenone, 900 for propiophenone,
and 905 for methyl benzoate, which was used as a secondary standard if only one ketone was reported).   (British Crown Copyright.
Courtesy of the Controller of Her Majesty's Stationary Office.)

In studies of polysulfur isomers by HPLC, Möckel et al. [141,141a] used retention indices based on the *n*-alkanes to record changes in elution with eluent composition. Further work has extended the study to selenium and tellurium chains [142].

D. Estimated Retention Indices

This chapter has concentrated on papers whose authors reported retention index values, but there are many publications that include sufficient index standard compounds for the retention indices to be calculated. These can form a valuable source for comparisons between results reported in different laboratories. Although there are major variations in the capacity factors, the measured and calculated retention indices are often very similar (Fig. 10) [81]. This method can also be applied to reports in which different proportions of the mobile phase were used, and again retention indices can largely compensate for the conditions.

In some cases it is possible to calculate the retention indices directly. Burda et al. [143] reported the capacity factors of a range of hydrocarbons and included the *n*-alkanes from pentane to undecane. In comparisons of the retentions of aromatic acids, Hanai and Hubert [144,145] also reported acetophenone, propiophenone, and butyrophenone and thus the retention indices based on the alkylaryl ketones can be calculated. In other studies an indirect method can be used as it has been found that the retention of methyl benzoate on an ODS-bonded silica column is virtually independent of the make of column material, although different values are found for different organic modifiers [52]. This is not unexpected, as from the Snyder selectivity triangle [135] esters and ketones fall in the same selectivity group and would be anticipated to behave a similar manner. This relationship enables approximate retention indices to be estimated for many compounds as acetophenone and methyl benzoate have frequently been included in sets of analytes [e.g., 136].

VI. CONCLUSION

Retention indices have been shown to be an effective method of recording retentions in HPLC. They have the advantage, compared to capacity factors or relative retention times, of providing a single uniform scale across a wide range of retentions and a robustness to small changes in operating conditions. A number of different scales of homologous retention standards have been used, but most of the work has used *n*-alkane (sometimes as derived alkylbenzenes) alkan-2-one, or alkylaryl ketone scales. In addition, a retention index scale based on the polycyclic aromatic hydrocarbons has been widely accepted for PAH. The retention indices have provided a standardization that has been used for analyte identification, comparisons of retention and physical or biological properties, and measurement and comparison of stationary phase/eluent combinations and separation conditions.

APPENDICES

Appendix 1   Groups of Compounds for Which Retention Indices Have Been Determined

Scale and compounds	Reference
n-Alkane scale	
Aromatic amines, phenols, and nitro derivatives	42
Olefins (argentation chromatography)	39
Selenium and tellurium	141
Sulfur polysulfides	141,141a, 142
n-Alkylbenzene scale	
PAH	125
Alkan-2-one scale	
3-Anilidopiperidines	118
Anthranilic acid derivatives	104
Arthritis drugs (glucocorticoids, anti-inflammatory, tranquilizers)	88
Azabicycloalkanes	105,106
Barbiturates	104
Bridged nitrogen compounds	110
Depsides (lichens)	89
Drugs (various)	47,87
Ergopeptines	111
Glucuronides of drugs	109
4-Hydroxyquinoline-3-carboxylic acids	117
Imipramine and metabolites	108
Narcotic analgesics	105
Nitrogen-containing aromatics	119,120
PCBs	129
Propranolol analogs	104

Appendix 1   (Continued)

Scale and compounds	Reference
[Alkan-2-one scale]	
N-substituted-3-propananilidonortropane analogs	105
Quininazolinesulfonamides	121
Quinazoline derivatives	122,123
Steroids	88
Urushiols (poison ivy and poison oak)	107
Alkylaryl ketone scale	
Alkan-2-ones	50
N-Alkylanilines	141
Alkylbenzenes	96
Aromatics (substituted)	50
Barbiturates	29,82,83
Curcuminoids (turmeric)	93
Diuretic/thiazides	86
Eugenol/methyleugenol (pimento)	92
Gingerols/shogaols/paradols (ginger)	90,91
Gliadin proteins (wheat)	95
Local anesthetics	84,85
Mycotoxins	94
Column test compounds (nitrobenzene, toluene, 2-phenylethanol, p-cresol)	52,54,56– 58,83–86 135
Ester scales	
Unsaturated fatty acids (argentation)	39
Polycyclic hydrocarbon scales	
PAH	68–70, 96–99, 112,136, 126–128

Appendix 2 Examples of Retention Indices of Column Test Compounds with Different Stationary Phase/Eluent Combinations, Based on Alkylarylketone Scale

Stationary phase	Mobile phase	Retention indices				Reference
		PhMe[a]	PhEtOH[a]	p-Cresol	PhNO$_2$[a]	
Zorbax ODS	MeOH-water 70:30	1040	702	703	843	52
Partisil ODS	MeOH-water 70:30	965	767	753	849	52
Hypersil ODS	MeOH-water 70:30	1062	783	810	862	52
	MeCN-water 50:50	1027	694	760	873	52
	THF-water 40:60	1088	727	885	913	52
Spherisorb-ODS	MeOH-water 60:40	999	737	746	851	53
	MeCN-water 50:50	1014	688	731	935	53
	THF-water 35:65	1053	747	891	935	53
Hypersil ODS	MeOH-water 30:70	998	772	798	815	55
Hypersil SAS	MeOH-water 30:70	892	745	779	816	55
Magnusil C-22	MeOH-water 30:70	867	733	741	793	55
Spherisorb-phenyl	MeOH-water 30:70	745	625	618	802	55
PL-RPS-1 (PS-DVB)	MeOH-buffer 90:10	937	481	494	881	57

[a]PhMe, toluene; PhEtOH, 2-phenylethanol; PhNO$_2$, nitrobenzene.

REFERENCES

1. E. Kováts, Helv. Chim. Acta *41*, 1915 (1958).
2. L. S. Ettre, Anal. Chem. *36*(July), 31A (1964).
3. J. K. Haken, Adv. Chromatogr. *14*, 367 (1976).
4. M. V. Budahegyi, E. R. Lombosi, T. S. Lombosi, S. Y. Mészáros, Sz. Nyiredy, G. Tarján, I. Timár, and J. M. Takács, J. Chromatogr. Chromatogr. Rev. *271*, 213 (1983).
5. *The Sadtler Standard Gas Chromatographic Retention Index Library*, Vols. I and II, Heyden & Son, London, 1986.
6. *Gas Chromatographic Retention Indices of Toxicologically Relevant Substances on SE-30 or OV-1*, 2nd ed., VCH, Weinheim, 1985.
7. K. Pfleger, H. Maurer, and A. Weber, eds., *Mass Spectral and GC Data of Drugs, Poisons and Their Metabolites*, VCH, Weinheim, 1985.
8. R. Gill, B. Law, C. Brown, and A. C. Moffat, Analyst *110*, 1059 (1985).
9. F. P. Woodward and C. M. van Gent, J. Lipid Res. *1*, 188 (1960).
10. R. G. Ackman, J. Chromatogr. Sci. *10*, 535 (1972).
11. S. J. Hawkes, J. Chromatogr. Sci. *10*, 536 (1972).
12. F. Pacholec and C. F. Poole, Anal. Chem. *54*, 1019 (1982).
13. T. R. Schwartz, J. D. Petty, and E. M. Kaiser, Anal. Chem. *55*, 1839 (1983).
14. L. S. Ettre, J. Chromatogr. Chromatogr. Rev. *220*, 65 (1981).
15. R. A. de Zeeuw, J. Pharm. Biomed. Anal. *1*, 435 (1983).
16. A. M. Krstulović, H. Colin, and G. Guiochon, Anal. Chem. *54*, 2438 (1982).
17. H. Engelhardt, H. Müller, and B. Dreyer, Chromatographia *19*, 240 (1984).
18. G. E. Berendsen, P. J. Schoenmakers, L. de Galan, Gy. Vigh, Z. Varga-Puchony, and J. Inczédy, J. Liq. Chromatogr. *3*, 1669 (1980).
19. M. S. Wainwright, C. S. Nieass, J. K. Haken, and R. P. Chaplin, J. Chromatogr. *321*, 287 (1985).
20. R. E. Kaiser and A. J. Rackstraw, *Computer Chromatography*, Vol. 1, Dr. Alfred Hüthig Verlag, Heidelberg, 1983.
21. E. J. Kikta and A. E. Stange, J. Chromatogr. *138*, 41 (1977).
22. E. J. Kikta and E. Grushka, Anal. Chem. *48*, 1098 (1976).
23. P. C. Rahn and I. Neitring, in *Abstracts, Pittsburg Conference*, Atlantic City, N.J., 1981, p. 465.
24. *Pierce 1985—86 Handbook and General Catalog*, Rockford, Ill., 1985.
25. *Aldrich 1986—1987 Catalogue Handbook of Fine Chemicals*, Gillingham, Dorset, U.K., 1986.

26. M. Verzele, L. Use, and M. Van Kerrebroeck, J. Chromatogr. *289*, 333 (1984).

27. L. S. Ettre, J. Chromatogr. *198*, 229 (1980).

28. A. H. Stead, R. Gill, T. Wright, J. P. Gibbs, and A. C. Moffat, Analyst *107*, 1106 (1982).

29. R. Gill, A. C. Moffat, R. M. Smith, and T. G. Hurdley, J. Chromatogr. Sci. *24*, 153 (1986).

30. K. E. Bij, Cs. Horváth, W. R. Melander, and A. Nahum, J. Chromatogr. *203*, 65 (1981).

31. H. J. Möckel and B. Masloch, Z. Anal. Chem. *290*, 305 (1978).

32. N. E. Hoffman and J. C. Liao, Anal. Lett. *A11*, 287 (1978).

33. Gy. Vigh and Z. Varga-Puchony, J. Chromatogr. *196*, 1 (1980).

34. W. R. Melander, B.-K. Chen, and Cs. Horváth, J. Chromatogr. *185*, 99 (1979).

35. W. R. Melander and Cs. Horváth, Chromatographia *15*, 86 (1982).

36. H. Colin and G. Guiochon, J. Chromatogr. Sci. *18*, 54 (1980).

37. H. Colin, A. M. Krstulović, M.-F. Gonnord, G. Guiochon, Z. Yun, and P. Jandera, Chromatographia *17*, 9 (1983).

38. A. Tchapla, H. Colin, and G. Guiochon, Anal. Chem. *56*, 621 (1984).

39. B. Vonach and G. Schomburg, J. Chromatogr. *149*, 417 (1978).

40. H. J. Möckel and T. Freyholdt, Z. Anal. Chem. *368*, 401 (1981).

41. N. Tanaka and E. R. Thornton, J. Am. Chem. Soc. *99*, 7300 (1977).

42. F. Morishita, H, Kakihana, and T. Kojima, Anal. Lett. *17*, 2385 (1984).

43. H. Figge, A. Deege, J. Köhler, and G. Schomburg, J. Chromatogr. *351*, 393 (1986).

44. P. Jandera, J. Chromatogr. *353*, 91 (1986).

45. P. Jandera, J. Chromatogr. *353*, 111 (1986).

46. J. K. Baker, R. E. Skelton, and C.-Y. Ma, J. Chromatogr. *168*, 417 (1979).

47. J. K. Baker and C.-Y. Ma, J. Chromatogr. *169*, 107 (1979).

48. *Analabs Chromatography Catalog*, New Haven, Conn., 1982.

49. M. G. Mates, A. C. Auguet, and S. I. M. González, Afinidad, *39*, 479 (1982).

50. R. M. Smith, J. Chromatogr. *236*, 313 (1982).

51. M. P. Henry, personal communication.

52. R. M. Smith, Anal. Chem. *56*, 256 (1984).

53. R. M. Smith, J. Chromatogr. *324*, 243 (1985).

54. R. M. Smith, G. A. Murilla, and C. M. Burr, J. Chromatogr. *388*, 37 (1987).

55. R. M. Smith, J. Chromatogr. *236*, 321 (1982).

56. R. M. Smith, J. Chromatogr. *237*, 144 (1982).

57. R. M. Smith, J. Chromatogr. *291*, 372 (1984).
58. R. M. Smith and D. R. Garside, in preparation.
59. B. J. Compton and W. C. Purdy, Anal. Chim. Acta *141*, 405 (1982).
60. P. E. Antle, A. P. Goldberg, and L. R. Snyder, J. Chromatogr. *321*, 1 (1985).
61. J. Wilken, Z. Anal. Chem. *320*, 696 (1985).
62. C. F. Simpson, personal communication.
63. M. J. M. Wells and C. R. Clark, J. Chromatogr. *235*, 31 (1982).
64. P. Dufek, J. Chromatogr. *281*, 49 (1983).
65. P. Dufek and E. Smolková, J. Chromatogr. *257*, 247 (1983).
66. M. Popl, V. Dolanský, and J. Mostecký, J. Chromatogr. *91*, 649 (1974).
67. M. Popl. V. Dolanský, and J. Mostecký, J. Chromatogr. *117*, 117 (1976).
68. D. Karlseky, D. C. Kelly, and I. M. Warner, J. Liq. Chromatogr. *6*, 471 (1983).
69. C. N. Ho, D. L. Karlesky, J. R. Kennedy, and I M. Warner, J. Liq. Chromatogr. *9*, 1 (1986).
70. L. C. Sander and S. A. Wise, Adv. Chromatogr. *25*, 139 (1986).
71. D. L. Vassilaros, R. C. Kong, D. W. Later, and M. E. Lee, J. Chromatogr. *252*, 1 (1982).
72. M. L. Lee, D. L. Vassilaros, C. M. White, and M. Novotny, Anal. Chem. *51*, 768 (1979).
73. S. Hara, T. Ohkuma, T. Watanabe, and A. Ohta, Yakugaku Zasshi *103*, 245 (1983); Chem. Abstr. *99*, 22429 (1983).
74. S. Hara and S. Ohnishi, J. Liq. Chromatogr. 7, 69 (1984).
75. S. Hara and S. Ohnishi, Yakugaku Zasshi *103*, 750 (1983); Chem. Abstr. *99*, 116294 (1983).
76. S. Hara and S. Ohnishi, Yakugaku Zasshi *103*, 756 (1983); Chem. Abstr. *99*, 116295 (1983).
77. H. Colin, G. Guiochon, and P. Jandera, Chromatographia *17*, 83 (1983).
78. K. Huovinen, R. Hiltunen, and M. Von Schantz, Acta Pharm. Fenn. *94*, 99 (1985).
79. R. M. Smith, T. G. Hurdley, R. Gill, and A. C. Moffat, Anal. Proc. *22*, 331 (1985).
80. R. M. Smith, T. G. Hurdley, R. Gill, and A. C. Moffat, Irish Chem. News *Spring*, 21 (1986).
81. R. M. Smith, T. G. Hurdley, R. Gill, and A. C. Moffat, LC.GC Magazine *4*, 314 (1986).
82. R. M. Smith, T. G. Hurdley, R. Gill, and A. C. Moffat, Chromatographia *19*, 401 (1984).
83. R. M. Smith, T. G. Hurdley, R. Gill, and A. C. Moffat, Chromatographia *19*, 407 (1984).

84. R. M. Smith, T. G. Hurdley, R. Gill, and A. C. Moffat, J. Chromatogr. *355*, 75 (1986).
85. T. G. Hurdley, R. M. Smith, R. Gill, and A. C. Moffat, Anal. Proc. *23*, 161 (1986).
86. R. M. Smith, G. A. Murilla, T. G. Hurdley, R. Gill, and A. C. Moffat, J. Chromatogr. *384*, 259 (1987).
87. J. K. Baker, L. A. Cates, M. D. Corbett, J. W. Huber, and D. L. Lattin, J. Liq. Chromatogr. *5*, 829 (1982).
88. J. K. Baker and E. K. Fifer, J. Pharm. Sci. *69*, 590 (1980).
89. C. F. Culberson, M. E. Hale, T. Tønsberg, and A. Johnson, Mycologia *76*, 148 (1984).
90. R. M. Smith, Chromatographia *16*, 155 (1982).
91. R. M. Smith, in *Electrochemical Detectors. Fundamental Aspects and Analytical Applications* (T. H. Ryan, ed.), Plenum, Elmsford, N.Y., 1984, pp. 83–88.
92. R. M. Smith and S. Beck, J. Chromatogr. *291*, 424 (1984).
93. R. M. Smith and B. A. Witowska, Analyst *109*, 259 (1984).
94. D. W. Hill, T. R. Kelley, K. J. Langner, and K. W. Miller, Anal. Chem. *56*, 2576 (1984).
95. J. A. Bietz and L. A. Cobb, Cereal Chem. *62*, 322 (1985).
96. S. A. Wise, S. N. Chester, H. S. Hertz, L. R. Hilpert, and W. E. May, Anal. Chem. *49*, 2306 (1977).
97. S. A. Wise, W. J. Bonnett, and W. E. May, in *Polynuclear Aromatic Hydrocarbons: Chemistry and Biological Effects* (A. Bjørseth, A. J. Dennis, eds.), Battelle, Columbus, Ohio, 1980, pp. 791–806.
98. J. Chmielowiec and A. E. George, Anal. Chem. *52*, 1154 (1980).
99. A. P. Murray, C. F. Gibbs, and P. E. Kavanagh, Int. J. Environ. Anal. Chem. *16*, 167 (1983).
100. C. Hansch and A. Leo, *Substituent Constants for Correlation Analysis in Chemistry and Biology*, Wiley, New York, 1979.
101. R. F. Rekker, *The Hydrophobic Fragmental Constant*, Elsevier, Amsterdam, 1977.
102. R. Kaliszan, J. Chromatogr. Chromatogr. Rev. *220*, 71 (1981).
103. J. K. Baker, presented at the Symposium on Analytical Methods in Forensic Chemistry, Division of Analytical Chemistry, ACS, Miami, April 1985.
104. J. K. Baker, Anal. Chem. *51*, 1693 (1979).
105. J. K. Baker, R. E. Skelton, T. N. Riley, and J. R. Bagley, J. Chromatogr. Sci. *18*, 153 (1980).
106. J. K. Baker, G. J. Hite, M. Reamer, and P. Salva, Anal. Chem. *56*, 2932 (1984).
107. C.-Y. Ma, M. A. Elsohly, and J. K. Baker, J. Chromatogr. *200*, 163 (1980).
108. C. D. Hufford, G. A. Capiton, A. M. Clark, and J. K. Baker, J. Pharm. Sci. *70*, 151 (1981).

109. J. K. Baker, J. Liq. Chromatogr. *4*, 271 (1981).
110. A. Shalaby, Zs. Budvári-Bárány, and Gy. Szász, J. Liq. Chromatogr. *7*, 1133 (1984).
111. H. Magg and K. Ballschmiter, J. Chromatogr. *331*, 245 (1985).
112. S. A. Wise, W. J. Bonnett, F. R. Guenther, and W. E. May, J. Chromatogr. Sci. *19*, 457 (1981).
113. M. N. Hasan and P. C. Jurs, Anal. Chem. *55*, 263 (1983).
114. H. Lamparczyk, Chromatographia *20*, 283 (1985).
115. H. Lamparczyk and A. Radecki, Chromatographia *18*, 615 (1984).
116. J. K. Baker, D. O. Rauls, and R. F. Borne, J. Med. Chem. *22*, 1301 (1979).
117. E. A. Coats, K. J. Shah, S. R. Milstein, C. S. Genther, D. M. Nene, J. Roesener, J. Schmidt, M. Pleiss, and E. Wagner, J. Med. Chem. *25*, 57 (1982).
118. T. N. Riley and J. R. Bagley, J. Med. Chem. *22*, 1167 (1979).
119. T. W. Schultz and B. A. Moulton, Environ. Toxicol. Chem. *4*, 353 (1985).
120. T. W. Schultz and F. M. Applehans, Ecotoxicol. Environ. Safety *10*, 75 (1985).
121. T. Tanaka, E. Yamada, T. Sone, and H. Hidaka, Biochemistry *22*, 1030 (1983).
122. T. Sekiya, S. Imada, S. Hata, and S.-I. Yamada, Chem. Pharm. Bull. *31*, 2779 (1983).
123. A. Shalaby, Zs. Budvári-Bárány, K. Hankó-Novák, and Gy. Szász, J. Liq. Chromatogr. *7*, 2493 (1984).
124. D. A. Brent, J. J. Sabatka, D. J. Minick, and D. W. Henry, J. Med. Chem. *26*, 1014 (1983).
125. W. A. Bruggeman, J. Van der Steen, and O. Hutzinger, J. Chromatogr. *238*, 335 (1982).
126. H. Govers, C. Ruepert, and H. Aiking, Chemosphere *13*, 227 (1984).
127. C. Ruepert, A. Grinwis, and H. Govers, Chemosphere *14*, 279 (1985).
128. N. Adler, D. Babić, M. Gotić, J. Mencer, and N. Trinajstić, Collect. Colloq. Semin. (Inst. Fr. Pet.) *40*, 406 (1984).
129. J. Brodsky and K. Ballschmiter, in *Abstracts, 15th International Symposium on Chromatography*, Nürnberg, October 1984. p. 3PF−40.
130. G. Schulze, M. Wachotsch, and H. J. Möckel, in *Abstracts, 15th International Symposium on Chromatography*, Nurberg, October 1984, p. 3PF−62.
131. L. Rohrschneider, J. Chromatogr. *22*, 6 (1966).
132. W. R. Supina and L. P. Rose, J. Chromatogr. Sci. *8*, 214 (1970).

133. W. O. McReynolds, J. Chromatogr. Sci. *8*, 685 (1970).
134. R. M. Smith, Trends Anal. Chem. *3*, 186 (1984).
135. L. R. Snyder, J. Chromatogr. Sci. *16*, 223 (1978).
136. P. J. Schoenmakers, H. A. H. Billiet, and L. de Galan, J. Chromatogr. *185*, 179 (1979).
137. J. J. Pesek and S. A. Swedberg, J. Chromatogr. *361*, 83 (1986).
138. J. J. Pesek and G. Guiochon, J. Chromatogr. in press (1987).
139. J. S. Thomson and J. W. Reynolds, Anal. Chem. *56*, 2434 (1984).
140. R. M. Smith, T. G. Hurdley, R. Gill, and A. C. Moffat, J. Chromatogr. *351*, 259 (1986).
141. H. J. Möckel, T. Freyholdt, J. Weiss, and I. Molnár, in *Practical Aspects of Modern High Performance Liquid Chromatography* (I. Molnár, ed.), de Gruyter, Berlin, 1983, pp. 162–185.
141a. H. J. Möckel, J. Chromatogr. *317*, 589 (1984).
142. H. J. Möckel, in *Practice of High Performance Liquid Chromatography* (H. Engelhardt, ed.), Springer-Verlag, Berlin, 1986, pp. 215–240.
143. Z. Burda, M. Kuráš, J. Kříž, and L. Vodička, Z. Anal. Chem. *321*, 549 (1985).
144. T. Hanai and J. Hubert, J. Chromatogr. *239*, 527 (1982).
145. T. Hanai, C. Tran, and J. Hubert, J. High Resolut. Chromatogr. Chromatogr. Commun. *4*, 454 (1981).

8

HPLC of Neurotransmitters and Their Metabolites

Emilio Gelpí   *Department of Neurochemistry, CSIC, Barcelona, Spain*

I. INTRODUCTION

In the late 1970s it was already evident that high-performance liquid chromatography (HPLC) could become a major technique for the determination of tissue and body fluid levels of catecholamines, indoleamines, and their metabolites. The rapid growth of HPLC as a convenient research method in neurochemistry promised to overshadow other sensitive analytical techniques such as spectrofluorescence, gas chromatography, gas chromatography-mass spectrometry (GC-MS), and radiochemical assays [1].

Historically, the practical application of HPLC in neurotransmitter assays stems mainly from two facts. First, the pellicular ion exchange and microparticulate reversed-phase (RP) sorbents were directly compatible with the aqueous nature of biological matrices such as tissue homogenates, CSF, urine, or plasma. This characteristic permitted more direct determinations without the need to resort to extensive and cumbersome sample purification steps. Second, the development of on-line detector technology resulted in the introduction of highly efficient low-volume flow through detector cells, for both fluorescence and electrochemical detection. These detectors were capable of maintaining chromatographic efficiency while providing very high levels of sensitivity, often comparable to those achieved by GC-MS methods [2].

The relatively large number of reports on HPLC procedures for the determination of neurotransmitters and metabolites up to 1982 have been covered in several reviews [3-8]. A few references from 1983 and 1984 are also included in two more recent chapters on HPLC of catecholamines [9] and HPLC with electrochemical detection in the neurosciences [10]. The present chapter attempts to cover in some depth the prolific period from 1982 to date and, while not claiming comprehensive coverage of all articles published on the HPLC of catecholamine and serotonin neurotransmitter systems, provides an updated overview of the scope and limitations of this technique in this particular field.

The increase in applications of HPLC to the determination of neurotransmitters in the past 4 to 5 years has been founded on practical considerations such as speed of analysis, versatility for profiling studies of the amines and their metabolites, high specificity and sensitivity for the assay of complex matrices, ease of automation, relatively good precision, and low cost. The sum of all of these characteristics makes the technique ideally suited for routine as well as difficult determinations of biological samples.

The approach followed here differs somewhat from the usual practice of reviewing the facts about the theory of a given topic and supporting them with selected references. In this case, the contents are dictated not by the specific interests of the author regarding the topic under consideration but by the direct evaluation of recent publications in the literature. The practical uses and common trends within the field are highlighted. For instance, although short HPLC columns of 3-μm particle size are supposed to yield faster analyses, are they actually used in common practice? Are neurochemists really using the more efficient coulometric detectors, or does the amperometric detector continue to be the most popular for catecholamine assays? Are column switching and gradient elution techniques being used at all?

Regarding the mode of chromatographic separation, the discussion will be almost exclusively centered on RP HPLC systems, as these have totally dominated this type of application in the past 3 to 4 years. Ion exchange HPLC was initially applied to the determination of catecholamines and serotonin in tissues and urine samples, but it was acknowledged in 1982 that this mode of HPLC has been almost totally supplanted by the RP systems [3]. This situation is still true today, and relatively few publications appear on the use of ion exchange HPLC methods to measure biogenic amines [11–14].

II. ION EXCHANGE HPLC

The number of reported applications of ion exchange (IE) in the determination of neurotransmitter amines within the past 4 years has clearly diminished compared to the use of reversed-phase HPLC columns. Table 1 summarizes some of the most recent applications, which, as indicated by the large particle size, in two cases would not even be representative of true HPLC runs.

It is to be noted that the three major catecholamines, adrenaline (A), noradrenaline (NA), and dopamine (DA), can be separated from serotonin (5HT) by using α-methyldopamine (MDA) as the internal standard [12]. MDA was used instead of 3,4-dihydroxybenzylamine (DHBA) because of the problems posed by separation of the latter from A in several ion exchange systems [3,12]. It has been claimed that the advantages derived from the use of pellicular cation exchangers easily offset those expected from the higher-efficiency RP systems [12]. These advantages are (a) minimum sample pretreatment, since these columns do not retain hydrophobic components that interfere with the assay of urinary catecholamines, as in the case of RP columns [11]; (b) no need to use ion pair reagents (IPR) as in reversed-phase systems in order to achieve optimum k' values for the catecholamines, which otherwise tend to elute too close to the solvent front [13], as we will see later; and (c) longer column lifetime and long-term stability, although this seems to depend on the particular

Table 1 Ion Exchange Chromatography of Neurotransmitter Amines

Amines[a]	Column	Particle size (μm)	Organic modifier	Buffer	pH	Flow (ml/min)	Reference
NA, A, DA, *MDA*	Nucleosil SA	5	7% THF	Citrate	5.0	1.0	11
A, DA, 5HT, *MDA*	Zipax SCX	25	—	Citrate/acetate	5.3	0.6	12
NA, A, DA, *MDA*, 5HT	Vydac-CX	35	5% MeOH	Citrate/acetate	5.3	0.6	12
NA, *DHBA*, A, DA	SCX	5	—	Phosphate	3.5		13
DA 3- and 4-sulfates	TSKgel DEAE 2SW	5	—	Phosphate	6.5	0.5	14

[a]Internal standards α-methyldopamine (MDA) and dihydroxybenzylamine (DHBA) are underlined.

batch used [11]. Presumably, the absence of organic solvent is an
important factor in the increased lifetime of the IE columns and detector cells as well [11]; this also results in higher sensitivities due
to the use of neat aqueous eluents, which are highly compatible with
electrochemical cells [11]. The most common buffer systems used in
IE HPLC have been citrate and citrate/acetate at about pH 5 (Table
1), although phosphate buffers have also been used. Solute retention, expressed by the capacity factor k', is controlled by the ionic
strength and pH of the mobile phase buffers, as shown by a plot of
the k' values for catecholamines as a function of the sodium ion concentration in neat aqueous citrate/acetate buffers [12]. The effect
is especially remarkable for the most highly retained solutes such as
5HT, where an increase in the concentration of sodium ions from
4.5×10^{-2} to 9×10^{-2} M cuts the k' for 5HT from 42 to 21. On the
other hand, additional reductions in k' can be brought about by the
same buffer containing 5% methanol (e.g., k' further reduced to 17).
This indicates that the mechanism of separation by IE is not only of
an electrostatic nature but also involves hydrophobic interactions,
as previously discussed [5].

Ion exchange HPLC on Partisil-SCX has also been used for the
separation of indolalkylamines and their methyl and tetrahydrocarboline
analogs [15,16]. In both cases, methanol or acetonitrile was used as
an organic modifier in phosphate buffer. More recently, anion exchange HPLC with postcolumn hydrolysis and fluorimetric detection
has been applied to the determination of plasma levels and urinary
excretion rates of dopamine-3- and 4-O-sulfate in normal subjects
[14].

Nevertheless, the use of IE in the analysis of neurotransmitter systems is seriously limited by practical considerations regarding reproducibility of retention times, column lifetime, and range of compounds
that can be determined simultaneously. Basic amines are compatible
with cation exchangers, neutral alcohols are not amenable to ion exchange chromatography, and practically no work is done on anion
exchange HPLC of acidic metabolites. In contrast, as illustrated below, all of these compounds can be efficiently separated on RP columns.

III. REVERSED-PHASE HPLC

The results of a noncomprehensive but rather detailed search of the
latest literature on RP isocratic HPLC assay methods for neutrans-
mitters and metabolites are summarized in Tables 2 to 6. These
tables include data from papers dealing with the determination of
neurotransmitter amines in brain (Table 2), plasma, CSF, and urine
(Table 3); acid metabolites in brain (Table 4), plasma, CSF, and
urine (Table 5); and neutral metabolites in brain, CSF, plasma, and
urine (Table 6). Publications describing work with standards and

Table 2   HPLC Reversed-Phase Separation Parameters for Brain Biogenic Amines[a]

Compound	Column type: length × i.d. (mm) packing	Particle size (μm)	Organic modifier (%)
5HT, T	100 × 3; Polygosil C_{18} + TBP	5	MeOH, 20
A, NA, DA, 5HT	200 × 4.6; Hypersil ODS	5	ACN, 25
NA, A, DA, NMN, 3MT, 5HT	250 × 4.0; Nucleosil C_{18}	7	—
NA, DA, 3MT, 5HT	100 × 4.6; Perkin Elmer HC-18	3	ACN, 7.5
	75 × 4.6; Ultrasphere ODS	3	ACN, 7.5
NA, DA	250 × 4.0; Lichrosorb RP 18	5	ACN, 35
DA, 5HT	300 × 3.9; μBondapak	10	MeOH, 12.5
NA, A, DA	250 × 4.6; Lichrosorb RP	7	MeOH, 5
NA, A, DA, 5HT	250 × 4.6; Zorbax ODS	5	MeOH, 5
DA, 5HT	300 × 3.9; μBondapak	10	MeOH, 12.5
5HT	300 × 4.6; Chromegabond	10	MeOH, 15
NA, A, DA	50 × 5.0; Little Champ	3	ACN, 1
NA, DA, 5HT, 3MT	250 × 4.6; Lichrosorb C_8	5	MeOH, 10
NA, DA, 5HT	250 × 4.6; Biophase ODS C_{18}	5	ACN, 7.3
NA, A	—; Biophase ODS	5	—

Buffer	mM; ion pair reagents	mM; additives	pH	Flow (ml/min)	Ref.
0.14 M PCA + 0.05 M P	—	TBP	1.95	1.0	18
10 mM Na_2HPO_4	20; SDS	0.5; EDTA	2.0	—	19
0.1 M NaP	—	—	2.2	0.74	20
0.1 M CA	0.25; SOS	0.05; EDTA + 0.06% DEA	2.45	1.85	21
0.1 M CA	0.17; SOS	0.05; EDTA	2.5	2.2	21
0.05 M MCA	—	—	2.5	1.0	22
0.1 M C-P	0.032%; SOS	0.1; EDTA	2.6	1.5	23
0.15 M NaH_2PO_4	5; HexSA	0.1; EDTA	2.7	1.0	24
0.2 M NaH_2PO_4	2.5; HSA	0.1; EDTA + 0.5 TEA	2.7	1.0	24
0.1 M C-P	SOS	20μm, EDTA	2.8	1.5	25
0.1 M MCA	—	—	3.0	2.0	26
0.1 M MCA	0.3; SOS	0.1; EDTA	3.0	1.0	27
0.04 M C-P	2; HSA	0.2; EDTA	3.0	1.2	28
0.15 M MCA	SOS	0.1; EDTA	3.0	1.5	8
MCA	SOS	EDTA	3.05	1.3	29

Table 2 (Continued)

Compound	Column type: length × i.d. (mm); packing	Particle size (μm)	Organic modifier (%)
NA, DT, 5HT	150 × 4.6; Ultrasphere IPC_{18}	5	ACN, 20; MeOH, 10
NA, DA, 5HT	300 × 3.9; μBondapak	10	MeOH, 10
NA, DA, 5HT	150 × 4.6; Ultrasphere ODS	5	MeOH, 3.4
NA, DA, 5HT	250 × 4.6; Lichrosorb RP18	10	MeOH, 16
NM, NA	—	5	MeOH, 5
NA, A, DA	250 × 4.6; Nucleosil C_{18}	5	2-Propanol
NA, DA, 5HT	75 × 4.0; Supelcosil C_{18}	3	—
DA, 5HT	100 × 8.0; Radial Pak C_{18}	10	MeOH, 15
NA, A, NMN, DA, 5HT	—; RCM 100 C_{18}	5	MeOH, 10
NA, DA	200 × 0.4; DMOS R	10	MeOH, 15
NA, DA, NMN	150 × 4.6; Nucleosil C_{18}	5	MeOH, 5−20
DA, 5HT	150 × 4.0; Bio Sil ODS	5	MeOH, 1.5
NA, A, DA, 5HT, 3MT	—; Supelcosil LC-18-DB	5	MeOH, 16
NA, DA, 3MT, 5HT	150 × 4.6; Ultrasphere ODS	5	ACN, 10.5
	75 × 4.6; Ultrasphere ODS	3	

Buffer	mM; ion pair reagents	mM; additives	pH	Flow (ml/min)	Ref.
0.02 M TCA + 0.07 M NaP	1.5; SDS	1.5 µM EDTA	3.1	1.0	30
0.066 M MCA + 0.046 M NaH$_2$PO$_4$	1.66; OSA	0.06; EDTA	3.2	1.2	31
0.02 M K$_2$HPO$_4$	NaHS	—	3.3	2.0	3.2
0.1 M NaH$_2$PO$_4$	2.6; OSA	0.1; EDTA + 0.25; TEA	3.35	1.0	24
NaAc	0.1 M TCA	0.1; ETDA	3.4	1.0	33
NaAc-CA	HSA	EDTA	3.25– 4.2	2– 2.5	34
7 mM CA + 11.5 mM Na$_2$PO$_4$	0.43; SOS	1.3; EDTA + 0.12% DEA	3.5	2.2	35
0.1 M KH$_2$PO$_4$	5; HSA	0.1; EDTA	3.6	1.0	36
0.1 M NaAc + 0.1 M CA	0.5; SOS	0.15; EDTA + 1; BA	3.7	1.0	37
0.1 M KH$_2$PO$_4$	NaOS 0.1%	5; EDTA	3.8	1.0	38
0.1 M TCA	—	—	3.9	1.0	39
400 mM NaAc	—	1; EDTA	4.0	1.0	40
100 mM NaH$_2$PO$_4$	1; SOS	0.1; EDTA + 0.4; TEA	3.85– 4.25	1.0	41
0.1 M K$_2$HPO$_4$	1; NaOS	1;EDTA	4.0	1.0	42

Table 2 (Continued)

Compound	Column type: length × i.d. (mm); packing	Particle size (μm)	Organic modifier (%)
NA, DA, 5HT, MT	250 × 4.6; Ultrasphere ODS	5	THF, 1
NA, DA, 5HT	150 × 4.6; Ultrasphere ODS	5	ACN, 7
NA, A, DA	250 × 4.6; Lichrosorb C_{18}	5	MeOH, 15
NA, A, DA, 5HT	250 × 4.6; Ultrasphere ODS C_{18}	5	MeOH, 20
5HT	250 × 4.6; Zorbax C_8	5	MeOH, 30
NA, A, DA	250 × 4.6; Biosil ODS	10	MeOH, 3.5
NA, DA, 3MT, 5HT	—; μBondapak	10	MeOH, 0.04
T	300 × 3.9; μBondapak	10	MeOH, 20
DA, 5HT	300 × 3.9; μBondapak	10	MeOH, 6
5HT, NAS	250 × 4.6; Spherisorb ODS	5	MeOH, 15
NA, A, DA, 5HT	45 × 4.6; Ultrasphere C_8	5	—
NA, A, DA, 5HT	250 × 4.6; Biophase ODS	5	ACN, 3.5; THF, 1.8
NA, DA, 5HT	See Ref. 81 in Table 4		

[a]Data are taken directly from the references. Where no mention is made of any given parameter, a dash is shown.

Buffer	mM; ion pair reagents	mM; additives	pH	Flow (ml/min)	Ref.
0.05 MNaC-CA	—	—	4.0	1.3	43
100 mM P + 87 mM acetic	10; NaOS	1; EDTA	4.0	—	44
0.1 M NaAc-CA	1.8; HSA	0.1; EDTA	4.1	1.0	28
35 mM CA + 12.5 mM Na$_2$HPO$_4$	0.25; SOS	0.05; EDTA	4.2	1.0	45
0.01 M NaAc	—	—	4.3	1.2	46
1 M NaAc-CA	SOS	EDTA	4.35	0.6–2.0	47
NaAc-acetic	SOS	EDTA	4.6	1.1	48
0.01 M NaAc	—	—	4.6	2.5	49
0.03 M CA-0.06 M P	—	0.1; EDTA	4.8	1.5	50
0.01 M NaAc	—	—	4.85	1.4	51
20 mM citrate	10; SDS	2; EDTA	5.0	1.0	52
MCA	SOS	EDTA		1.6	53

Table 3   HPLC Reversed-Phase Separation Conditions for Biogenic Amines

Column	Column type: length × i.d. (mm); packing	Particle size (μm)	Organic modifier (%)
			A. In Plasma
5HT	250 × 4.6; ODS	5	MeOH, 12
5HT, 5MT	100 × 4.6; RP-18 Spheri	5	MeOH, 35 AcN, 6
5HT	250 × 4.6; Spherisorb ODS 1	5	MeOH, 15
5HT	250 × 4.6; Analytical C_8	10	MeOH, 10
5HT	300 × 3.9; μBondapak C_{18}	10	MeOH, 8
5HT	300 × 3.9; μBondapak C_{18}	10	MeOH, 4 ACN, 2
NA, αMeDA	150 × 4.6; Sperisorb ODS 2	5	—
NA, A, DA	250 × 4.0; Yanaco ODS-T	10	ACN, 6
NA, A	150 × 4.6; Ultrasphere ODS	5	MeOH, 12
NA, A, DA	100 × 4.6; Hypersil ODS	3	MeOH, 10
NA, A	100 × 4.6; Hypersil ODS	3	—
A, NA	—; RCM 100 C_{18}		MeOH, 4
NA, A, DA	150 × 4.6; TSK Gel ODS	5	MeOH, 10 ACN, 50
NA, αMNA	250 × 4.0; Spherisorb	5	—
NA, A, DA, 5HT, 3MT	See Ref. 41 in Table 2		
NA, A, DA	See Ref. 47 in Table 2		

Buffer	mM; ion pair reagents	mM; additives	pH	Flow (ml/min)	Ref.
0.1 M CA-Na_2HPO_4	—	0.1; EDTA	3.8	1.0	54
0.85 M NaAc	SDS	—	4.0	1.0	55
0.1 M NaAc	—	—	4.2	1.2	56
0.1 M NaAc	—	1; EDTA	4.5	1.0	57
9 mM CA + 82 mM NaAc	—	—	4.7	1.1	58
80 mM NaAc	—	EDTA	4.75	1.2	59
0.1 M NaH_2PO_4	2; NaHS	0.001% EDTA	5.0	1.5	60
0.1 M NaC	6; NaOS	2; EDTA	5.0	1.0	61
NaC-CA	NaOS	EDTA	5.2	1.0	62
40 mM KH_2PO_4-CA	0.5; OSA	4; EDTA	6.0	1.2	63
C-P	—	—	6.0	1.0	64
50 mM NaAc-CA	2; NaOS	0.1; EDTA + 1; DBA	—	—	65
50 mM Tris-HCl	—	—	7.0	1.0	66
70 mM NaH_2PO_4	1.85; OSA	13.4; EDTA	3.0	1.0	67

Table 3   (Continued)

Column	Column type: length × i.d. (mm); packing	Particle size (µm)	Organic modifier (%)
		B. Cerebrospinal Fluid (CSF)	
NA, DA	150 × —; Spherisorb C$_{18}$	5	MeOH, 3
5HT, 5MT	100 × 4.6; RP-18 Spheri	5	MeOH, 35 ACN, 6
	—; RP-2	—	MeOH, 10
NA, DA	300 × 3.9; µPorasil	—	ACN, —
DA conjugates	Radial Pak C$_{18}$	—	ACN, 8
NA, A, DA, 5HT, 3MT	See Ref. 41 in Table 2		
NA, DA	See Ref. 50 in Table 2		
		C. In Urine	
NA, A, DA	75 × 4.6; Hypersil	3	MeOH, 1
MN, NMN	—; Radial Pak C$_{18}$	10	ACN, 6
NA, A, DA	250 × 4.6; Ultrasphere	5	MeOH, 10
NMN	300 × 3.9; µBondapak C$_{18}$	10	ACN, 2–6
NA, A, DA	—; Radial Pak C$_{18}$		MeOH, 7
Melatonin	250 × 4.6; Spherisorb ODS2	5	ACN, 18
NMN, MN, 3MT	250 × 4.6; µBondapak	5	ACN, 3
NA, A, DA	250 × 4.6; Hypersil ODS	5	—
A, NA	—; RCM 100 C$_{18}$		MeOH, 4

Buffer	mM; ion pair reagents	mM; additives	pH	Flow (ml/min)	Ref.
0.07 M P	SOS	EDTA	3.0	1.0	68
0.085 NaAc	1.1; SDS	—	4.0	1.0	55
0.05 NaAc	—	—	4.0	1.0	55
NaAc	OSA	EDTA	4.88	—	69
0.6% H_3PO_4	HSA	EDTA + TEA	—	—	70
$(NH_4)_2SO_4$-acetic	DS	EDTA	2.9	1.8	71
H_3PO_4	—	—	3.0	2.0	72
0.04 M NaH_2PO_4	OSA	EDTA	3.0	0.9	73
0.01 H_3PO_4	2; HSA	0.27; EDTA	3.1	1.0	74
0.1 M NaH_2PO_4	OSA	0.1; EDTA	3.5	2.0	75
0.05 M $(NH_4)_3PO_4$	—	—	3.9	1.0	76
0.1 M NaP	NA S	—	4.5	1.0	77
$(NH_4)_2SO_4$-acetic	OSA	EDTA	—	1.4	78
50 mM NaAc-CA	NaOS	EDTA + DBA	—	—	65

Table 3 (Continued)

Column	Column type: length × i.d. (mm); packing	Particle size (μm)	Organic modifier (%)
MN, NMN, 3MT	300 × 3.9; μBondapak C$_{18}$	10	MeOH, 10
5HT, 5MT	See Ref. 55 in Table 3B		
NA, A, DA, 5HT, 3MT	See Ref. 41 in Table 2		
NA, A, DA	See Ref. 47 in Table 2		

Table 4 HPLC Reversed-Phase Separation Conditions for Acid Metabolites of Catecholamines and Serotonin in Brain

Column	Column type: length × i.d. (mm); packing	Particle size (μm)	Organic modifier (%)
5HIAA	150 × 4.6; Ultrasphere ODS	5	MeOH
DOPAC, 5HIAA, HVA	250 × 4.6; Lichrosorb RP18	5	MeOH, 16
DOPAC, HVA, 5HIAA	250 × 4.6; Ultrasphere ODS	5	MeOH, 10 THF, 1
IAA	300 × 3.9; μBondapak C$_{18}$	10	MeOH, 40
HVA, DOPAC	100 × 3.0; Polygosil TBP	5	—
DOPAC, HVA 5HIAA	150 × 4.0; Biosil ODS	5	—
VMA, DOPAC, 5HIAA, HVA	250 × 4.6; Nucleosil C$_{18}$	7	—
VMA, DOPAC, 5HIAA, HVA	200 × 4.6; Hypersil ODS	5	ACN, 25

Buffer	mM; ion pair reagents	mM; additives	pH	Flow (ml/min)	Ref.
0.002 M CA-NaHPO$_4$	2.5; NaOS	0.05; EDTA	3.2	1.5	79

Buffer	mM; ion pair reagents	mM; additives	pH	Flow (ml/min)	Ref.
0.02 M K$_2$HPO$_4$	HSA	—	3.3	2.0	32
0.1 M NaH$_2$PO$_4$	2.6; OSA	0.1; EDTA + 0.25 TEA	3.35	1.0	24
0.075 M NaC-CA	—	12% acetic	3.5	0.8	43
0.01 M NaAc	—	—	4.0	—	80
0.05 M CA	0.2 M ClO$_4^-$	TBP	4.9	0.6	81
500 mM NaAc	—	1; EDTA	5.0	1.0	40
0.1 M NaAc	—	—	5.2	1.0	20
10 mM Na$_2$HPO$_4$	30; TEAm	0.5; EDTA	7.0	—	19

338 / *Gelpi*

Table 4 (Continued)

Compound	Column type: length × i.d. (mm); packing	Particle size (μm)	Organic modifier (%)
DOPAC, 5HIAA, HVA	250 × 4.6; Spherisorb ODS$_2$	5	MeOH, 16
HVA, 5HIAA	See Refs. 35, 39 in Table 2		
DOPAC, HVA 5HIAA	See Refs. 8, 21, 23, 24, 25, 28, 31, 37, 40, 42, 43, 53		
5HIAA, IAA	See Ref. 46 in Table 2		
5HIAA	See Refs. 18, 26, 32, 45, 51 in Table 2		
VMA, DOPAC, 5HIAA, HVA	See Refs. 19, 20, 36, 41, 50 in Table 2		
5HIAA, DOPAC	See Ref. 44 in Table 2		
DOPAC	See Refs. 27, 47 in Table 2		

Table 5 HPLC Reversed-Phase Separation Parameters for Acid Metabolites of Catecholamines and Serotonin

Metabolite	Column type: length × i.d. (mm); packing	Particle size (μm)	Organic modifier (%)
			A. In Plasma
HVA	300 × 4.0; Lichrosorb RP-18	5	MeOH, 20
IAA	300 × 3.9; μBondapak C$_{18}$	10	MeOH, 35
IAA	300 × 3.9; μBondapak C$_{18}$	10	MeOH, 6
DOPAC, HVA 5HIAA	250 × 4.6; Yanapak ODS-A	7	MeOH, 18
HVA, 5HIAA	250 × 4.6; Analytical C$_8$	10	MeOH, 10

Buffer	mM; ion pair reagents	mM; additives	pH	Flow (ml/min)	Ref.
0.1 M Na_2HPO_4	0.1% OSA	0.1% EDTA	—	1.0	82

n Table 2

Buffer	mM; ion pair reagents	mM; additives	pH	Flow (ml/min)	Ref.
0.05 M KH_2PO_4	—	—	2.2	0.8	83
—	5; PSA	—	3.1	1.0	84
—	PSA	—	3.1	1.0	85
0.1 M P	—	10 μm; EDTA	3.2	1.2	86
0.1 M NaAc	—	1; EDTA	4.5	1.0	87

Table 5 (Continued)

Metabolite	Column type: length × i.d. (mm); packing	Particle size (μm)	Organic modifier (%)
5HIAA	300 × 3.9; μBondapak C$_{18}$	10	MeOH, 6
VMA, DOPAC, 5HIAA, HVA	See Refs. 41, 56 in Table 2		
			B. In CSF
HVA, 5HIAA	250 × 5.0; Hypersil ODS	5	—
HVA, 5HIAA	250 × 4.6; Ultrasphere ODS	5	MeOH, 8
5HIAA, HVA	250 × 4.1; Ultrasphere-IP	5	MeOH, 4
HVA, 5HIAA	250 × 4.6; Biophase ODS	5	THF, 0.9
DOMA, VMA, DOPAC, 5HIAA, HVA 5HICA	250 × 4.6; R Sil C$_{18}$ HL	10	MeOH, 10
VMA, DOPAC, HVA, 5HIAA	See Refs. 41, 50 in Table 2		
DOPAC, 5HIAA HVA	See Ref. 23 in Table 3		
			C. In Urine
HVA	150 × 4.6; ODS	5	ETOH, PropOH
VMA	150 × 4.6; ODS	5	MeOH, 1.6
VMA, HVA	150 × 4.6; Ultrasphere IP	5	MeOH, 10
HVA, VMA DOPAC	2 × 300 × 4.6; μBondapak C$_{18}$	10	MeOH, 10
5HIAA, HVA	300 × 3.9; μBondapak C$_{18}$	10	MeOH, 20

Buffer	mM; ion pair reagents	mM; additives	pH	Flow (ml/min)	Ref.
CA-Na$_2$HPO$_4$	—	—	4.8	1.5	85
0.1 M NH$_4$Ac	—	0.27; EDTA	5.15	1.5	89
0.1 M NaAc 0.01 M CA	—	0.25; EDTA	5.22	1.3	87
78 mM P + 75 mM NaC	—	—	5.25	—	91
0.05 M NH$_4$Ac	—	0.1; EDTA	5.3	1.5	90
0.07 M NaH$_2$PO$_4$	—	0.1; EDTA	5.4	1.5	88
0.05 M P	—	—	2.3	1.5	92
0.05 M P	—	—	3.0	2.0	92
0.04 M NaH$_2$PO$_4$	OSA	EDTA	3.0	0.9	73
100 mM KH$_2$PO$_4$	—	—	3.3	1.0	93
0.1 M NaH$_2$PO$_4$	—	25 mg/l EDTA	3.5	2.0	94

Table 5 (Continued)

Metabolite	Column type: length × i.d. (mm); packing	Particle size (μm)	Organic modifier (%)
5HIAA	300 × 3.9; μBondapak C_{18}	10	MeOH, 14
VMA	300 × 3.9; μBondapak C_{18}	10	—
HVA, 5HIAA	100 × 8; Radial Pak C_{18}	10	ACN, 3
5HIAA	250 × 4.6; Biophase ODS	5	MeOH, 15
VMA, DOPAC, 5HIAA, HVA	See Ref. 41 in Table 2		

Table 6 HPLC Reversed-Phase Separation Conditions for Neutral Metabolites of Catecholamines and Serotonin

Compound	Column type: length × i.d. (mm); packing	Particle size (μm)	Organic modifier (%)
			A. In Brain
MHPG	250 × 4.6; Lichrosorb RP 18	10	MeOH, 11
MHPG	—; μBondapak C_{18}	10	MeOH, 10
DHPG, MHPG	—	—	—
MHPG	250 × 4.6; Nucleosil C_{18}	7	—
DHPG, MHPG	150 × 4.6; Nucleosil C_{18}	5	MeOH, 0–3
MHPG	200 × 4.6; Hypersil ODS	5	MeOH, 25
DHPG	300 × 3.9; μBondapak C_{18}	10	MeOH, 10
MHPG	See also Refs. 28, 35, 41 in Table 2		
DHPG, MHPG	See also Refs. 36, 37 in Table 2		
TOL	See Ref. 46 in Table 2		

Buffer	mM; ion pair reagents	mM; additives	pH	Flow (ml/min)	Ref.
0.1 M NaAc	—	—	4.5	1.4	95
NaH_2PO_4	OTEAP	EDTA	5.0	—	96
NaC	—	—	5.0	2.0	97
NH_4OH + Acetic	—	EDTA	5.1	—	98

Buffer	mM; ion pair reagents	mM; additives	pH	Flow (ml/min)	Ref.
0.15 M NaH_2PO_4	—	0.13; EDTA	3.5	1.0	99
0.05 M C-P	—	—	3.5	1.0	100
Acetate-P	—	0.1; EDTA	4.8	1.0	33
0.1 M NaAc	--	—	5.2	1.0	20
0.35 M C-P	—	—	5.8	1.0	39
10 mM Na_2HPO_4	30; TEAm	0.5; EDTA	7.0	—	19
70 mM NaH_2PO_4	2; HSA	0.05%; EDTA	3.8	1.0	101

Table 6 (Continued)

Compound	Column type: length × i.d. (mm); packing	Particle size (μm)	Organic modifier (%)
			B. In CSF
DHPG	250 × 4.0; Spherisorb	5	—
MHPG	—; Biophase ODS	5	MeOH, 3
MHPG	See also Refs. 51, 50 in Table 2 and 87, 90, 91 in Table 5		
DHPG, MHPG, DOPET	See Ref. 88 in Table 5		
			C. In Plasma
MHPG	150 × 4.6; Chemcosorb 50DSH	5	MeOH, 9
MHPG	300 × 3.9; μBondapak C_{18}	10	MeOH, 2
MHPG	—; μBondapak C_{18}	10	MeOH, 9
MHPG	250 × 4.6; Ultrasphere ODS	5	MeOH, 4
MHPG	100 × 4.6; Spherisorb ODS-2	3	—
MHPG	250 × 4.6; Yanapak ODS-A	—	MeOH, 10
MHPG	250 × 4.5; Ultrasphere ODS	5	MeOH, 10
MHPG	100 × 4.6; ODS	3	MeOH, 10
MHPG	See Ref. 41 in Table 2		
			D. In Urine
MHPG	250 × 4.6; Ultrasphere ODS	5	ACN, 12
MHPG	300 × 3.9 μBondapak C_{18}	10	ACN, 2
MHPG	—; μBondapak C_{18}	10	MeOH, 9

Buffer	mM; ion pair reagents	mM; additives	pH	Flow (ml/min)	Ref.
70 mM NaH_2PO_4	1.85; OSA	13.4; EDTA	3.0	1.0	67
CA-NaAc	—	—	5.1	—	102
C-P	—	0.1; EDTA	3.0	1.0	103
0.05 M Na_2HPO_4	—	1.34; EDTA	3.0	1.0	104
0.05 M C-P	—	—	3.5	1.0	100
0.1 M NaAc + 0.09 M CA	—	0.25; EDTA	3.75	1.5	105
0.1 M NaAc	—	—	4.0	1.5	106
0.1 M P	—	10 μM; EDTA	4.8	1.1	107
0.009M NaAc-CA	—	—	5.0	1.0	108
0.1 M NaAc	—	—	5.0	1.0	109
50 mM KH_2PO_4	—	—	2.5	—	110
0.1 M P	—	0.27; EDTA	3.1	1.0	111
0.5 M C-P	—	—	3.5	1.0	100, 112

Table 6   (Continued)

Compound	Column type: length × i.d. (mm); packing	Particle size (μm)	Organic modifier (%)
MHPG	300 × 3.9; μBondapak C_{18}	10	—
MHPG	300 × 3.9; μBondapak C_{18}	10	MeOH, 4
MHPG	See Refs. 41 in Table 2 and 106 in Table 6C		
TOL	See Ref. 76 in Table 3C		

not involving actual measurements on biological samples are not included in these tables and will be referred to as the need arises.

The evaluation of all of these data is useful to establish the most common trends in the assay of these compounds in RP systems. Of the various possible elution techniques described for RP HPLC [3,17], the ion suppression and ion pairing modes, or a combination of both, are most commonly used. Simple ionization control with neat aqueous buffers is not very popular as it does not provide for fine tuning of the separation and retention of amines and it often leads to peak asymmetry. In this case, the nature and characteristics of the stationary phase and the analyte are more important for the retention than the characteristics of the eluent, so in practice there is not much room available for modification of the elution pattern. Overall, ion suppression reversed-phase HPLC is in theory a more efficient way to optimize column retention of basic compounds such as the neurotransmitters and especially of acidic ones such as their metabolites. In the ion suppression mode, the separation achieved is a function of the changes in hydrophobicity of the neutral or quasi-neutral species induced by appropriate changes in the pH of the eluent buffer. At low pH values the amines, which would be protonated due to their higher pK values, elute rapidly, whereas the acidic components of the sample are more retained. In this case, the hydrophobic character of the acids is increased by the excess population of undissociated species predominating below their pK_a values. On the other hand, at high pH values of the buffer the retention of amines increases, with a corresponding decrease in that of the acids. However, at pH values >7 both the catecholamines and the RP sorbents are unstable. Thus, the mechanism of ion suppression is mostly applied only to acid compounds. Because of their

Buffer	mM; ion pair reagents	mM; additives	pH	Flow (ml/min)	Ref.
CA-NaAc	—	EDTA	3.75	2.0	113
?	—	EDTA	6.0	—	113

pK_a values, true ion suppression for biogenic amines on RP systems would start at mobile phase pH values greater than 6, but as indicated by the data in Tables 2 and 3 this is not common practice. Rather, enhanced retention of the catecholamines and the indolamine serotonin is achieved instead by ion pairing with a counterion in opposite charge. For this purpose, the eluent is buffered at low pH to protonate the amines, which then associate with the counterion added to the mobile phase. The resulting neutral ion pairs are selectively retained by the sorbent due to their enhanced hydrophobicity.

As indicated in Tables 2 and 3, the most common ion-pairing reagents are sodium octylsulfate, heptane, and octanesulfonic acid. Sodium docecylsulfate and trichloroacetic acid have been used occasionally. Regarding buffers for ion-pairing RP HPLC of amines, the most often used have been acetate, citrate, and phosphate or combinations of acetate and citrate.

Concerning the best choice of buffer, column efficiency is better with acidic amine buffers than with phosphate buffers. Ammonium acetate is an excellent masking agent for effectively suppressing residual silanol groups on the surface of the bonded phase [114]. For instance, it has been reported that acetate, phosphate, and citrate-phosphate buffers at pH 5.1 to 5.3 give lower resolution for CSF metabolites than does ammonium acetate [90]. It is also well known that retention in ion suppression or ion-pairing reversed-phase HPLC can be further controlled by the addition of an organic modifier such as methanol or acetonitrile, as indicated in Tables 2 to 6. This effectively decreases both retention and selectivity and can be used for fine adjustment of retention times. However, the absence of organic modifier (see Tables 2 to 6) may be useful in practice, as possible concentration changes during degassing when using a single pump would be avoided in this way.

A. Effects of Eluent Modifications

As indicated above, the acidic metabolites of neurotransmitter amines can be selectively retained by ionization suppression at pH values below their corresponding pK_a values. For instance, Table 7 illustrates the behavior of some of these acids in a given eluent system variously adjusted within the pH range 3.5 to 6.5 [50]. It can be seen that the retention of acids, such as DOPAC, 5HIAA, and HVA, steadily decreases from pH 3.5 to 6.5. At pH 3.5 both DOPAC and 5HIAA elute well after their parent amines DA and 5HT, but at pH 6.5 the two acids have moved ahead of these amines. This can be explained by the solvophobic interaction theory of reversed-phase HPLC, which states that hydrophobic solute molecules are excluded from the polar solvent into the hydrocarbon chains of the bonded stationary phase [17]. When the pH increases, the acid molecules gain a net electrostatic charge as a result of their dissociation. This effectively

Table 7   Retention Times (min) of Neurotransmitters and Metabolites under Different Eluent Conditions[a,b]

Compound	A	B	C	D	E	F
VMA	3.8	2.5	2.4	2.4	2.4	2.4
DA	3.8	3.6	3.4	3.4	3.4	3.8
DOPAC	12.6	5.0	4.3	4.7	4.8	2.9
MHPG	6.7	6.6	6.6	6.6	6.7	6.7
5HT	9.5	8.2	8.3	8.2	8.3	9.4
5HIAA	15.2	11.6	10.0	11.1	11.1	5.6
HVA	27.2	13.0	10.9	12.4	12.4	5.8
pH	3.5	4.8	4.8	5.0	5.0	6.5

[a]A, 0.02 M citric acid/0.02 M hydrogen phosphate; B, 0.02 M citric acid/0.05 M hydrogen phosphate; C, same as B + 0.1 M EDTA; D, 0.04 M citric acid/0.1 M hydrogen phosphate; E, same as D + 0.1 mM EDTA; F, 0.01 M citric acid/0.05 M hydrogen phosphate.
[b]Organic modifier 6% MeOH in all cases.
Source: Data from Ref. 50.

reduces their hydrophobicity and thus increases their tendency to stay in solution in the mobile phase. The result is a decrease in their retention. On the contrary, at low pH values the amines are protonated and thus tend to elute as a block ahead of the acids, which are more hydrophobic as their dissociation is suppressed.

The relatively small changes in retention observed for VMA Table 7 as well as for the amines DA and 5HT can be explained on the same basis. In the case of VMA, its pK_a is probably much lower than those of the other acids, so at pH 3.5 its dissociation would be practically complete. On the other hand, the same data (Table 7) indicate that the pK values for the amines must be over 6.5 so that they already exist as protonated species within the pH range considered in this table. Likewise, MHPG, being a neutral species, would not be expected to be much affected by pH changes in the eluent, as deduced from the data in Table 7. Thus, retention of MHPG is not as easily manipulated [104]. The effect of eluent pH is considered in some detail in several of the reports summarized in Tables 2 to 6 as well as other publications [18,25,32,36,37,42,50,81, 90,114—117].

While Table 7 provides a direct comparison of elution patterns with changing pH, Table 8 introduces the effect of other eluent parameters such as the content of organic modifier and the use of ion-pairing reagents. These data are taken directly from the recent literature and include separations performed under different conditions of flow rate and particle size column packing. Therefore, no direct comparison of absolute retention times should be attempted, but only a comparative appraisal of relative changes in the elution of these compounds under the different eluent conditions.

An in-depth evaluation of the data in Table 8 shows, as expected, that when no ion-pairing reagents are used the amines elute together within the first five min, with the possible exception of 5HT at eluent pH values over 3.2. Also, although the elution order does not change, an increase in the methanol content from 9 to 18% (Table 8, B and C) at practically the same pH brings about a reduction in total chromatographic time from more than 40 min to less than 20 min. Along the same lines, allowing for the concomitant effect of the different proportions of organic modifier, changes observed in the relative positions of acids and amines are those expected in going from pH 3.0 to 4.85. Note the inversion of the retention times for 5HT and 5HIAA observed at the same pH value (pH 4.8) (see E and F in Table 8). As would be predicted, the increased methanol (10 vs. 15%) content at slightly higher flow rate (1.1 vs. 1.4 ml/min) decreases the retention of 5HIAA by a factor of 4. However, in HPLC system F, 5HT is selectively retained. It is interesting to consider that in this case a sodium acetate buffer was used instead of potassium phosphate. A comparison of ammonium acetate- and phosphate based

Table 8 Relative Changes in Retention Times (Min) Induced by Modifications of Eluent Composition and pH

Compound	A	B	C	D	E	F
DHPG		2.79	3.38		4.65	
VMA		4.28		3.8		
NA	2.5		2.79			
MHPG		5.36	4.80	6.7	9.24	
A		1.92	2.79			
DHBA						
DOPAC	8.3	12.94	8.06	12.6	8.54	
NMN	2.30					
DA		2.63	3.08	3.8	4.67	
MN		2.68				
5HIAA	17.8	23.91	11.14	15.2	19.65	5.8
HVA	30.1	41.55	18.20	27.2	22.74	
5HT		5.19	3.87	9.5	10.70	12.2
Reference	92	103	86	50	107	51

mobile phases for the separation of catecholamines has been reported [114], with the conclusion that the ammonium acetate system is superior in performance to the phosphate buffers. Incidentally, this has important practical implications for the use of thermospray HPLC-MS equipment for the analysis of neurotransmitters, since the operation of the thermospray ion source requires ammonium acetate in the eluent for optimum performance [118]. Work in the literature [114] indicates that the effect of pH on the capacity factors is much greater in the ammonium acetate system and that the system does not require the addition of small amounts of organic modifier. However, no explanation is given for the greater pH effect with ammonium acetate, except for stating that some factor other than pH influences the separation. On the other hand, the effect on 5HT retention observed in columns E and F (Table 8) cannot be explained solely by the 0.05 unit difference in pH. A possible alternative explanation is formation of an ion pair between 5HT and the acetate ion at a pH of 4.8. This could also be the explanation for the effects described in Ref. 114.

G	H	I	J	K	L	M
		0.75				
		1.00		4.05		
2.91	7.49	1.16		3.50	5.3	4.65
3.85		1.33				5.6
3.65		1.92		3.65	8.1	
		2.67		3.84	11.1	
6.67	10.10	3.00	5.3	9.12	3.9	6.1
4.85		3.50				
5.89	17.56	4.67	3.6	4.47	18.1	10.4
7.22		5.75		5.17		
13.15	14.69	6.58	8.8	16.59	6.4	12.3
18.67	21.98	9.50	11.2	20.96		14.0
16.33	40.96	16.00	6.8		44	30.0
121	24	35	23	122	44	48

As indicated above, the formation of ion pairs at low pH values constitutes an effective means of retarding the elution of basic amines while avoiding an unduly fast elution of acids, which would happen by the mechanism of ion suppression at pH 6. This is illustrated in columns G to M in Table 8. In ion-pairing systems the basic amines are more retained and spread out; this means that they can be more readily determined because interferences due to their elution close to or under the solvent front are prevented. For instance, in the presence of OSA, even with 16% methanol as an organic modifier, DA takes 18 min to elute whereas 5HT takes a full 41 min.

Retention can also be controlled by changing the concentration of the ion pair reagent so that increased concentration results in increased values of the capacity ratio. Likewise, an increase in the size of the nonpolar moiety of the counterion results in increased retention of the corresponding ion pairs. The usual concentrations of counterion reported for neurochemical applications of HPLC range from 0.01 to 150 mM (Tables 2 and 5), with 100 mM being the concentration

Table 8A Eluent Characteristics

Eluent	Buffer	Ion pair reagent
A	KH_2PO_4, 0.05 M	—
B	Citrate-phosphate, 0.03 ml/liter	—
C	Potassium phosphate, 0.1 M	—
D	Citric acid, 0.02 M + H_3PO_4, 0.02 M	—
E	Potassium phosphate, 0.1 M	—
F	Sodium acetate, 0.01 M	—
G	Citric acid, 0.1 M + DEA, 0.06%	HSA, 225 mg/ml
H	NaH_2PO_4, 0.1 M	OSA, 2.6 mM
I	Citric acid, 7.0 mM + NaH_2PO_4, 11.5 mM + DEA, 0.12%	SOS, 1.3 mM
J	Citrate-phosphate, 0.1 M	SOS, 0.004%
K	NaH_2PO_4, 0.1 M	SOS, 0.025 mM
L	Potassium phosphate, 100 mM + Acetic acid, 87 mM	SOS, 10 mM
M	Sodium acetate	SOS, 125 mg/liter

most often reported. Selection of the proper counterion is important for optimizing separation. It has been reported that replacing OSA with sodium dodecyl sulfate (SDS) resulted in improved separation of biogenic amines and also provided a less noisy baseline when using the electrochemical detector at high oxidation potentials. This was the result of the lower concentration of SDS ion pair reagent needed for the same separation [30]. According to the authors, the increased resolution obtained with SDS at 2 mM can be explained by its strong adsorption to the sorbent and the absence of micelle formation, as this concentration is below the critical micelle concentration for a mobile phase containing 32% methanol.

Micelle formation in the eluent should be avoided in order to prevent the formation of aggregates as a result of the hydrophobic nature of the counterion and the hydrophobic character of its hydrocarbon chain. The strategy in this case was based on the use of a relatively high concentration of SDS (1.5 to 2.0 mM) to retard the

EDTA	Organic modifier, (%)	pH	Flow (ml/min)	Temperature (°C)
—	MeOH, 1.6	3.0	2.0	·35
0.1 mM	MeOH, 9	3.0	1.0	—
10 µM	MeOH, 18	3.2	1.2	40
—	MeOH, 6	3.5	1.5	—
10 µM	MeOH, 10	4.8	1.1	—
—	MeOH, 15	4.85	1.4	—
0.05 mM	ACN, 3	2.55	1.5	—
1 mM	MeOH, 16	3.35	1.0	25
1.3 mM	—	3.35	2.2	—
—	MeOH, 12.5	3.5	1.5	—
0.08 mM	ACN, 8	3.5	0.3	—
1 mM	ACN, 7	4.0	—	—
15 mg/liter	MeOH, 4	4.6	1.1	—

amines together with a high proportion of organic modifier (∿30%) to speed elution.

In addition to considering in some detail the effects of eluent pH on retention, several publications on HPLC of neurotransmitter amines and metabolites discuss the role of selected modifications of other operating parameters such as the nature and concentration of ion pair reagents and organic modifiers and the flow rate [18,19,23,42, 50,53,73,81,89,93,116,117,119].

The role of EDTA, which, as shown in Tables 2 to 6, is commonly added to the HPLC eluent, is mostly that of complex formation with the metal ions on the silica surface. This helps to reduce tailing of amines on particularly active columns. Citrate buffers may act in the same manner because of their complexing properties. Complexation of metal ions released from the syringes, injector, or pump surfaces also contributes to suppressing both baseline noise and early-eluting peaks due to the oxidation of these ions in the electrochemical detector cells [50]. However, care must be taken when using eluents containing THF above a given concentration, as EDTA may precipitate [53].

Considering the data in these tables, we can also conclude that there has not been any major advance in the preparation of eluents for reversed-phase HPLC of neurotransmitters. In all cases there is a straightforward use of modifiers, buffers with or without ion-pairing reagents, and additives. A variant mode of operation worth mentioning in this context is the so-called reversed-phase ion pair partition chromatography, first described in 1980 [120]. This HPLC system is based on the use of tri-n-butylphosphate (TBP) as the stationary phase and a buffered aqueous medium with or without perchlorate ions as the mobile phase. Polygosil C_8 or C_{18} and Nucleosil C_8 (5 μm) have been used as chromatographic supports, on which TBP is loaded by pumping mobile phase saturated with TBP until the pores of the stationary phase are filled and the system stabilizes. Serotonin, its acid metabolite (5HIAA), tryptamine, and the precursor tryptophan were separated in such a system [18]. More recently, the addition of boric acid to the system to improve the separation of catecholamines and corresponding metanephrines as well as acid and neutral metabolites has been described [115]. The effect of boric acid stems from its characteristic reversible complexation of catechols.

In this system the increase in concentration of perchlorate ion results in longer retention times for the amines, as expected, whereas the increase in pH in the range 3 to 6 only affects retention of the acids, decreasing it substantially. On the other hand, addition of borax to the mobile phase at pH 5 was found to afford complete separation of catecholamines and their metabolites [115]. The influence of boric acid on the retention of the catecholamines depends on the pH the mobile phase. Relative to the metanephrines and MHPG, which are not affected by pH changes, the retention of the catecholamines (NA, A, DA) and of the acid metabolites of DA (HVA, DOPAC) is substantially decreased by lowering the pH. This is due to the formation of a boric acid-catecholamine complex that will not extract into the stationary TBP organic phase.

In an approach similar to the TBP ion pair partition mode described above, the adsorption of cetyltrimethylammonium (CTMA) ions on a bare silica surface, rather than on a C_8 or C_{18} bonded phase, provides a dynamically modified sorbent for HPLC. This system has been demonstrated as an alternative to reversed-phase systems on chemically bonded phases for the separation of catecholamines [121]. The eluents used were mixtures of organic modifier and phosphate buffer with CTMA bromide added.

Finally, it must also be considered that eluent modifications are often implemented with the aim of permitting concurrent determination of the basic neurotransmitter amines and their neutral and acid

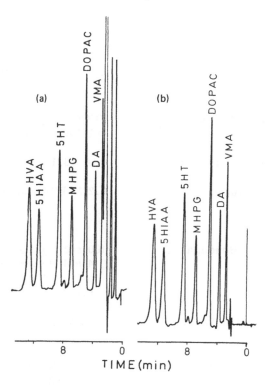

Fig. 1   HPLC profile from a sample containing amines and metabolites belonging to the noradrenergic, dopaminergic, and serotonergic pathways.   (a) Column, μBondapak C_{18} eluted with 0.03 M citric acid-0.06 M hydrogen phosphate and 6% methanol at 1.5 ml/min. Electrochemical detector at +0.7 V.   (b) Same conditions except for the addition of 0.1 mM EDTA to the mobile phase.   (From Ref. 50).

metabolites; this is a definite advantage of RP systems in the sense that they can provide true metabolic profiles as illustrated in Fig. 1 [18,21,24,36–39,50,53,81,82].   Most of these methods are based on the retention of basic amines by ion pairing at low pH, which concomitantly enhances the retention of the acidic components of the sample by ionization suppression.   The overall elution profile, including the neutral metabolites, is then adjusted by appropriate modifications of the content of the organic modifier [122].   This approach is a very useful alternative to gradient elution for metabolic profiling as it avoids the drawbacks related to the use of eluent gradients, discussed below.

Table 9 Gradient Elution Conditions for Catecholamines, Indoleamines, and Metabolites[a]

Column	Column : length × i.d. (mm); packing	Eluent A
5HT, T, 5HIAA, IAA	300 × 3.9; μBondapak C_{18}	HCOOH/H_2O 1/1000 (2.7)
5HT, T, 5HIAA, IAA	300 × 3.9; μBondapak C_{18}	PSA (3.1)
MHPG, A, NA, VMA, DA, HVA, DOPAC, 5HT, 5HIAA	100 × 3.0; Nucleosil C_8	0.2 M PCA (2.15)
DHPG, NA, MHPG, A, DOPAC, DA, 5HIAA HVA, 5HT	300 × 3.9; μBondapak C_{18}	100 mM K_2PO_4 + 4 mM HSA + 1% ACN + 100 μM EDTA
VMA, 5HIAA, HVA, VMA	100 × 8; Radial Pak C_{18}	0.05 M CA (3.55)
VMA	150 × 4.6; Hypersil ODS	100 mM $NaHPO_4$ + PCA (2.7) + EDTA
VMA, HVA, 5HIAA	250 × 4.0; Yanapak ODS-T	0.2 M P (3.0)
DOPAC, NA, A, DA	ODS	NaAc + OSA + EDTA (3.5)
MHPG	250 × 4.6; Ultrapshere ODS	0.05 M NaC/CA (4.0) + 1% THF
NA, A, DA, NMN, MN	150 × 4.6; Cosmosil C_{18}-P	0.05 M P (3.1)
3-O-Me-DA, 4-O-Me-DA	250 × 4.0; Yanapak ODS	0.05 M P (3.1)
VMA	300 × 3.9; μBondapak C_{18}	0.1 M PA

[a]Compounds are given by order of elution. Precursor amino acids have been omitted.

(pH) B	Program	Detection	Reference
MeOH/H_2O 7/3	20 $\xrightarrow{20\ min}$ 100% B	F 278/335	123
MeOH/H_2O 7/3	20 $\xrightarrow{20\ min}$ 100% B	F 278/335	124
0.02 M PCA (6.0)	0 $\xrightarrow{9\ min}$ 99% B	Dual ECD F 284/318	117
A + 20% ACN	A $\xrightarrow{25\ min}$ B	ECD	125
ACN/CA (5.25)	1 $\longrightarrow$ 99% B	F 285/315	97
H_2O/propanol	A + B	ECD	126
P/ACN 9/1	A $\xrightarrow{6\ min}$ B $\xrightarrow{20\ min}$	ECD	127
A + ACN 40 ml/liter	0% B $\xrightarrow{6\ min}$ 100% B	ECD	128
0.75 NaC/CA (3.5) + 1% THF + 10 % MEOH + 12% acetic acid	A $\xrightarrow{9\ min}$ B $\longrightarrow$	ECD	43
MeOH	A $\xrightarrow{18\ min}$ A $\xrightarrow{8\ min}$ B $\longrightarrow$	Dual ECD	129
MeOH	O $\xrightarrow{30\ min}$ 15% B	F 282/322	130
ACN/H_2O (3.1)		ECD	74

B. Gradient and Flow Programming in Reversed-Phase HPLC of Neurotransmitters

The technique of gradient elution to reduce the k' of long-retained components (peak compression effect), while maintaining satisfactory k' values for early-eluting components, has various practical drawbacks such as the extra time of column reequilibration to initial conditions and baseline drift and noise, which affect detection limits especially in high-sensitivity electrochemical (EC) detection. The EC detector, now the most widely used detection system for the HPLC of neurotransmitters, is very sensitive to changes in the conductivity of the mobile phase (pH and ionic strength). Consequently, gradient elution has been receiving much less attention in this field, although a few papers on this topic have appeared within the past 4 years. A summary is given in Table 9. To our knowledge, the two papers from this laboratory [123,124] represent two pioneering approaches to comprehensive metabolic profiling of the tryptophan indoleamine pathway, including tryptophan itself and 5-hydroxy-tryptophan (not shown in the table), as illustrated in Figs. 2 and 3. The concurrent determination of all of these compounds had been reported previously only by GC-MS in the selected ion mode [131]. By capitalizing on the high natural fluorescence of indoles, the limits of detection were as with GC-MS in the picogram range [123], allowing the determination of these compounds in brain extracts (Figs. 2 and 3).

One of the advantages of fluorimetric detection in gradient elution HPLC is that it can be relative insensitive to changes in the eluent buffer (pH, ionic strength) [117] so that baselines do not show the same degree of drift as with the EC detectors. On the other hand, the problem of baseline shift correction in gradient elution with EC detectors has been approached by resorting to the use of dual detection systems [117] placed in series toward the flow direction so that the downstream detector is used to follow changes in the mobile phase. Also as can be deduced from the data in Table 9 (Ref. 117), the increase from eluent B of 0 to 99% is equivalent in effect to performing a pH gradient, which can be considered more effective than a counterion or modifier gradient. Unfortunately, reproduction of some of the reported gradients would be difficult because of the lack of precision in reporting the actual conditions used. This may be blamed on excessively casual refereeing of such manuscripts. Another possible way to reduce the overall analysis time in metabolic profiling studies covering a wide range of compound types would be based on the use of flow rate gradients. However, this possibility is more a theory than a reality, and there is practically no work done in this direction. In this way the analysis time for the separation of DOPA, NA, DOPAC, A, and DA was cut from 40 to less than 20 min. This was achieved by varying the flow rate from 0.5 to 2.0

Tryptophan in Rat Brain

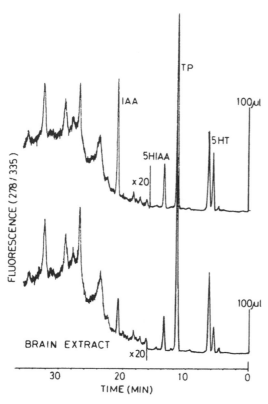

Fig. 2   HPLC separation of the components of the indolaminergic pathway in a rat brain extract prepared as described in Ref. 131. Upper trace, sample spiked with known amounts of 5HT, 5HIAA, and IAA; lower trace, profile of endogenous components. Column, μBondapak C_{18} eluted under a 20-min linear gradient from 20 to 100% of methanol-water (7/3, v/v) with initial postinjection delay of 1 min at a flow of 1 ml/min; low-strength eluent, formic acid-water (1/1000, v/v) at pH 2.7.   Fluorescence detection at 278 nm (excitation) and 335 nm (emission). (From Ref. 123.)

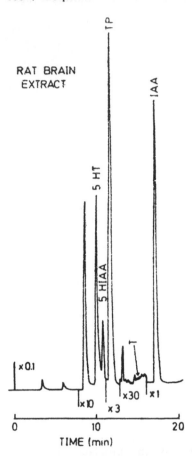

Fig. 3 HPLC profile from a brain extract (cortex plus striatum) of a rat treated with 150 ng/kg tryptamine intraperitoneally. Separation was performed on a μBondapak C₁₈ reversed-phase system eluted under the same gradient profile as in Fig. 2. Low-strength eluent was a 0.005 mol/liter solution of pentanesulfonic acid (PIC-B5) at pH 3.1. High-strength eluent was a mixture of methanol and water (7/3, v/v). Elution was monitored by fluorescence detection as in Fig. 2. (From Ref. 124.)

ml/min in increments of 0.1 ml/min [47]. However, a much simpler way to cut retention would be by an increase in the column temperature. The capacity ratio can be halved at each 10 to 15°C increase in temperature [18]. Nevertheless, this is not standard practice, although a few reports have described the use of temperatures ranging from 25 to 50°C [18,22,24,40,79,82].

In general, the role of gradient or flow programming in metabolic profiling studies of amines, acids, and neutral components has been rather limited, as indicated by the data in the most recent literature on the subject. In contrast, the work published to date on the reversed-phase isocratic elution of catecholamines, indolamines, and metabolites (see Tables 2 through 6) shows that a judicious selection of column operating conditions results in separations in many ways comparable to or even faster than those achieved by eluent programming techniques. For instance, on standard 25 to 30-cm-long columns packed with materials of 5- or 10-μm particle size an isocratic separation of several catecholamines and indoles similar to that reported by gradient elution in Refs. 117, 124, 125, and 127 (Table 9) can be obtained in 10 to 40 min [24,50,53]. By means of a reduction in column length to 10 cm and using 10-μm particles, 13 amines and metabolites were resolved in 28 min [36]. Also, the same indoles as in Ref. 124 (Table 9) can be separated in 10 min on a column 10 cm long × 3.0 mm inner diameter (i.d.) packed with 5-μm particles [18].

An interesting way to achieve good resolution for several compounds in relatively short times with isocratic elution as an alternative to gradient programming is by the use of multiple organic modifiers. By means of this approach, seven neurochemicals were resolved in 15 min with a mixture of THF and MeOH (Ref. 53; see Table 2). However, examples of the use of various solvents in the separation of neurotransmitters and metabolites are very scarce, as can be seen from the data summarized in Tables 2 to 6 [30,53,55, 59,66].

IV. HIGH-SPEED HPLC SYSTEMS

The use of shorter columns packed with 3-μm particles affords a more efficient way to speed up analysis than gradient elution methods. For example, a total of 18 catecholamines and indoleamines plus their corresponding metabolites have been resolved in only 5 min on a 7.5 cm × 4.6 mm i.d. column packed with 3-μm reversed-phase particles and eluted at a flow rate of 2.2 ml/min [21]. The high efficiencies associated with such a small particle size compensates for the loss of efficiency determined at the higher than optimum flow rates and reduced column lengths required for very fast analysis. However, to get the most out of these high-speed systems, judicious use must be made of elution parameters. In this regard, and allowing for differences in the characteristics of Supelcosil C_{18} [35] and Ultrasphere ODS [21] stationary phases, an interesting comparison can be made of the elution times of the components listed in Table 8, column 1, and the same components as shown in Ref. 21. The compositions of the eluting solvents in both cases are compared in Table 10. As indicated in Table 8, column I, elution of the last peak (5HT) takes 16 min in the Supelcosil LC-18 column under the conditions summarized in Table 10. However, the analysis of the same compounds can be completed in 3.5 or 5.5 min under the conditions given in Table 10 for the Ultrasphere and HC-18 columns, respectively.

Evidently, it woud be advantageous in terms of resolution per unit time to maintain a low concentration of ion pair reagents adding ACN as organic modifier. Apparently, using phosphate and temperature would be unnecessary to realize the full capabilities of the high-speed system. Thus, inadequate selection of eluent composition may be responsible for the time of analysis for 3-μm columns still being in the range of 15 to 20 min [35].

One of the practical problems related to the use of small particle packings is their restricted loading capacity. In this regard, it has been stated that injection volumes should be kept below 5 μl for early-eluting components, whereas up to 50 μl is allowable for later-eluting components [21]. In contrast, other authors have injected

Table 10   Comparison of Eluting Solvent Composition in Three Different High-Speed HPLC Systems

	Ultrasphere 3 μm	PE HC-18 3 μm	Supelcosil LC-10 3 μm
Citric acid	100 mM	100 mM	7 mM
NaH_2PO_4	0	0	7 mM
Diethylamine	0	0.06%	0.12%
SOS	0.175 mM	0.255 mM	0.43 mM
ACN	7.5%	7.5%	0
pH	2.5	2.45	3.5
Temperature	∿21	∿21	40−50°C
Column dimensions	7.5 cm × 4.6 mm	10 cm × 4.6 mm	7.5 cm × 4.0 mm
Reference	21	21	35

Source: Adapted from Refs. 21 and 35.

up to 70 μl before the resolution for early-eluting peaks and column efficiency show signs of being affected [27].

A few other papers have described the use of high-speed systems in this field [42,106,109,128,132,133].

V. MULTIDIMENSIONAL HPLC

To optimize the resolution of the components of the sample with lower k' values without increasing total analysis time, limited use has been made of column switching techniques. The systems described range from the most simple, in which a sample enrichment column is connected to the analytical column through a six-way valve [134], to systems comprising three or four columns interconnected by an electro-pneumatically programmed valve switching unit [135,136]. The columns can be all of the same material but different lengths, as in the separation of DOPA, NA, A, DA, and isoproterenol (IP) in urine samples. The total analysis time was reduced from 60 to 35 min by eluting DOPA, NA, and A through four columns, DA through two columns, and IP through one column. All columns were packed with Nucleosil C_{18}, 10 μm, and were eluted with a mobile phase consisting of 0.06 M KH_2PO_4, 50 mg/liter Na_2EDTA, 100 mg/liter SOS, and 12.5% MeOH at pH 3.5 [136]. Direct injection of serum, plasma, and urine samples is made possible by the automated multidimensional approach. For instance, the determination of A and NA in human serum has been achieved by direct injection of the sample into a microprecolumn of alumina. The catecholamines in the appropriate buffer solution are thus enriched until, by switching a six-way valve, the adsorbed catecholamines are eluted by the mobile phase from the microprecolumn and simultaneously separated by the analytical column [134]. A similar approach to the direct injection of plasma or urine for the determination of catecholamines involves complexation of the amines on columns of phenylboronic acid bound to a polyacrylamide gel. The catecholamines are eluted with phosphoric acid and enriched as ion pairs on an ODS column, from which they are then eluted into an analytical reversed-phase column for their separation [135].

Aside from other occasional publications describing either similar approaches [137,138] or off-line collection of fractions and reinjection on analytical systems, as reported for HVA and 5HIAA [84] or MHPG in human urine [113], no extensive use has been made of these techniques in neurochemical work to date.

VI. DETECTION SYSTEMS IN THE HPLC OF NEUROTRANSMITTERS

A wide range of detectors is available for the HPLC detection of different analytes. However, in practice, only two types of detector have been providing consistent and reliable data on neurotransmitters and related products. These are the fluorescence (F) and electrochemical detectors, with the latter clearly dominating in the field in terms of frequency of use and scope of applications, as indicated in Table 11. A few recent reviews have addressed the history and principles of operation of fluorescence and electrochemical detectors [3,9,10], so the discussion here will be focused on current trends and developments within the past 3 or 4 years regarding the use of these detectors, in the neurosciences. In this context, a relatively wide literature survey shows overwhelming use of the EC detector for biogenic amines (Tables 11A and 11B) and especially catechol-related neurochemicals, since these readily undergo a two-electron oxidation to the corresponding orthoquinone. A list of electroactive neurochemicals has been compiled [10].

Evaluation of Table 11 indicates that the EC detector is being applied routinely to the determination of catecholamines (NA, A, DA) and their basic (NMN, MN), acid (DOPAC, HVA, VMA), and neutral (MHPG, DHPG) metabolites, as well as serotonin (5HT) and its metabolites (5HIAA). Precursor amino acids such as tyrosine, DOPA, or tryptophan have also been determined simultaneously although they are not included in this review. These compounds have been assayed in brain, CSF, plasma, blood, and urine samples with sensitivities claimed to be in the picomole to femtomole range, such as the 51 fmol reported for 5HIAA in brain, equivalent to about 0.2 pg [8], or the 15 to 100 fmol reported for NA, A, and DA in urine [71]. As seen in Table 11, limits of detection are variously reported by different authors, although in most cases they fall in the low picogram range for absolute amounts injected on column or in the range of nanograms to picograms per unit weight or volume of sample. From the data in Table 11A it is clear that there are no standardized criteria for reporting and comparing overall assay performance in terms of sensitivity and precision. For instance, the limits of detection (LOD) for a given assay are often obtained with solutions of authentic standards, but this kind of information is meaningless with regard to the practical detectabilities of the same compounds in their corresponding biological matrix. The values reported in the table refer mostly to sensitivities given for the endogenous compounds when these are quoted by the authors. More than anything, the picogram range sensitivities for electroactive endogenous neurochemicals have been the determining factor in the rapid rise in popularity of the EC detector.

The small dimensions of the thin-layer EC cells are fully compatible with the use of columns packed with 3-μm particles. A sevenfold increase in sensitivity for biogenic amines and metabolites has recently been reported with this approach [35]. Sensitivity becomes an important consideration when analyzing minute samples such as a few milligrams of tissue or microliters of blood or CSF. Detector noise adversely affects assay sensitivity, so both electrical and chemical background noise must be minimized for optimum performance. In this sense, as already mentioned, the increase in noise observed especially at low pH values [39] can be attenuated with agents capable of complexing metal ion released from the system. For this purpose EDTA has proved very useful [39,50,125]. As illustrated in Fig. 1, EDTA helps to eliminate early-eluting peaks that interfere with the assay of VMA [50]. An influence of mobile phase composition on detector noise has also been observed, with noise increasing in the following order: phosphate/citrate buffer = phosphate/acetate buffer, TCA-containing eluent, and alkylsulfonate-containing eluent [39].

As Table 11 illustrates, of the two carbon-based electrode materials available for HPLC work, glassy carbon is the more commonly used. Both carbon paste and glassy carbon are used as the working electrodes in amperometric detectors, in which they are kept at a constant potential difference relative to the reference electrode. The current generated by the oxidation of the electroactive analyte on the working electrode is dependent on its concentration. Although the carbon paste electrode should provide more sensitivity, the glassy carbon electrodes seem to be favored in routine use as they do not require as much care in handling, preparation, and maintenance. Also, because of the binder (wax, silicone grease, etc.) needed to hold the graphite paste in place on the Plexiglas electrode housing assembly, mobile phases should not contain more than 10 to 15% organic modifier when using the amperometric detector with carbon paste working electrodes. Likewise, the rotating-disk electrode, a design in which the glassy carbon working electrode is rotated to promote diffusion of the analytes, resulting in increased sensitivity, has been used in this field only by the group who first introduced it in 1980 [33,39].

The dynamic ranges of amperometric detectors (Table 11) in applications to real samples are sufficiently wide to cover the concentrations of the various analytes. Also, in most cases, the reported precision within and between runs stays below 10%, depending on the concentration range assayed [95]. However, values up to 22.5% for the interday coefficient of variation have been reported [109].

A. Dual Electrochemical Detectors

Although not as extensively used as the single-electrode EC detectors, the dual EC systems have been very useful in the determination of electroactive neurochemicals in biological samples, as summarized in Table 11B. These systems were introduced as early as 1976 but were not commercially available until the early 1980s [9].

The dual EC detectors simultaneously monitor oxidative and/or reductive processes occurring at two different potentials. For these

Table 11A Detection of Catecholamines and Serotonin: Electrochemical Detection

Metabolite	Sample[a]	Electrode[b]	Potential (V)
NA, DA, 5HT, DOPAC, HVA, 5HIAA	B	GC	0.95
A, NA, DA, 5HT, VMA, DOPAC, 5HIAA, HVA, MHPG	B	GC	0.7
NA, NMN, DA, 5HT, A, VMA, DOPAC, 5HIAA, HVA	B	—	0.65
DA, 5HT, NA, DOPAC, HVA, 5HIAA	B	CP	0.75
5HT, 5HIAA	B	CP	0.65
NA, DA, 5HT	B	GC	0.65
NM/DHPG, MHPG	B	RD	0.85/0.55
NA, A, DA	B	GC	0.75
NA, A, NMN, DA, 5HT, DOPAC, 5HIAA, HVA, MHPG	B	GC	0.8
NA, DA	B	—	0.75
NA, DA, NMN, HVA, DOPAC, 5HIAA	B	RD	0.75
DA, 5HT, HVA, 5HIAA	B	CP	0.7
NA, DA, 5HT	B	GC	0.8
NA, DA, 5HT	B	GC	0.65
NA, A, DA, 5HT, 5HIAA	B	GC	0.72
NA, A, DA	B, P, U	GC	0.7
NA, DA, 5HT	B	GC	0.85

Linear range	LOD	Precision (%) intra/interassay	Ref.
1—400 pmol	51 fmol HIAA	<2/—	8
—	21—33 ng/ml	2/—	19
-5 nmol	0.1 pmol	—/—	20
1—30 ng	1—5 pmol	—/—	23
—	20 pg, 40 pg	10/—	26
5—80 ng	15 pg NA	—/—	30
—	100—200 pg/mg	—/—	33
—	20—40 pg	<0.6/—	34
0.5—9.4 pmol	0.15—0.75 pmol	3.1—5.4 /—	37
200 pg—1.6 ng	40 pg	3—5.3/5.0—7.7	38
—	3—7 pg	—/—	39
—	1—50 amp/V	3—6/—	40
—	20—150 pg	5/—	42
5—80 ng	15 pg NA	—/—	44
10—320 ng	—	<6.9/<6.7	45
100—500 pg	—	—	47
—	20—60 pg	—	48

Table 11A   (Continued)

Metabolite	Sample[a]	Electrode[b]	Potential (V)
DA, 5HT, VMA, DOPAC, 5HIAA, HVA, MHPG	B, CSF	GC	0.7
NA, DA	CSF	—	0.61
5HT, 5HIAA	B	GC	0.72
5HT	P	GC	0.6
5HT	P	GC	0.6
5HT, HVA, 5HIAA	P	GC	0.95
5HT	Blood	CP	0.5
NA	P	GC	0.5
NA, αMNA	P	GC	0.66
NA, A, DA	U	GC	0.55
MN, NMN	U	GC	0.85
NA, A, DA	U	GC	0.7
NMN	U	GC	0.85
NA, A, DA	U	GC	0.67
NMN, MN	U	GC	0.82
NA, A, DA	U	GC	0.73
NMN, MN	U	GC	0.9
HVA, 5HIAA	CSF	GC	0.75
MHPG	P	GC	0.80
5HIAA, HVA	CSF	GC	0.85
HVA	P	GC	0.7

Linear range	LOD	Precision (%) intra/interassay	Ref.
10 pg—15 ng	5—18 pg	—	50
—	0.1—0.3 pmol/liter	12.3—16.6/—	69
0.1—50 ng	1 ng/sample	6.3/—	129
—	200 pg/ml	5.3/5	54
5—10,000 nmol/liter	250 fmol	5.6—7.7/3.5—14.9	56
0.25— 1 ng	—	6.8/—	57
—	—	1.1—1.5/—	58
90—1200 pg/ml	30 pg/ml	7—8/—	60
—	30 pg	3.4/4.6	67
—	15—100 fmol	—	71
—	10 µg/liter	2.9;1.4/4.9;5.3	72
0.3—1250 mg/liter	20—32 pg	2—12/—	73
0.5—10 ng	0.2 ng	4.5/6.8	74
—	—	1.9—3.2/8.1—12.9	75
1—20 ng	1 ng; <10 ng/ml	7/5—9	77
0.3—2000 µg/liter	1—5 µg/liter	—	78
0—0.4 nmol	—	—	79
0.5 pmol—0.5 nmol	0.5 pmol	3.6—9/—	70
0.4—200 pmol	0.4 pmol	6.8—7.5/8.8	70
5—35 nmol/liter	40—50 pg	—	89
5—60 ng/ml	2 ng/ml	5.3/11.9	83

Table 11A (Continued)

Metabolite	Sample[a]	Electrode[b]	Potential (V)
5HIAA	P	GC	0.55
VMA, HVA	U	GC	0.9
HVA, VMA, MHPG	U	GC	0.8
HVA, 5HIAA	U	GC	0.7
5HIAA	U	GC	0.55
MHPG	CSF	GC	
DHPG	P	GC	0.66
MHPG	B,P,U	GC	0.85
MHPG	P	GC	0.75
MHPG	P,U	GC	0.9
MHPG	P	GC	0.7
NHPG	P	GC	0.75
MHPG	U	GC	0.85
MHPG	U	GC	0.75

[a]B, Brain; P, plasma; U, urine.
[b]GC, glassy carbon; CP, carbon paste.

Linear range	LOD	Precision (%) intra/interassay	Ref.
—	—	5.3/—	85
0.1—220 mg/liter 0.5—950 mg/liter	22; 140 pg	5.8/—	73
—	20—40 pg	5.7—6.4/—	90
—	—	5—6/4.4—9.5	94
0—10 µg	1 mg/liter	4.8/7.1	98
	1.6 ng/ml	3.8/7.2	102
—	30 pg	4.1/8.9	67
5—20 ng/ml	30 pg/ml	3/—	100
—	30 pg; 2—3 ng/ml	6.5/8.9	104
0.1 ng—40 mg	0.1 ng	~0.1/—	106
0.5—20 ng/ml	—	2.6—5.4/—	107
—	0.18 ng/ml	4.1/5.6	108
2—32 ng	2 ng	5.8/—	111
—	20—40 ng	6.4/6.9	113

Table 11B Detection of Catecholamines and Serotonin: Dual Electrochemical Detection

Metabolite	Sample[a]	Electrode[b]	Potential (V) E_1	E_2
NA, DA, 5HT, DOPAC	B	2GC	0.55	0.8
HVA, 5HIAA, MHPG			0.6	0.85
NA, A, DA, 5HT, HVA, 5HIAA, DOPAC	B	2GC	0.8 0.85	0.65 -0.165
NA, A, DA	P	2GC	0.5	0.55
NA, DA	CSF	2GC	0.75	-0.1
NA, A, DA	P	2GC	0.5	0.8
NA, A, DA	P	2GC	0.3;0.15	-0.35
DA	P	2Clm	0.05	0.35
NA, DA, 5HT	B	2Clm	0.5	0.75
NA, DA, 5HT, HVA, 5HIAA, MHPG	SC	2GC	0.65	0.8
NA, A, DA, 5HT, DOPAC, HVA, 5HIAA, MHPG	B	2Clm	0.3	-0.2
HVA, 5HIAA, MHPG	CSF	2Clm	0.23	0.40
DOPAC, 5HIAA, HVA	P	2GC	0.6	0.75
VMA	U	2GC	0.18	0.30
HVA	U	2GC	0.2	0.46
MHPG	P	2Clm	0.1	0.4

[a]SC, spinal cord.
[b]2GC, Dual glassy carbon; 2Clm, dual coulometric.

Linear range	LOD	Precision (%) intra/interassay	Ref.
—	—	—/—	43
100 pg–100 μg	200 pg	3/—	43
1.5–87 ng/ml	100 pg	—/—	53
10–500 pg	5 pg; 10 pg/ml	ᷪ5/—	61
10 pg–1 ng	3; 5 pg	6.8–5.8/—	68
0.1–500 pml	75 fmol	7.7–9.8/—	63
50–2500 pg	—	<20/—	128
1 ng/ml	—	—	130
—	—	—	30
5/20 pg–20 ng	5–20 pg	3.5/5.2	35
—	0.01 pmol	—	28
—	20 pg	— —	91
1–50 ng/ml	200 pmol	2.1–4.3/—	86
—	—	1.5/3.9	92
—	—	0.9/4.6	92
1–200 nmol/liter	0.5 nmol/1; 2 pg	9/22.5	109

Table 11C   Detection of Catecholamines and Serotonin: Fluorescence Detection

Metabolite	Sample	Derivative[a]	$\lambda_{exc}/\lambda_{em}$ (nm)
NA, DA	B	OPA(p)	351,363/450
NA, DA, 5HT, 5HIAA	B	—	290/330
5HT, 5HIAA, IAA	B	—	280/350
5HT, NAS, 5HIAA	B	—	285/345
5HT	CSF, P, U	—	290/338
NA, A, DA	P	DPE(p)	350/480
Melatonin, TOL	U	—	283.5/333
CA-S	U	ED	410/520
IAA	B	—	280/340
IAA	B	—	280/340
IAA	P	—	280/340
5HIAA	U	—	295/345
VMA, HVA, 5HIAA	U	—	285/315
TOL	B	—	280/350

[a](p) Precolumn derivative; DPE, 1,2-diphenylethylenediamine; ED, ethylene-diamine.

Linear range	LOD	Precision (%) intra/interassay	Ref.
0.2—1.6 ng NA 0.4—3.2 ng DA	5.16 pg	—	22
	1—5 pmol	6.1—7.9/—	32
—	5—15 pg	1—4/—	46
	75—140 pg	1.3/5.8	51
—	0.5 nmol/liter	—/10	55
—	7—10 fmol/ml	1.8—3.7/—	66
50—1000 pg	30 pg/ml	2.8,4.1/—	76
10—200 ng	1—2 pmol	2.3/4—7	139
—	—	—	84
—	80 pg	7.1/—	80
—	65 fmol	5.3/—	85
—	0.2 μg/ml	9.2/6.7	95
—	—	3.2—5.5/—	97
—	10 pg	1.4/—	46

Table 11D   Detection of Catecholamines and Serotonin: Dual Fluorescence-Electrochemical Detection

Metabolite	Sample	$\lambda_{exc}/\lambda_{em}$	Electrode	Potential (V)
5HT, 5HIAA, HVA	CSF, SC	225/	GC	0.75
NA, A, NMN, DA, MN	P	282/322	GC	0.6/0.9
VMA	U	285/325	GC	0.75
DOMA, DOPAC	U	282/314	GC	0.72

purposes they can be operated in basically two configurations [53, 116]: (a) the parallel mode, in which the detector provides two chromatograms in a manner resembling dual-wavelength UV detection; peak homogeneity can thus be ascertained through the corresponding peak height ratios of standards compared to those of biological samples; and (b) the series configuration, which is analogous to fluorescence detection as it depends on specific detection at the downstream working electrode (equivalent to emission wavelength in fluorescence) of the reaction product generated in the upstream working electrode (excitation wavelength). Overall selectivity can thus be significantly improved. In Table 11B the potentials of the two working electrodes are given under E_1 and E_2, although in many cases the authors do not specify whether they are working in parallel or in series and the reader must figure it out from the data presented. One of the important parameters in the operation of a dual EC detector in the series configuration is the so-called collection efficiency; the ratio of downstream peak current to the upstream peak current [53]. The ratio indicates the fraction of upstream oxidation products which are converted at the downstream working electrode and it provides an indication of peak purity and types of functional groups being converted by oxidative or reductive processes. Theoretically, this value can only approach 0.4 [53] because of design considerations, but it was demonstrated that it could be as high as 0.78 for catecholamines in a microelectrolytic cell at a flow rate of 8.3 μl/min [116].

Response for any given electroactive species can be optimized by selection of the most appropriate detection potential for each electrode according to the current-voltage curve. These curves are highly

Linear range	LOD	Precision (%) intra/interassay	Ref.
—	—		140
ED; 0.25–100 ng/ml FD; 1–100 ng/ml	0.1–0.2 µg/ml 0.5–1 ng/ml	—	130
—	FD; 400 pg EC; 50 pg	2.75/7.9	96
—	FD; 9.6 ng/ml ED; 2.5 ng/ml	—	136

dependent on the nature and composition of the mobile phase [141]. Nevertheless, the dual-electrode operation, while affording higher selectivities, does not increase sensitivity (Table 11B). Quantitation is carried out at the more selective electrode, which could be the downstream electrode if interferences can be eliminated at the first electrode, provided they are irreversibly oxidized at potentials lower than those of the analytes. Also, it must be taken into account that EC detector response decreases with increasing content of organic modifier in the eluent [11] because of the resulting reduction in overall dielectric constant. Likewise, detector response is dependent on flow rate [47].

In theory, sensitivity can be substantially improved by the use of coulometric detectors, providing 100% efficiencies as compared to the efficiency of $\simeq 10\%$ for amperometric detectors. In other words, less than a 10% of the molecules going through an amperometric thin-layer cell are oxidized or reduced, so the coulometric EC detector should be 10 times more sensitive provided the expected increase in background noise can be minimized. This can be achieved by placing a guard cell between the pump delivering the mobile phase and the injector [9,28,30,109,129] with a setting of the oxidation potential higher than that of the detecting cell. This eliminates most of the noise generated by the solvent. Accordingly, a recent comparison of amperometric and coulometric sensitivities for noradrenaline showed the latter to be 10 times more sensitive [28]. This type of EC detector is most often used in the dual series configuration and, as shown in Table 11B, the oxidation potentials required for biogenic amines are much lower than those required in amperometric detection

[30]. The detector can be operated in either the screen, redox, or differential mode. Operation in the screen mode offers increased selectivity. In this case the first detector is set at a lower potential at which all easily oxidized interferences are removed and thus are screened from the downstream detector [28,30,109,142]. In the redox mode oxidized compounds are reduced at the second electrode, whereas in the differential mode the difference between the two currents is monitored.

A remarkable gain in selectivity for the HPLC assay of neurotransmitters has been realized by the introduction of series multielectrode coulometric EC detectors [142]. For instance, the use of three electrodes in series increases selectivity and resolution for direct analysis of neurochemicals. In this mode, the first electrode oxidizes the analytes and the second and third provide a ratio of two reduction potentials. This capability has been expanded into "gate" cells of four or more coulometric cells and array cells of up to 15 electrodes which increase resolution to the point that coeluting compounds with differences in half-wave potentials of only 30 to 40 mV can be separated [142]. Although the use of this approach for the determination of neurotransmitters and metabolites in brain and CSF samples has been demonstrated, it has been very restricted in practice.

Additional advantages of the flow-through porous graphite coulometric detectors are their insensitivity to changes in eluent flow rate [47] and temperature in terms of peak areas [9] and the fact that, because of their larger surface area, their response will be less affected by electrode contamination. It is known that detector response in amperometric detectors with glassy carbon or carbon paste working electrodes is affected by contamination [18,40] and a gradual decay of detector sensitivity is to be expected with time [8]. This can be corrected by repolishing the electrode surfaces. In contrast, the coulometric electrode requires less maintenance. For instance, it has been claimed that it only requires cleaning with 35% nitric acid every 4000 to 5000 samples [28].

VII. FLUORESCENCE DETECTION

This kind of detection system can be very selective and sensitive, but in the field of neurochemistry it is not as widely used as the GC detectors. As Table 11 indicates, it is used mostly for the determination of indolic compounds such as serotonin and its metabolites, where it can be almost as sensitive to EC detection. Advantage is taken in this case of the strong natural fluorescence of the indole nucleus [32,143]. However, EC detection is more sensitive for catecholamines, which, when analyzed in an HPLC system equipped with a fluorescence detector, usually must be preoxidized to trihydroxyindole derivatives or derivatized to suitable fluorophores with phthaladehyde, OPA (Table 11C, Ref. 22), DPE, or ED (Table 11C, Refs. 66 and 139), fluorescamine, or dansyl chloride [22]. In any case, fluorescence detection for underivatized catecholamines can be an order of magnitude less sensitive than for the indoles [18]. Nevertheless, direct HPLC assays of catecholamines by fluorescence monitoring have been described [32,130], because the total levels of catecholamines in the samples analyzed lie within the response range of the detector [96,130]. Comparison of typical detection limits in nanograms per milliliter for underivatized catecholamines in plasma shows that the EC detector can be five to seven times more sensitive [130]. Similar results have been reported for urine, although in this case the response of the acidic metabolite DOPAC is 50 times more sensitive in the EC detector [136]. Also, the sensitivity of detection has been increased by using an argon laser as the excitation source. This system was applied to the determination of NA and DA in rat brain [22]. A comparison of the values thus obtained with values obtained with various HPLC detection systems shows that laser fluorescence can be as sensitive for catecholamines as the trihydroxyindole or EC detection and more than 20 times as sensitive as OPA derivatives and native fluorescence.

Postcolumn derivatives of catecholamines with glycylglycine or 2-cyanoacetamide yields fluorescence derivatives suitable for HPLC determination [144,145]. Overall, not much work has been done on pre- or postcolumn derivatization procedures for neurotransmitter amines and related metabolites. Applications based on the natural fluorescence of the indole ring seem to reflect the most common use of the fluorescence detection systems.

In a few applications both F and EC detectors have been connected in series, with the EC acting as the downstream detector. This is summarized in Table 11D. In one case, comparison of the F and EC values for VMA [96] showed a high correlation.

VIII. MICROBORE HPLC

It is well known that a decrease in the particle size of the HPLC packing material from 10 to 3 μm results in increased sensitivity and faster analyses in shorter columns.

Recently, another approach to further improving detectability has become available with the advent of the technology necessary to reduce the cross section of the columns from the standard 4 to 5 mm i.d. (Table 2) to less than 2 mm i.d. These are the so-called microbore columns. However, specific applications of these new columns to neurotransmitters are still scarce. Only a handful of papers describe practical applications in this field [116,134,140,146,147].

Columns are usually 10 to 20 cm long × 0.5 to 1.2 mm i.d., and their presumed advantages over conventional systems are centered on their lower detection limits and lower solvent consumption. Typical flow rates range from <10 μl/min [116,134] to 200 to 250 μl/min [147] and detection limits can be an order of magnitude lower than in conventional systems, such as the 280 fg reported for NA [147]. This may require a modification of the time constant of the detector.

Because of their small void volumes, the efficiency of these columns is very much dependent on extracolumn band broadening effects due to the use of inadequate hardware. For instance, a given column tested with an HPLC setup with negligible dead volumes gave a plate number of 15,000 to 17,000. However, the plate number was reduced to 8000 in another setup [146]. This level of efficacy would be equivalent to using a conventional bore 4.6 mm i.d. column, yet the nonoptimal use of the microbore column would be advantageous from the point of view of the lesser dilution of the analyte consistent with an increased mass sensitivity. Nevertheless, this is often not realized in practice because of the reduced sample capacity of these columns, as maintenance of their high efficiencies is contingent on the injection of small sample volumes, though in some cases direct injection of up to 200 μl of sample has been described [134]. Unfortunately, no study was made of the effect on column efficiency.

An interesting approach to the problem of injection of large sample volumes without influencing efficiency is based on concentration of the sample on top of the column [146]. This can be achieved by dissolving the sample in a noneluting liquid, such as water in reversed-phase HPLC. The effect of perchlorate and hydrogen ion on retention of solutes on top of an HPLC column resulting in peak sharpening has also been discussed in the literature [81]. However, with regard to efficiency, it is surprising that only one of the references on applications of microbore columns to the analysis of neurochemicals used 3-μm particles. This may reflect the practical limitations of very high efficiency setups due to excessive pressure buildup from clogging of lines and columns if samples are not adequately filtered.

IX. PEAK IDENTIFICATION AND INTERNAL STANDARDS

It is acknowledged that identifications based on retention times alone are not reliable because of the possibility of misidentification of interfering peaks for the analyte. Identifications of neurochemicals detected by F or EC detectors can be confirmed by (a) the chromatographic behavior in different mobile phases, (b) the responses obtained at different excitation/emission wavelengths or electrode potentials according to previously run voltammograms for both the sample and standards solutions, and (c) the selective use of pharmacological manipulations that enhance or inhibit the formation of the neurochemical of interest [8,23,26,29,37,55,60,86,91].

Whereas deuterated internal standards (IS) are commonly used in GC-MS [148] for precise quantitative work, no such possibility exhits in HPLC, where one must resort to the use of analogs or standard addition techniques. Regarding analogs, various compounds have been used. The most common for catecholamines has been di-hydroxybenzoic acid (DHBA) [21,22,27,29,38,44,52,61,62]. The monohydroxy analog (MHBA) has also been used [77]. MHPG has been quantitated by using the ethoxy (EHPG) or *iso*-MHPG analogs as standards [74,103,108]. α-Methyldopamine (α-MDA) has also been substituted for DHBA [12,60]. 3-Methoxy-4-hydroxyphenylacetic acid (MHPLA) has served as an IS for MHPG, HVA, and 5HIAA [91]. Catecholamines have been determined with isoproterenol or dihydroxy-phenylpropionic acid (DOPPA) [136]. As to indoles, 5-methoxy-tryptamine [55], 6-fluoro-5HT [59], and *N*-methyl-5HT [54,119] have been used as internal standards for serotonin determinations and 5-hydroxyindole-3-propionic acid (5HIPA) for 5HIAA [98]. DOPAC, HVA, and 5HIAA have been determined with 5-hydroxyindolecarboxy-lic acid as the IS [86,95], and *p*-hydroxymandelic acid (POMA) has been used for HVA and VMA [73].

Obviously, in cases where the IS selected can be an endogenous constituent of the sample, its absence from the sample must be certified [55].

X. OTHER DETECTION SYSTEMS

Consideration has been given here only to detection systems that have been routinely applied in the determination of neurotransmitter amines and metabolites in biological samples and that have been in use to date in most of the laboratories working in this field. However, a limited use of other detection systems has also been described in the literature. For instance, catecholamine sulfoconjugate isomers were detected after HPLC separation by a photochemical fluorogenic reac-tion. The fluorescent compounds generated at the photochemical re-actor were monitored with a spectrofluorometer [139]. More recently, on-line HPLC radiochemical detection of radiolabeled reaction products from catechol-O-methyltransferase activity in brain tissue have been reported [149]. Chemiluminiscence detection using the peroxyoxalate reaction has provided picomole sensitivity for the detection of acetyl-choline and choline [150]. For this purpose the eluent from the analytical column was fed to an inmobilized enzyme column containing 80 units of acetylcholinesterase and 12 units of choline oxidase. The resulting H_2O_2 was detected by the peroxyoxlate chemiluminiscence reaction. A similar procedure was reported previously, but in that case the hydrogen peroxide generated was detected with an electro-chemical detector [151].

HPLC has been coupled with success to mass spectrometers, but whereas the GC-MS combination has provided a wealth of information on neurotransmitter metabolism and turnover [148], the LC-MS equipment is not available in many laboratories at present and the number of reported applications is very small. Nevertheless, although no significant data have been reported on HPLC-MS assay of neurotransmitters, it may become an important tool in the near future, as evidenced by recent data from this laboratory. For instance, the selected-ion current traces depicted in Fig. 4 illustrate the potential of this new technique in the neurosciences [153]. In this case an extract obtained from the hypothalamus of rat brain was injected directly onto a reversed-phase HPLC column connected to the thermospray ion source of a mass spectrometer. Elution of endogenous

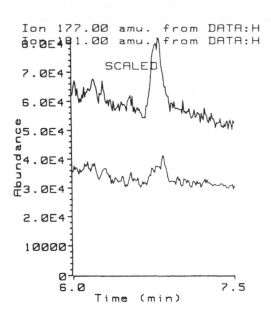

Fig. 4   Selected monitoring of the ion at m/z 177 corresponding to the $(M + H)^+$ species from serotonin in rat hypothalamus. Extraction as described in Ref. 131. Lower trace is from equivalent ion in deuterated serotonin added as an internal standard. Extract was injected into the reversed-phase HPLC column and eluted at 1 ml/min with a 10-min linear gradient of 10 to 50% ammonium acetate at pH 4.0. The HPLC-MS interface was of the thermospray type with stem, tip, and vapor temperatures set at 116, 205, and 210°C, respectively. The ion source was at 270°C.

serotonin was monitored by the response at mass-to-charge ratio (mz/) 177. Quantitation was achieved by the isotope dilution technique, using a deuterated internal standard [131]. The same determination by GC-MS would require previous derivatization of the sample to render it suitable for analysis in the gas phase [131]. The HPLC-MS technique might become a very useful adjunct in the direct assay of brain perfusates, thus simplifying sample preparation procedures.

XI. CONCLUSIONS

Considerations of space have prevented the inclusion of other neurotransmitters such as the amino acids, histamine, and acetylcholine, although the existing body of literature in this case is much more restricted and thus easier to gather and evaluate. Also, no discussion has been given of the preparation of the sample itself for HPLC determination, which will be the subject of a separate review with regard to sample stability, deproteinization, homogenization, extraction, etc. For those interested, a brief outline of cleanup and liquid-solid extraction method in use for catecholamine assays has been presented [9]. A detailed general account of liquid-solid sample preparation methods is also available [152].

In conclusion, the body of data taken directly from recent literature sources can be considered an accurate reflection of the state of the art in the HPLC of neurotransmitters. No assumptions have been made as to what could have been achieved according to the theory of HPLC. Instead, the chapter has attempted to describe what has been realized in practice.

Overall, there seems to be an excessive description of HPLC methods which for the most part do not add anything new since they are based on minute variations of operating parameters, as illustrated in Tables 2 to 6. The basic concepts underlying the mechanism of separation of neurotransmitters and their metabolites are clear, as discussed here (see Sec. III.A), and there is no point in cluttering the literature with methods that differ only in a small percentage of organic modifier or in the molarity of the buffer used.

The use of nonisocratic methods in this field has been relatively very limited, and no extensive use of multiple solvent capabilities to enhance selectivity has been made, either. Likewise, the smaller 3-μm particle columns are not very popular, possibly because of plugging problems and stability considerations when analyzing complex brain extracts. The new microbore column technology has not been adopted yet for routine applications whenever sensitivity is a determining factor, possibly because of the limitations of sample size, which may require additional preconcentration steps.

Surprisingly, neuroscientists have been rather conservative with regard to the utility and potential of the latest HPLC advances. Could this be justified by the quality of the data they have been obtaining with standard HPLC procedures? On reflection, the achievements of the HPLC-ECD procedures in neurochemistry have been impressive, but there is still plenty of room for more demanding applications, such as direct determination of neurotransmitters in CSF and brain perfusates for the study of neurotransmitter dynamics in vivo. Evidently this will call for use of the latest HPLC developments in terms of both selectivity and sensitivity.

ABBREVIATIONS

ACN	acetonitrile
BA	butylamine
CA	citric acid
CA-S	catecholamine sulfates
C-P	citrate-phosphate
DBA	dibutylamine
DS	decylsulfonate
ECD	electrochemical detector
GC-MS	gas chromatography-mass spectrometry
HexSA	Hexanesulfonic acid
HSA	heptanesulfonic acid
IPR	ion pair reagent
MCA	monochloroacetic acid
NaAc	sodium acetate
NaC	sodium citrate
NaHS	sodium heptanesulfonate
NaOS	sodium octanesulfonate
NaP	sodium phosphate
NH_4Ac	ammonium acetate
OSA	octanesulfonic acid
OTEAP	octyltriethylammonium phosphate
P	phosphate

PA	phosphoric acid
PCA	perchloroacetic acid
PSA	pentanesulfonic acid
RP	reversed phase
SDS	sodium dodecyl sulfate or lauryl sulfate
SOS	sodium octylsulfate
TBP	tributylphosphate
TEA	triethylamine
TEAm	tetraethylammonium

REFERENCES

1. J. M. P. Holly and H. L. J. Makin, Anal. Biochem. *128*, 257 (1983).
2. E. Martinez, A. Adell, C. Suñol, F. Artigas, J. M. Tusell, and E. Gelpí, *Progress in Tryptophan and Serotonin Research*, de Gruyter, Berlin, 1984, p. 83.
3. J. J. Warsh, A. S. Chiu, and D. D. Goodse, *Techniques in Instrumentation in Analytical Chemistry*, Part A, *Analysis of Biogenic Amines* (G. B. Baker and R. T. Coutts, eds.), Elsevier, Amsterdam, 1982, p. 203.
4. G. M. Anderson and J. G. Young Life Sci. *28*, 507 (1981).
5. P. T. Kissinger, C. S. Bruntlett, and R. E. Shoup, Life Sci. *28*, 455 (1981).
6. S. Allenmark, J. Liq. Chromatogr. *5*, 1 (1982).
7. A. M. Krustulovic and H. Colin, Trends Anal. Chem. *2*, 42 (1983).
8. S. M. Lasley, I. A. Michaelson, R. D. Greenland, and P. M. McGinnis, J. Chromatogr. Biomed. Appl. *30*, 27 (1984).
9. R. C. Causon, Res. Methods Neurochem. *6*, 211 (1985).
10. C. A. Marsden, *Drug Determination in Therapeutic and Forensic Contexts* (E. Reid and I. D. Wilson, eds.), Plenum, New York, 1984, p. 319.
11. B. M. Eriksson, S. Gustafsson, and B. A. Persson, J. Chromatogr. Biomed. Appl. *29*, 255 (1983).
12. M. Patthy and R. Gyenge, J. Chromatogr. *286*, 217 (1984).
13. A. H. B. Wu and T. G. Gornet, Clin. Chem. *31*, 298 (1985).
14. T. Yamamoto, A. Yamatodani, M. Nishimura, and H. Wada, J. Chromatogr. Biomed. Appl. *43*, 261 (1985).
15. B. R. Sitaram, G. L. Blackman, W. R. McLeod, and G. N. Vaughan, Anal. Biochem. *128*, 11 (1983).

16. B. R. Sitaram, R. Talomsin, G. L. Blackman, W. R. McLeod, and G. N. Vaughan, J. Chromatogr. *275*, 21 (1983).

17. A. M. Krstulovic and P. R. Brown, *Reversed-Phase High-Performance Liquid Chromatography: Theory, Practice and Biomedical Applications*, Wiley, New York, 1982.

18. J. De Jong, U. R. Tjaden, W. Van't Hof, and C. F. M. Van Valkenburg, J. Chromatogr. *282*, 443 (1983).

19. R. B. Taylor, R. Reid, K. E. Kendle, C. Geddes, and P. F. Curle, J. Chromatogr. Biomed. Appl. *28*, 101 (1983).

20. K. Oka, K. Kojima, A. Togari, T. Nagatsu, and B. Kiss, J. Chromatogr. Biomed. Appl. *33*, 43 (1984).

21. P. Y. T. Lin, M. C. Bulawa, P. Wong, L. Lin, J. Scott, and C. L. Blank, J. Liq. Chromatogr. *7*, 509 (1984).

22. H. Todoriki, T. Hayashi, H. Naruse, and A. Y. Hirakawa, J. Chromatogr. *276*, 45 (1983).

23. J. A. Nielsen and C. A. Johnston, Life Sci. *31*, 2847 (1982).

24. J. Wagner, P. Vitali, M. G. Palfreyman, M. Zraika, and S. Huot, J. Neurochem. *38*, 1241 (1982).

25. W. H. Lyness, Life Sci. *31*, 1435 (1982).

26. T. Di Paolo, A. Duppont, P. Savard, and M. Daigle, Can. J. Physiol. Pharmacol. *61*, 530 (1983).

27. J. F. Reinhard and J. A. Perry, J. Liq. Chromatogr. *7*, 1211 (1984).

28. G. Achilli, C. Perego, and F. Ponzio, Anal. Biochem. *148*, 1 (1985).

29. F. M. Siri and C. D. Kauer, Life Sci. *37*, 1923 (1985).

30. R. J. Martin, B. A. Bailey, and R. G. H. Downer, J. Chromatogr. Biomed. Appl. *29*, 265 (1983).

31. G. B. Freeman, P. Nielsen, and G. E. Gibson, J. Chromatogr. Biomed. Appl. *374*, 239 (1986).

32. M. A. Peat and J. W. Gibb, Anal. Biochem. *128*, 275 (1983).

33. B. H. C. Westerink, J. Neurochem. *42*, 934 (1984).

34. F. Ehrenstrom and P. Johansson, Life Sci. *36*, 867 (1985).

35. S. L. Salzman, C. Llados Eckman, and E. Hirofuji, J. Liq. Chromatogr. *8*, 345 (1985).

36. E. Morier and R. Rips, J. Liq. Chromatogr. *5*, 151 (1982).

37. M. Warnhoff, J. Chromatogr. Biomed. Appl. *32*, 271 (1984).

38. K. Koike, T. Aono, F. Chatani, T. Takemura, and K. Kurachi, Life Sci. *30*, 2221 (1982).

39. B. H. C. Westerink, J. Liq. Chromatogr. *6*, 2337 (1983).

40. W. E. Wilson, S. W. Mietling, and J. S. Hong, J. Liq. Chromatogr. *6*, 871 (1983).

41. R. F. Seegal, K. O. Brosh, and B. Bush, J. Chromatogr. Biomed. Appl. *377*, 131 (1986).

42. C. F. Salier and A. I. Salama, J. Chromatogr. Biomed. Appl. *34A*, 287 (1984).

43. K. Ishikawa and S. Shibanoki, Anal. Biochem. *147*, 441 (1985).
44. V. M. Gregory, B. Larsen, and B. Benson, J. Chromatogr. Biomed. Appl. *345*, 140 (1985).
45. C. Kim, C. Campanelli, and J. M. Khanna, J. Chromatogr. *282*, 151 (1983).
46. J. Yaneda, Y. Sugimoto, and K. Horisaka, Anal. Biochem. *129*, 460 (1983).
47. Z. L. Rossetti, G. Mercuro, and C. A. Rivano, Life Sci. *33*, 2387 (1983).
48. D. L. Sparks and J. T. Slevin, Life Sci. *36*, 449 (1985).
49. A. A. Larson and N. L. Dalo, J. Chromatogr. Biomed. Appl. *48*, 37 (1986).
50. J. M. Tusell, C. Suñol, F. Artigas, E. Martinez, and E. Gelpí, Chromatographia *16*, 112 (1982).
51. W. A. Wolf and D. M. Kuhn, J. Chromatogr. *275*, 1 (1983).
52. W. D. Ruwe, A. M. Naylor, L. Bauce, and W. L. Eale, Life Sci. *37*, 1749 (1985).
53. G. S. Mayer and R. E. Shoup, J. Chromatogr. *255*, 533 (1983).
54. M. Picard, D. Olichon, and J. Gombert, J. Chromatogr. Biomed. Appl. *42*, 445 (1985).
55. E. Kwarts, J. Kwarts, and H. Rutgers, Ann. Clin. Biochem. *21*, 425 (1984).
56. P. C. Tagari, D. J. Boulin, and C. L. Davies, Clin. Chem. *30*, 131 (1984).
57. J. I. Javaid, T. S. Liu, J. W. Maas, and J. M. Davis, Anal. Biochem. *135*, 326 (1983).
58. G. Santagostino, P. Frattini, S. Schinelli, M. L. Cucchi, and G. L. Corona, Farmaco Ed. Prat. *38*, 229 (1983).
59. E. R. Korpi, Clin. Chem. *30*, 487 (1984).
60. G. P. Jackman, C. J. Oddie, H. Skews, and A. Bobik, J. Chromatogr. Biomed. Appl. *33*, 301 (1984).
61. K. Maruta, J. Fujita, S. Ito, and T. Nagatsu, Clin. Chem. *30*, 1271 (1984).
62. C. C. T. Smith and J. Betteridge, Clin. Chem. *30*, 1432 (1984).
63. P. Bouloux, D. Perrett, and G. M. Besser, Anal. Clin. Biochem. *22*, 194 (1985).
64. R. C. Causon and M. J. Brown, Clin. Chem. *29*, 737 (1983).
65. H. Weicker, M. Feraudi, H. Haegele, and R. Pluto, Clin. Chim. Acta *141*, 17 (1984).
66. A. Mitsui, H. Nohta, and Y. Ohkura, J. Chromatogr. Biomed. Appl. *344*, 61 (1985).
67. L. G. Howes, S. Miller, and J. L. Reid, J. Chromatogr. Biomed. Appl. *39*, 401 (1985).
68. S. A. McClintock, W. C. Purdy, and S. N. Young, Anal. Chim. Acta *166*, 171 (1984).
69. T. Seppala, M. Scheinin, A. Capone, and M. Linnoila, Acta Pharmacol. Toxicol. *55*, 81 (1984).

70. M. Scheinin, T. Seppala, M. Koulu, and M. Linnoila, Acta Pharmacol. Toxicol. *55*, 88 (1984).

71. J. Odink, H. Sandman, and W. H. P. Schreurs, J. Chromatogr. Biomed. Appl. *377*, 145 (1986).

72. P. J. Orsulak, P. Kizuka, E. Grab, and J. J. Schildkraut, Clin. Chem. *29*, 305 (1983).

73. E. Gerlo and R. Malfait, J. Chromatogr. Biomed. Appl. *343*, 9 (1985).

74. R. T. Brown, K. L. Kirk, and J. Oliver, J. Liq. Chromatogr. *9*, 831 (1986).

75. D. F. Davidson and J. Fitzpatrick, Ann. Clin. Biochem. *22*, 297 (1985).

76. M. H. Mills, M. G. King, N. G. Keats, and R. A. McDonald, J. Chromatogr. Biomed. Appl. *377*, 350 (1986).

77. N. T. Buu, M. Angers, D. Chevalier, and O. Kuchel, J. Lab. Clin. Med. *104*, 425 (1984).

78. A. J. Speek, J. Odink, J. Schrijver, and W. H. P. Schreurs, Clin. Chim. Acta *128*, 103 (1984).

79. J. Jouve, N. Mariotte, C. Sureau, and J. P. Muh, J. Chromatogr. Biomed. Appl. *25*, 53 (1983).

80. J. M. Tusell, F. Artigas, C. Suñol, E. Martinez, and E. Gelpí, J. Chromatogr. Biomed. Appl. *31*, 338 (1984).

81. C. Van Valkenburg, U. Tjaden, J. Van der Krogt, and B. Van der Leden, J. Neurochem. *39*, 990 (1982).

82. G. L. Diggory and W. R. Buckett, Pharmacol. Methods *11*, 207 (1984).

83. A. Yoshida, Y. Ichihashi, and M. Yoshioka, J. Chromatogr. Biomed. Appl. *343*, 155 (1985).

84. F. Artigas, E. Martinez, J. M. Tusell, C. Suñol, and E. Gelpí, Biochem. Pharmacol. *32*, 3251 (1983).

85. E. Martinez, F. Artigas, C. Suñol, and E. Gelpí, and J. M. Tusell, Clin. Chem. *29*, 1354 (1983).

86. A. Minegishi and T. Ishizaki, J. Chromatogr. Biomed. Appl. *33*, 55 (1984).

87. M. Scheinin, W. Chang, K. L. Kirk, and M. Linnoila, Anal. Biochem. *131*, 246 (1983).

88. M. Van Bockstade, L. Dillen, M. Claeys, and W. P. de Potter, J. Chromatogr. Biomed. Appl. *275*, 11 (1983).

89. T. Bottiglieri, C. L. Lim, and T. J. Peters, J. Chromatogr. Biomed. Appl. *36*, 354 (1984).

90. L. Elrod and G. L. Mayer, J. Liq. Chromatogr. *8*, 1817 (1985).

91. M. A. Javors, C. L. Bowden, and J. W. Maas, J. Chromatogr. Biomed. Appl. *37*, 259 (1984).

92. S. R. Binder and G. Sivorinovsky, J. Chromatogr. Biomed. Appl. *336*, 173 (1984).

93. R. F. Seegal, K. O. Brosh, and B. Bush, J. Chromatogr. Biomed. Appl. *273*, 253 (1983).
94. G. Anderson, K. R. Schlicht, and D. J. Cohen, Anal. Biochem. *144*, 27 (1985).
95. V. Skrinska and S. Hahn, J. Chromatogr. Biomed. Appl. *36*, 380 (1984).
96. G. M. Anderson, F. C. Feibel, and D. J. Cohen, Clin. Chem. *31*, 819 (1985).
97. J. P. M. Wielders and C. J. K. Mink, J. Chromatogr. Biomed. Appl. *35*, 379 (1984).
98. P. P. Chou and P. K. Jaynes, J. Chromatogr. Biomed. Appl. *341*, 167 (1985).
99. J. M. Hornsperger, J. Wagner, J. P. Hinkel, and M. J. Jung, J. Chromatogr. Biomed. Appl. *31*, 364 (1984).
100. F. Karege, J. Chromatogr. Biomed. Appl. *36*, 361 (1984).
101. L. G. Howes, R. J. Summers, P. R. Rowe, and W. J. Louis, J. Chromatogr. *287*, 133 (1984).
102. R. K. Yang, J. P. Edasery, and K. L. Davis, J. Liq. Chromatogr. *6*, 1997 (1983).
103. J. Semba, A. Watanabe, and R. Takahashi, Clin. Chem. Acta *152*, 185 (1985).
104. N. S. Sharpless, V. Italbreich, and H. Feldfogel, J. Chromatogr. Biomed. Appl. *377*, 101 (1986).
105. M. Scheinin, W. Chang, D. C. Jimerson, and M. Linnoila, Anal. Biochem. *132*, 165 (1983).
106. P. A. Shea and J. B. Howell, J. Chromatogr. Biomed. Appl. *31*, 358 (1984).
107. A. Minegishi and T. Ishizaki, J. Chromatogr. Biomed. Appl. *311*, 51 (1984).
108. S. Schinelli, G. Santagostino, P. Frattini, and M. L. Cucchi, J. Chromatogr. Biomed. Appl. *39*, 396 (1985).
109. S. G. Molyneux and M. Franklin, J. Chromatogr. Biomed. Appl. *42*, 160 (1985).
110. J. R. Shipe, J. Savory, and M. R. Willis, Clin. Chem. *30*, 140 (1984).
111. R. T. Brown, J. Oliver, K. L. Kirk, and I. J. Kopin, Life Sci. *34*, 2313 (1984).
112. F. Karege, J. M. Gaillard, P. Bovier, R. Tissot, and D. Pringuey, Clin. Chem. *30*, 1416 (1984).
113. G. M. Anderson, K. R. Schlicht, and D. J. Cohen, Anal. Chem. *55*, 1399 (1983).
114. W. A. Bartlett, J. Liq. Chromatogr. *8*, 719 (1985).
115. J. De Jong, C. F. M. Van Valkenburg, and U. R. Tjaden, J. Chromatogr. *322*, 43 (1985).
116. M. Goto, E. Sakurai, and D. Ishii, J. Liq. Chromatogr. *6*, 1907 (1983).

117. U. R. Tjaden, J. De Jong, and C. F. M. Van Valkenburg, J. Liquid Chromatogr. *6*, 2255 (1983).

118. D. J. Liberato and A. L. Yergey, Anal. Chem. *58*, 6 (1986).

119. R. B. Mailman and C. D. Kilts, Clin. Chem. *31*, 1849 (1985).

120. H. J. L. Jansen, V. R. Tjaden, H. J. de Jong, and K. G. Wahlund, J. Chromatogr. *202*, 203 (1980).

121. R. Helboe, J. Pharm. Biomed. Anal. *3*, 293 (1985).

122. H. G. Hedfield, P. Crane, M. E. King, E. A. Nugent, C. Milio, and M. Naresimhachari, J. Liq. Chromatogr. *8*, 2689 (1985).

123. E. Martinez, J. M. Tusell, C. Suñol, N. Mahy, F. Artigas, and E. Gelpí, in *Biological/Biomedical Applications of Liquid Chromatography IV* (G. L. Hawk, ed.), Chromatographic Science Series, Vol. 20, New York, Marcel Dekker, 1982, pp. 305–317.

124. A. Adell, J. M. Tusell, F. Artigas, E. Martinez, C. Suñol, and E. Gelpí, J. Liq. Chromatogr. *6*, 527 (1983).

125. W. A. Hunt and T. K. Dalton, Anal. Biochem. *135*, 269 (1983).

126. P. Moleman and J. J. M. Borstrok, Clin. Chem. *29*, 878 (1983).

127. K. Fujita, K. Maruta, S. Ito, and T. Nagatsu, Clin. Chem. *29*, 876 (1983).

128. D. S. Goldstein, R. Stull, R. Zimlichman, P. D. Levinson, H. Smith, and H. R. Keiser, Clin. Chem. *30*, 815 (1984).

129. T. Ishimitsu and S. Hirose, Anal. Biochem. *150*, 300 (1985).

130. T. Ishimitsu and S. Hirose, J. Chromatogr. Biomed. Appl. *337*, 239 (1985).

131. F. Artigas and E. Gelpí, Anal. Biochem. *92*, 233 (1979).

132. G. Eisenhofer, J. Chromatogr. Biomed. Appl. *377*, 328 (1986).

133. J. M. Di Bussolo, M. W. Dong, and J. R. Grant, J. Liq. Chromatogr. *6*, 2353 (1983).

134. M. Goto, G. Zou, and D. Ishii, J. Chromatogr. Biomed. Appl. *275*, 271 (1983).

135. P. O. Edlund and D. Westerlund, J. Pharm. Biomed. Anal. *2*, 315 (1984).

136. C. Julien, C. Rodriguez, G. Cuisinaud, N. Bernard, and J. Sassard, J. Chromatogr. Biomed. Appl. *344*, 51 (1985).

137. L. Hanson, M. Glad, and C. Hanson, J. Chromatogr. *265*, 37 (1983).

138. A. Niederwieser, W. Standenmann, and E. Netzel, J. Chromatogr. *290*, 237 (1984).

139. Y. Arakawa, K. Imai, and Z. Tamura, Anal. Biochem. *132*, 389 (1983).

140. C. G. Honegger, R. Burri, H. Langeman, and A. Kempf, J. Chromatogr. Biomed. Appl. *34*, 53 (1984).

141. B. A. Bailey, R. J. Martin, and R. G. H. Downer, J. Liq. Chromatogr. *5*, 2435 (1982).
142. W. R. Matson, P. Langlais, L. Volicer, P. H. Gamache, E. Bird, and K. A. Mark, Clin. Chem. *30*, 1477 (1984).
143. C. Suñol, J. M. Tusell, F. Artigas, E. Martinez, A. Adell, and E. Gelpí, in *Neurobiology of the Trace Amines* (A. A. Boulton, G. B. Baker, W. G. Dewhurst, and M. Sandler, eds.), Humana Press, N.J., 1984, p. 77.
144. S. Honda, M. Takahashi, Y. Araki, and K. Kakehi, J. Chromatogr. Biomed. Appl. *25*, 45 (1983).
145. T. Seki and Y. Yamaguchi, J. Chromatogr. *287*, 407 (1984).
146. G. Kamperman and J. C. Kraak, J. Chromatogr. Biomed. Appl. *38*, 384 (1985).
147. T. A. Durkin, E. J. Caliguri, I. N. Mefford, D. M. Lake, I. A. McDonald, E. Sundstrom, and Gosta Jonsson, Life Sci. *37*, 1803 (1985).
148. E. Gelpí, in *Advances in Mass Spectrometry 1985* (J. F. J. Todd, ed.), Wiley, New York, 1986, p. 397.
149. E. Nissinen, Anal. Biochem. *144*, 247 (1985).
150. K. Honda, K. Miyeguchi, H. Nishine, H. Tanaka, T. Yao, and K. Imai, Anal. Biochem. *153*, 50 (1986).
151. C. Eva, M. Hadjiconstantinou, N. H. Neff, and J. L. Meek, Anal. Biochem. *143*, 320 (1984).
152. R. D. McDowald, J. C. Pierce, and G. S. Murkitt, J. Pharm. Biomed. Anal. *4*, 3 (1986).
153. F. Artigas and E. Gelpí, J. Chromatogr. 1987 (in press).

Index

Printed and bound by CPI Group (UK) Ltd, Croydon, CR0 4YY

17/10/2024

01775696-0016